Condensed
Matter Theories

Volume 25

Condensed Matter Theories

Volume 25

Eduardo V Ludeña
Venezuelan Institute for Scientific Research, IVIC, Venezuela

Raymond F Bishop
The University of Manchester, UK

Peter Iza
SENACYT, Ecuador

Editors

World Scientific

NEW JERSEY · LONDON · SINGAPORE · BEIJING · SHANGHAI · HONG KONG · TAIPEI · CHENNAI

Published by

World Scientific Publishing Co. Pte. Ltd.

5 Toh Tuck Link, Singapore 596224

USA office: 27 Warren Street, Suite 401-402, Hackensack, NJ 07601

UK office: 57 Shelton Street, Covent Garden, London WC2H 9HE

British Library Cataloguing-in-Publication Data
A catalogue record for this book is available from the British Library.

CONDENSED MATTER THEORIES
Volume 25
Proceedings of the 33rd International Workshop

ISBN-13 978-981-4340-78-6
ISBN-10 981-4340-78-2

Printed in Singapore by B & Jo Enterprise Pte Ltd

PREFACE

The *Thirty-Third International Workshop on Condensed Matter Theories* (CMT33) was held in the city of Quito, the beautiful Andean capital of Ecuador, during the period 16-22 August 2009. Workshops in the CMT Series have taken place annually since 1977, when the inaugural meeting was held in São Paulo, Brazil. Subsequent meetings have been held thereafter in Trieste, Italy (1978), Buenos Aires, Argentina (1979), Caracas, Venezuela (1980), Mexico City, Mexico (1981), St. Louis, USA (1982), Altenberg, Federal Republic of Germany (1983), Granada, Spain (1984), San Francisco, USA (1985), Argonne, USA (1986), Oulu, Finland (1987), Taxco, Mexico (1988), Campos do Jordão, Brazil (1989), Isle of Elba, Italy (1990), Mar del Plata, Argentina (1991), San Juan, Puerto Rico (1992), Nathiagali, Pakistan (1993), Valencia, Spain (1994), Caracas, Venezuela (1995), Pune, India (1996), Luso, Portugal (1997), Nashville, USA (1998), Isle of Ithaca, Greece (1999), Buenos Aires, Argentina (2000), Canberra, Australia (2001), Luso, Portugal (2002), Toulouse, France (2003), St. Louis, USA (2004), Kyoto, Japan (2005), Dresden, Germany (2006), Bangkok, Thailand (2007), and Loughborough, UK (2008). The present volume contains the Proceedings of the 2009 Workshop, CMT33.

The primary aims of the CMT Workshops have remained unchanged since the inception of the Series, namely: to encourage cross-fertilization between different approaches to many-body systems; to promote continuing collaborative efforts involving groups of Workshop participants; and to foster communication and cooperation between scientists in developed and developing nations. An important objective throughout has been to work against the ever-present trend for physics to fragment into increasingly narrow fields of specialization, between which communication is difficult. The CMT Workshops have traditionally sought to emphasize the unity of physics, and CMT33 fully subscribed to that ideal, as the present volume of the Proceedings will amply attest.

From their Pan-American origins, the Workshops in the CMT Series rapidly developed into the significant scientific meetings that they have now become, and where work at the forefronts of the fields covered is presented by leading experts from around the world. This last Workshop, CMT33 in Ecuador, for example, had participants from more than 15 different countries. The Workshops have thus successfully fostered truly international collaborations between scientists divided not only by interdisciplinary barriers but also by geographical and political boundaries.

As we have already pointed out, the orientation and the physical context of the CMT Series of Workshops have always been cross-disciplinary, but with an emphasis placed on the common concerns of theorists applying many-particle concepts and formalisms in such diverse areas as solid-state, low-temperature, atomic, nuclear, particle, chemical, statistical and biological physics, as well as in quantum field theory, quantum optics, quantum information theory, strongly correlated electronic systems and the theory of complex systems. While adhering strongly to this tradition as an over-arching principle, the organizers of each particular CMT Workshop have traditionally been encouraged to highlight specific areas of topical and/or local interest. In this spirit CMT33 chose to focus special attention on exotic fermionic and bosonic systems, quantum magnets and their quantum and thermal phase transitions, novel condensed matter systems for renewable energy sources, the physics of nanosystems and nanotechnology, and applications of molecular dynamics and density functional theory.

The Series Editorial Board, on behalf of all the participants, wishes to express its gratitude to the many people who contributed to the choice of scientific programme and in other ways to the success of the CMT33 Workshop. The main sponsor of CMT33, to whom a huge debt of gratitude is owed for its very generous financial support, for its gracious hospitality and for the provision of excellent secretarial and organizational support, was the Secretaría Nacional de Ciencia y Tecnología (SENACYT) of Ecuador. Particular thanks are due to our SENACYT colleagues Pedro Montalvo (National Secretary) and Peter Iza (Director of Scientific Research). We are grateful to Pedro for opening the meeting and for welcoming delegates on behalf of SENACYT, and to Peter for acting as Chair of the Local Organizing Committee. Huge thanks are also due to Catalina Díaz of SENACYT, who acted with great skill and diplomacy as the Secretary to the Workshop and to its Local Organizing Committee. It was largely due to her capable hands and cool head that the meeting ran as smoothly as it undoubtedly did. Perhaps most thanks of all, however, are due to Eduardo V. Ludeña for suggesting Quito in the first place as a venue for CMT33. Without his dogged perseverance and his formidable negotiating skills, however, the Workshop would not have taken place. It is fitting, therefore, that he was also chosen to act as Chair of the Scientific Programme Committee (SPC), a job which he performed with his customary enthusiasm and skill, as the final list of speakers and attendees clearly demonstrates. He was aided in that task by Raymond Bishop. Khandker Quader and Bilal Tanatar, who acted as the other members of the SPC. Without the active participation of all of them, the meeting would not have been the undoubted success it so clearly was. Our grateful thanks are due to each of them, as well as to all of the speakers and participants, for making CMT33 such a high-quality and memorable Workshop.

R.F. Bishop   Manchester, U.K.
(On behalf of the Editorial Board   31 March 2010
for the Series of International Workshops on
Condensed Matter Theories)

ORGANIZING COMMITTEES

Series Editorial Board
for the Series of International Workshops on
Condensed Matter Theories

R. F. Bishop   – The University of Manchester, Manchester, UK
J. W. Clark   – Washington University, St. Louis, USA
M. de Llano   – UNAM, Mexico City, Mexico
F. B. Malik   – Southern Illinois University, Carbondale, USA

International Advisory Committee

H. Akai (Japan)
G. Anagnostatos (Greece)
R. F. Bishop (United Kingdom)
L. Blum (Puerto Rico)
C. E. Campbell (USA)
M. Ciftan (USA)
J. W. Clark (USA)
J. Dabroski (Poland)
J. da Providencia (Portugal)
M.P. Das (Australia)
M. de Llano (Mexico)
D. Erns (USA)
R. Guardiola (Spain)
E. V. Ludeña (Venezuela)
F. B. Malik (USA)
J. Navarro (Spain)
A. Plastino (Argentine)
A. N. Proto (Argentine)
S. Rosati (Italy)
E. Suraud (France)
P. Vashishta (USA)
C. W. Woo (China)

Scientific Program Committee
for the XXXIII International Workshop on
Condensed Matter Theories

Raymond F. Bishop	– The University of Manchester, Manchester, UK
Eduardo V. Ludeña	– IVIC, Caracas, Venezuela
Khandker F. Quader	– Kent State University, Kent, USA
Bilal Tanatar	– Bilkent University, Ankara, Turkey

Local Organizing Committee
for the XXXIII International Workshop on
Condensed Matter Theories

| Peter Iza (Chair) | – SENACYT, Quito, Ecuador |
| Catalina Díaz (Secretary) | – SENACYT, Quito, Ecuador |

Editorial Committee
for the XXXIII International Workshop on
Condensed Matter Theories

Eduardo V. Ludeña	– IVIC, Caracas, Venezuela
Raymond F. Bishop	– The University of Manchester, Manchester, UK
Peter Iza	– SENACYT, Quito, Ecuador

HOST INSTITUTION AND MAIN SPONSOR
for the XXXIII International Workshop on
Condensed Matter Theories

Secretaría Nacional de Ciencia y Tecnología, SENACYT
(National Secretariat for Science and Technology, SENACYT)
National Secretary: Econ. Pedro Montalvo
Quito, Ecuador

Other Sponsoring Institutions

- Centro Latinoamericano de Física, CLAF
  (Latin American Physics Center, CLAF)
  CBPF-Rio de Janeiro, Brazil
- IOP Publishing
  Dirac House, Temple back
  Bristol BS1 6BE, UK
- Ministerio de Turismo del Ecuador
  (Ministry of Tourism, Ecuador)
  Quito, Ecuador
- Escuela Superior Politécnica del Ejército, ESPE
  (High Polytechnic Army School, ESPE)
  Quito, Ecuador
- Escuela Politécnica Nacional, EPN
  (National Polytechnic School, EPN)
  Quito, Ecuador
- Universidad Andina Simón Bilívar, UASB
  (Simon Bolivar Andean University, UASB)
  Quito, Ecuador
- Ecuadorian Paxi & Rupay Travel
  Quito, Ecuador

LIST OF PARTICIPANTS

XXXIII International Workshop on Condensed Matter Theories

Quito, Ecuador, August 16-22, 2009

(1) ALBARRACIN, Patricia
CFA, Quito, Ecuador

(2) ALDAS, Oswaldo
Escuela Politécnica Nacional,
Quito, Ecuador;
oswaldo.aldas@epn.edu.ec

(3) ALVARADO, Juan
SENPLADES, Quito, Ecuador;
jalvarado@senplades.gov.ec

(4) ANDREI, Eva
Rutgers University,
New Jersey, USA;
eandrei@physics.rutgers.edu

(5) ARMAS, Pablo
Escuela Politécnica Nacional,
Quito, Ecuador

(6) ARMIJOS, Eduardo
COMCIEC-ANC,
Guayaquil, Ecuador;
eduarmijos@hotmail.com

(7) AVENDAÑO, Maribel
Escuela Politécnica Nacional,
Quito, Ecuador

(8) AYALA, Luis Alfonso
Universidad Laica Eloy Alfaro,
Manta, Ecuador;
alfonsoayalacastro@yahoo.es

(9) AYERS, Paul
McMaster University,
Hamilton, Ontario, Canada;
ayers@mcmaster.ca

(10) BISHOP, Lev
Yale University,
New Haven, Connecticut, USA;
lev.bishop@gmail.com

(11) BISHOP, Raymond F.
University of Manchester,
Manchester, United Kingdom;
raymond.bishop@manchester.ac.uk

(12) BOEHM, Helga
Johannes Kepler University,
Linz, Austria;

helga.boehm@jku.at

(13) BURGOS, Armando
Escuela Politécnica Nacional,
Quito, Ecuador;
armandoburgos14@hotmail.com

(14) BURGOS, Meri Fernanda
Escuela Politécnica Nacional,
Quito, Ecuador.

(15) CABEZAS, Roberto
Colegio San Gabriel, Quito, Ecuador

(16) CAICEDO, Pal
Escuela Politécnica Nacional,
Quito, Ecuador;
poulmack@hotmail.com

(17) CALDERON, Juan
Universidad de Guayaquil,
Guayaquil, Ecuador;
jtcalderon@gmail.com

(18) CAMACHO, Ana Cecilia
Colegio 23 de Junio,
Manab, Ecuador;
anace_1960@hotmail.es

(19) CAMPBELL, Charles E.
University of Minnesota,
Minneapolis, Minnesota. USA;
campbell@umn.edu

(20) CAMPOZANO, Lenín
Universidad del Azuay,
Cuenca, Ecuador;
lenin_camp@yahoo.com

(21) CAMPUZANO, Bayardo
Universidad Politécnica Salesiana,
Quito, Ecuador;
gbcampuzano@yahoo.com

(22) CARDENAS, Luisa Mara
Escuela Politécnica Nacional,
Quito, Ecuador.

(23) CARDENAS, Víctor
Escuela Politécnica Nacional,
Quito, Ecuador;
victor.cardenas@epn.edu.ec

(24) CHIMBORAZO, Johnny
Universidad Politécnica Salesiana,

Quito, Ecuador
jchimborazo@yahoo.com.mx

(25) CIFUENTES, Jos
Escuela Politécnica Nacional,
Quito, Ecuador.

(26) CLARK, John
University of Washington,
St. Louis, Missouri, USA;
jwc@wuphys.wustl.edu

(27) COLE, Milton
University of Pennsylvania,
University Park, Pennsylvania,
USA;
miltoncole@aol.com

(28) CUEVA, Diego
SENACYT, Quito, Ecuador;
dcueva@senacyt.gov.ec

(29) DAVILA, Richard
EPR Travel, Quito, Ecuador;
infor@eprtravel.com

(30) DAS, Mukunda
Australian National University,
Canberra, Australia;
mpd105@physics.anu.edu.au

(31) DE LA CRUZ, Gonzalo
Galpagos 439 y Venezuela,
Quito, Ecuador.

(32) DE LlANO, Manuel
Universidad Autnoma de Mxico,
Mexico City, Mexico;
dellano@servidor.unam.mx

(33) DIAZ, Catalina
SENACYT, Quito, Ecuador;
cadiaz@senacyt.gov.ec

(34) DIEZ, Natasha
IOP Publishing, Mxico;
diez@ioppubusa.com

(35) EISFELD, Alexander
Max Planck Institute for the Physics
of Complex Systems,
Dresden, Germany;
eisfeld@pks.mpg.de

(36) ESCOBAR, Carlos

Colegio Fiscal Seis de Octubre,
Ventanas, Los Ros, Ecuador

(37) ESTEVES, Carlos
Colegio 23 de Junio,
Manab, Ecuador;
carlosesteves@hotmail.com

(38) FAITHFUL, Melanie
IOP Publishing, Mxico;
faithful@ioppubusa.com

(39) FAZZIO, Adalberto
Universidad de Sao Paulo,
Sao Paulo, Brazil;
fazzio@if.usp.br

(40) FLORES, Andrea
Escuela Politécnica Nacional,
Quito, Ecuador.

(41) GONZALEZ, Silvia
Universidad Técnica Particular,
Loja, Ecuador
sgonzalez@utpl.edu.ec

(42) GRANDA, Juan
Colegio Nacional 11 de Octubre,
Babahoyo, Los Ros, Ecuador.

(43) GUAÑO, Sonia
Universidad Politécnica Salesiana,
Quito, Ecuador;
sguano@ups.edu.ec

(44) GUARDERAS, Pedro
Escuela Politcnica Nacional,
Quito, Ecuador;
nampfgo@hotmail.com

(45) GUZMAN, Angela
Florida Atlantic University,
Miami, Florida, USA;
guzman@physics.fau.edu

(46) HALLBERG, Karen
Centro Atómico Bariloche,
Bariloche, Argentine;
karenhallberg@gmail.com

(47) HERNANDEZ, Guillermo
Universidad San Francisco de Quito,
Cumbayá, Ecuador;

ghernandez@usfq.edu.ec

(48) HERNANDEZ, Susana
Universidad de Buenos Aires,
Buenos Aires, Argentine;
shernand@df.uba.ar

(49) HUREL, Jorge
Escuela Superior Politcnica del
Litoral, Guayaquil, Ecuador;
jhurel@espol.edu.ec

(50) IZA, Peter
SENACYT, Quito, Ecuador;
peter.iza@senacyt.gov.ec

(51) JERVES, Alex
Universidad del Azuay,
Cuenca, Ecuador;
ajerves@uazuay.edu.ec

(52) KOINOV, Zlatko
University of Texas,
San Antonio, Texas, USA;
zlatko.koinov@utsa.edu

(53) KONNO, Rikio
Kinki University, 2800 Arima-cho
Kumano-shi, Mie 519-4395,
Japan; r-konno@ktc.ac.jp

(54) KROTSCHECK, Eckhard
Johannes Kepler University,
Linz, Austria;
kro@grizzly.tphys.uni-linz.ac.at

(55) LAGOS, Daniel
Escuela Politécnica Nacional,
Quito, Ecuador;
daniellagos@hotmail.com

(56) LEE, M. Howard
University of Georgia,
Athens, Georgia, USA;
mhlee@uga.edu

(57) LEMA, Jenny
EPR Travel, Quito, Ecuador;
info@eprtravel.com

(58) LUDEÑA, Eduardo V.
Centro de Qumica, IVIC,
Caracas, Venezuela;
popluabe@yahoo.es

(59) MALIK, F. Bary
University of Illinois at Carbondale,
Carbondale, Illinois, USA;
fbmalik@physics.siu.edu

(60) MAMEDOV, Tofik
Baskent University,
Ankara, Turkey;
tmamedov@baskent.edu.tr

(61) MENDEZ, Guillermo
Escuela Superior Politcnica del
Ejrcito, Quito, Ecuador.
guimencam@hotmail.com

(62) MONTENEGRO, Luis
Colegio Fiscal Tcnico Agr.;
Mocache, Los Ríos, Ecuador;
arturo-montenegro@yahoo.com

(63) MORALES, Leonardo
Pontificia Universidad Catlica,
Quito, Ecuador.

(64) MORALES, Eddie
Escuela Superior Politécnica del
Ejrcito, Quito, Ecuador.
raulmorales52@hotmail.com

(65) MORENO, Carlos
Escuela Superior Politécnica del
Litoral, Guayaquil, Ecuador;
cmoreno@espol.edu.ec

(66) MORETA, Alfonso
Privada de Construcciones,
Quito, Ecuador;
alfonsomoreta@yahoo.es

(67) NEILSON, David
University of Camerino,
Camerino, Italy;
david.neilson@unicam.it

(68) OCAMPO, Galo
SENACYT, Quito, Ecuador;
gocampo@senacyt.gov.ec

(69) OLA, Nelson
OLACORPORATION CEO,
Quito, Ecuador;
olacorp@email.com

(70) OÑA, David Fernando
Escuela Politécnica Nacional,
Quito, Ecuador;
dfoo788@hotmail.com

(71) OÑA, Jorge
Louisiana State University,
Department of Chemistry,
Baton Rouge, LA. USA;
jona1@lsu.edu

(72) ORTIZ, Vincent
Auburn University,
Auburn, Alabama, USA;
jvo0001@auburn.edu

(73) PATIÑO, Edgar Javier
Universidad de Los Andes,
Bogotá, Colombia;
epatino@uniandes.edu.co

(74) PEÑA, Cristina
BIOSAVIA, Quito, Ecuador;
crispena1@gmail.com

(75) PEREZ, Luis Antonio
Universidad Autnoma de Mxico,
Mexico City, Mexico;
lperez@fisica.unam.mx

(76) PINTO, Henry
Universidad Técnica Particular,
Loja, Ecuador;
pavlvs.pinto@gmail.com

(77) PLASTINO, Angelo
Universidad de La Plata,
La Plata, Argentine;
plastino@fisica.unlp.edu.ar

(78) PLAZA, Franklin
Escuela Politécnica Nacional,
Quito, Ecuador.

(79) PONCE, Daniel
Escuela Politécnica Nacional,
Quito, Ecuador;
draconix16@hotmail.com

(80) PROTO, Araceli N.
Universidad de Buenos Aires,
Buenos Aires,

Argentine; aproto@fi.uba.ar

(81) QUADER, Khandker
Kent State University,
Kent, Ohio, USA;
quader@kent.edu

(82) RANNINGER, Julius
Neel Institute, Grenoble, France;
julius.ranninger@grenoble.cnrs.fr

(83) RISTIG, Fred
University of Köln,
Köln, Germany;
ristig@thp.uni-koeln.de

(84) REINHOLZ, Heidi
University of Rostock,
Rostock, Germany;
heidi.reinholz@uni-rostock.de

(85) REINOSO, Miguel
Universidad Estatal de Milagro,
Milagro, Ecuador;
miguelreinoso@yahoo.es

(86) ROEPKE, Gerd
University of Rostock,
Rostock, Germany;
gerd.roepke@uni-rostock.de

(87) ROMAN, Vernica
Biogroups, Quito, Ecuador;
veronik_veronik@hotmail.com

(88) ROSERO, Franklin
ro_2001_ro@hotmail.com

(89) SANTISTEBAN, Ana
Universidad Vladimir Ilich Lenin,
Las Tunas, Cuba;
polett_0@hotmail.com

(90) SEVILLA, Francisco Javier
Universidad Autnoma de Mxico;
Mexico City, Mexico;
fjsevilla@fisica.unam.mx

(91) SILVERS, Mnica
NEOQUIM, Quito, Ecuador;
mxsilvers@hotmail.com

(92) SOBNACK, M. Binoy
Loughborough University,
Loughborough, United Kingdom;

m.b.sobnack@lboro.ac.uk

(93) SOLANO, Marco Vinicio
MEER, Quito, Ecuador;
marcosolano28@latinmail.com

(94) SOTOMAYOR, Ulbio
SENACYT, Quito, Ecuador;
usotomayor@senacyt.gov.ec

(95) TOBAR, Mara Antonieta
SENACYT, Quito, Ecuador;
mtobar@senacyt.gov.ec

(96) TORRES, Fernando Javier
Universidad San Francisco de Quito,
Cumbayá, Ecuador;
jtorres@usfq.edu.ec

(97) TORRES, Carlos
FENACA, Quito, Ecuador;
hc1iuf@hotmail.com

(98) TRUGMAN, Stuart
Los Alamos National Laboratoy,
Los Alamos, New Mexico, USA;
sat@lanl.gov

(99) UGALDE, Jesus M.
University of the Basque Country,
Donostia, Spain;
jesus.ugalde@ehu.es

(100) VALENCIA,Martha
Quito, Ecuador;
serprofin7@hotmail.com

(101) VALENCIA, Carlos
Empresa Metropolitana de Turismo,
Quito, Ecuador;
carelins@hotmail.com

(102) VALLEJO, Andrés
Escuela Politécnica Nacional,
Quito, Ecuador.

(103) VASHISHTA, Priya
University of South California,
Los Angeles, California, USA;
priyav@email.usc.edu

(104) VERA, Pilar
Colegio Fiscal 6 de Octubre,
Babahoyo, Los Ros, Ecuador;
pilarverav@yahoo.es

(105) VILLACRES, Juan Carlos
Escuela Politécnica Nacional,
Quito, Ecuador;
hombrepollo18@hotmail.com

(106) VIZCAINO, Gustavo
RELOGIS, Quito, Ecuador;
gustavovizca@yahoo.com

(107) WEXLER, Carlos
University of Missouri,
Columbia, Missouri, USA;
wexlerc@missouri.edu

(108) WIDOM, Michael
Carnegie Mellon University,
Pittsburgh, Pennsylvania, USA;
widom@andrew.cmu.edu

(109) YACELGA, Olga Cristina
DIRACSA, Quito, Ecuador;

(110) YAÑEZ, Estela
Universidad de Guayaquil,
Guayaquil, Ecuador;
estelita_del_rocioy@hotmail.com

(111) YAÑEZ, Alcbar
FAE, Quito, Ecuador;
alchi_y@hotmail.com

CONTENTS

Part A   Fermi and Bose Fluids, Exotic Systems

Part B   Quantum Magnets, Quantum Dynamics and Phase Transitions

Part C   Physics of Nanosystems and Nanotechnology

Part D   Quantum Information

Part E   Theory and Applications of Molecular Dynamics and Density Functional Theory

Part F   Superconductivity

Part G   Statistical Mechanics, Relativistic Quantum Mechanics

Part A
Fermi and Bose Fluids, Exotic Systems

REEMERGENCE OF THE COLLECTIVE MODE IN ³HE AND ELECTRON LAYERS

HELGA M. BÖHM*, ROBERT HOLLER, ECKHARD KROTSCHECK and
MARTIN PANHOLZER

*Institute of Theoretical Physics, Johannes Kepler University, Altenbergerstr. 69
A-4040 Linz, Austria*
* helga.boehm@jku.at

HENRY GODFRIN and MATHIAS MESCHKE

Institut Néel, CNRS et Université Joseph Fourier,
Grenoble Cedex 9, BP 166, F-38042 France

HANS-JOCHEN LAUTER

Institut Laue-Langevin,
Grenoble Cedex 9, BP 156, F-38042 France

Received 24 December 2009

Neutron scattering experiments on a ³He layer on graphite show an unexpected behavior of the collective mode. After having been broadened by Landau damping at intermediate wave vectors, the phonon-roton mode resharpens at large wave vectors and even emerges from the particle-hole continuum at low energies. The measured spectra cannot be explained by a random phase approximation with any static interaction. We show here that the data are well described if dynamic two-pair fluctuations are accounted for. We predict similar effects for electron layers.

Keywords: Fermi liquids; ³He; electron liquid; neutron scattering; dynamic correlations.

1. Motivation

1.1. *An experimentum crucis*

Understanding the excitations of quantum fluids has been a major goal of condensed matter physics for decades. For the helium liquids, a prime example of strongly correlated systems, Pines[1] argued long ago that the phonon-maxon-roton mode in ⁴He and ³He has a common physical origin in their strong and identical interatomic interaction. Quantum statistics, though quantitatively relevant, plays a less important role for the general features of the collective mode. The long wavelength plasmon in an electron liquid,[2,3] caused by a completely different interaction, is explained by the same theories that proved successful for the neutral systems.

There is, however, a fundamental difference in bosonic and fermionic spectra: whereas for bosons the collective mode remains well-defined over a wide range of

3

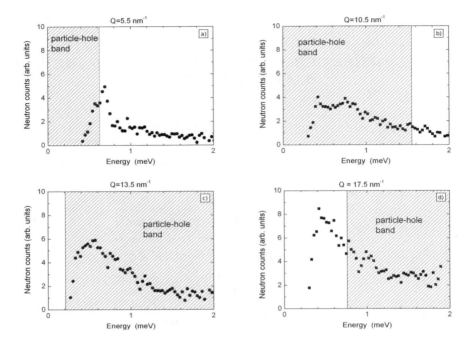

Fig. 1. Dynamic structure factor of two-dimensional ^{3}He, obtained from inelastic neutron scattering at 4 different wave vectors Q and an areal density of $4.9\,\mathrm{nm}^{-2}$. The shaded area is the particle-hole continuum. Note the strong peak below it in the lower right figure.

wave vectors q, it is rapidly damped when it enters the domain of incoherent single-particle-hole (PH) excitations ("Landau damping"). Now our refined measurements reveal a pronounced excitation at atomic wave vectors (Fig. 1). At the Institut Laue-Langevin (ILL) we determined the dynamic structure factor $S(q, E)$ of a mono layer of liquid ^{3}He ($E = \hbar\omega$ is the excitation energy). It is clearly seen that the collective mode, sharp at small q and damped at intermediate q (Fig. 1 a) and b)), recollects strength (Fig. 1c) and even re-emerges at the lower end of the PH continuum (Fig. 1d).

1.2. Effective static interactions

These unexpected findings put the existing theories to a severe test. In the long wavelength limit the common paradigm of quasi-particle excitations[2,4−5] works well. For the electron liquid it has led to the random phase approximation (RPA), which gives the density-density response function χ of a system as

$$\chi^{\mathrm{RPA}} = \frac{\chi^0(q, \omega)}{1 - v(q)\,\chi^0(q, \omega)}\ ,\tag{1}$$

where χ^0 is the Lindhard function and $v(q)$ is the Fourier transform of the inter-action. For many systems in nature (e.g. with hard-core or $1/r^3$ potentials) this transform does not exist. How the RPA can be extended to such strong interactions was shown in Ref. 6; it amounts to replacing the bare $v(q)$ in (1) by an appro-priately defined effective static interaction. Requiring consistency of $\Im m\,\chi$ and the static structure via the fluctuation-dissipation theorem unambiguously identifies this interaction as the "irreducible particle-hole interaction" $V_{\text{p-h}}$

$$\chi^{\text{cRPA}} = \frac{\chi^0(q,\omega)}{1 - V_{\text{p-h}}(q)\,\chi^0(q,\omega)} \;.\tag{2}$$

Different choices of static interactions (formulated via "local field corrections") were presented for the electron liquid,[3,7] based on different self–consistency require-ments. None of these, however, can provide an excitation spectrum *qualitatively* different from a low q collective mode, vanishing into the PH continuum.

1.3. *Effective mass*

Before entering the PH continuum, measured collective modes visibly deviate from the (c)RPA predictions based on (2). A common cure[8,9] is to introduce an effective mass m^*, replacing the bare m in the Lindhard function. Again, such an approach cannot give the newly observed re-emergent zero-sound mode. Furthermore, m^* in ^{3}He is strongly q-dependent with a peak around the Fermi vector k_{F} and additi-onally leads to a wrong density dependence of the mode.[10] We do *not* claim that correcting for effective mass effects is unimportant. Rather we here concentrate on clarifying the physical reason for the sharpening of the collective mode at high q.

2. Theory: Overview

2.1. *Ground state theory*

For strongly interacting fermions the variational Jastrow–Feenberg[11] *ansatz* has highly successfully described most ground state properties:

$$\left|\Psi_{\text{GS}}\right\rangle \equiv \left|\Psi_{\mathbf{0}}\right\rangle = \frac{1}{\sqrt{\mathcal{N}_{\text{GS}}}}\, e^{\frac{1}{2}\widehat{U}}\left|\Phi_{\mathbf{0}}\right\rangle \;.\tag{3}$$

Here, $\mathcal{N}_{\text{GS}}$ is the normalization, $\Phi_{\mathbf{0}}$ a Slater determinant and $\widehat{U}$ contains, in principle, correlation functions of arbitrarily high order n

$$\widehat{U} = \sum_i u^{(1)}(\mathbf{r}_i) + \sum_{i<j} u^{(2)}(\mathbf{r}_i,\mathbf{r}_j) + \sum_{i<j<k} u^{(3)}(\mathbf{r}_i,\mathbf{r}_j,\mathbf{r}_k) + \dots \;.\tag{4}$$

These functions are determined by *functionally minimizing* the ground state energy $E_{\text{GS}} \equiv H_{\mathbf{0},\mathbf{0}}$, thus providing a parameter-free and robust ab-initio theory. The essential physics contained in Eq. (3) is to intuitively account for core exclusion; in systems like ^{3}He, where no Fourier transform of the interaction exists, $\exp\{u^{(n)}\}$ is well-behaved. For the practical evaluation of the energy expectation value, Fermi hypernetted chain (FHNC) theory has the advantage of being consistent with the optimization procedure at any order n, summing ladder diagrams exactly and rings in a local approximation. This way both, short as well as long range correlations are very well described.[12]

If necessary, the nodal surface problem inherent in Eq. (3) can be overcome by using correlated basis function (CBF) theory.[11,12] Here the correlation operator

$$F_{\mathrm{GS}} \equiv \exp\{\tfrac{1}{2}\widehat{U}\} \tag{5}$$

is applied not to the ground state but to determinants describing free excitations

$$\left|\Psi_{\mathbf{m}}\right\rangle \equiv \tfrac{1}{\sqrt{N_{\mathbf{m}}}} F_{\mathrm{GS}} \left|\Phi_{\mathbf{m}}\right\rangle . \tag{6}$$

This creates a complete set of non-orthogonal functions. Conventional basis functions, *e.g.* plane waves for uniform systems with hard cores, require infinite summations of large (often divergent) terms: by contrast, CBF incorporates important aspects of the correlations right from the starting point.

The matrix elements of unity and of the Hamiltonian $\widehat{H}$ in the basis (6)

$$\begin{aligned} M_{\mathbf{m},\mathbf{m}'} &\equiv \left\langle\Psi_{\mathbf{m}}\middle|\Psi_{\mathbf{m}'}\right\rangle &\equiv \delta_{\mathbf{m},\mathbf{m}'} + N_{\mathbf{m},\mathbf{m}'} \\ H_{\mathbf{m},\mathbf{m}'} &\equiv \left\langle\Psi_{\mathbf{m}}\middle|\widehat{H}\middle|\Psi_{\mathbf{m}'}\right\rangle \equiv H'_{\mathbf{m},\mathbf{m}'} + H_{0,0} M_{\mathbf{m},\mathbf{m}'} \end{aligned} \tag{7}$$

constitute the essential building blocks of CBF ground state theory. Together with the matrix elements of the density operator $\widehat{\rho}(\mathbf{r})$

$$\rho_{\mathbf{m},\mathbf{m}'}(\mathbf{r}) \equiv \left\langle\Psi_{\mathbf{m}}\middle|\widehat{\rho}(\mathbf{r})\middle|\Psi_{\mathbf{m}'}\right\rangle \tag{8}$$

they are also key ingredients of our dynamic approach.

2.2. *Boson dynamics*

A natural generalization of the Jackson-Feenberg wave function to excited states is

$$\left|\Psi_t\right\rangle = \tfrac{1}{\sqrt{N_t}} e^{-\frac{i}{\hbar}t E_{\mathrm{GS}}} F_{\mathrm{GS}} e^{\frac{1}{2}\delta\widehat{U}(t)} \left|\Phi_{\mathbf{0}}\right\rangle , \tag{9}$$

where $\left|\Phi_{\mathbf{0}}\right\rangle$ is unity for bosons. The fluctuation operator

$$\delta\widehat{U}(t) = \sum_i \delta u^{(1)}(\mathbf{r}_i, t) + \sum_{i<j} \delta u^{(2)}(\mathbf{r}_i, \mathbf{r}_j, t) + \sum_{i<j<k} \delta u^{(3)}(\mathbf{r}_i, \mathbf{r}_j, \mathbf{r}_k, t) + \dots \tag{10}$$

is of a similar form as (4) but is now time dependent. Again, the $\delta u^{(n)}$ are determined by functional optimization, now based on the action principle corresponding to Schrödinger's equation.

Campbell *et al.*[13] investigated ^{4}He, including fluctuations up to the pair level (*i.e.*, $n=2$). A recent study[14] demonstrates i) that additional formal approximations to simplify the numerical treatment (the "uniform limit approximation" in Ref. 13) yield quite accurate results, and, ii) that pair fluctuations are highly relevant for correctly explaining large q dynamics. The state of the art for boson dynamics is to include triplet fluctuations,[15] which correct the dynamic pair excitations in a self-consistent manner.

2.3. *Fermion dynamics*

The logical extension of the formalism to fermions is to express $\delta\widehat{U}(t)$ of (10) by

$$\delta\widehat{U}(t) \;=\; \sum_{\mathbf{p}_1\mathbf{h}_1} \delta u^{(1)}_{\mathbf{p}_1\mathbf{h}_1}(t)\, a^\dagger_{\mathbf{p}_1} a_{\mathbf{h}_1} \;+\; \tfrac{1}{2} \sum_{\mathbf{p}_1\mathbf{h}_1\,\mathbf{p}_2\mathbf{h}_2} \delta u^{(2)}_{\mathbf{p}_1\mathbf{h}_1\,\mathbf{p}_2\mathbf{h}_2}(t)\, a^\dagger_{\mathbf{p}_1} a_{\mathbf{h}_1} a^\dagger_{\mathbf{p}_2} a_{\mathbf{h}_2} \;+\dots \quad (11)$$

with $\mathbf{p}_i$ and $\mathbf{h}_i$ denoting states which are occupied and unoccupied in the ground state ("*particles*" and "*holes*"); spin-dependencies are not explicitly spelled out. The sheer increase in the number of variables ($\delta u^{(2)}$ depends on 4 vectors for fermions compared to 2 for bosons), prevents a solution on the same level of sophistication.

So-called "local approximations" assume that a quantity depends only on the momentum-*transfer* $\mathbf{q}_i = \mathbf{p}_i - \mathbf{h}_i$ of each particle-hole pair. If applied to the fluctuation amplitudes $\delta u^{(n)}$, the approach is of the same complexity as the bosonic theory. Though necessary for some quantities to obtain numerical tractability, making this simplification for a specific $\delta u^{(n)}$ squeezes the corresponding n-pair continuum into a single mode, and cannot give a proper explanation of the data in Fig. (1).

2.4. *The cRPA*

Omitting all $n \geq 2$ fluctuations for bosons yields the Bijl-Feynman spectrum,[16,17] where the static structure factor $S(q)$ determines the collective mode dispersion

$$\chi^{\text{BF}}(q,\omega) \;=\; -\frac{\hbar^2 q^2/m}{\omega^2 - \varepsilon_q^2} \quad \text{with} \quad \varepsilon_q \equiv \frac{\hbar^2 q^2}{2m\,S(q)} \;. \qquad (12)$$

For fermions, not even the case $n=1$ can be solved analytically for χ. If exchange is neglected and the collective approximation is used for χ^0 (known as "plasmon pole approximation" in charged systems) the PH continuum shrinks into a single mode. This results again in the form (12). With the full particle-hole structure (but still neglecting exchange), Eq. (2), the cRPA, is obtained.[6]

For systems with a weak $v(q)$ the RPA is equivalent to time-dependent Hartree theory and the cRPA may also be interpreted as such, with an *effectively* weak interaction. Therefore our theory gives a systematic way to microscopically derive such interactions.[18,19]

3. Fermion Pair Fluctuation Theory

3.1. *Linear response theory*

For better clarity of the structure of the theory, we subsume the variables describing a particle-hole pair[a] into a single number, $i \equiv (\mathbf{p}_i, \mathbf{h}_i)$. The system is subject to a weak external perturbation $h^{\text{ext}}(\mathbf{r}, t)$; this implies small deviations of the wave function from the ground state

$$\left|\Psi_t\right\rangle \approx \left|\Psi_{\text{GS}}\right\rangle + \left|\delta\Psi_t\right\rangle + \dots \quad . \tag{13}$$

$\delta\Psi_t$ is obtained by expanding $\exp\{\delta\widehat{U}/2\}$ in (9) linearly in the fluctuations $\delta u^{(n)}$. This, in turn, gives the induced density $\delta\rho(\mathbf{r})$ as deviation from the ground state density ρ_{GS} in terms of CBF matrix elements:

$$\frac{1}{\mathcal{N}_t}\left\langle\Psi_t\middle|\widehat{\rho}(\mathbf{r})\middle|\Psi_t\right\rangle \approx \rho_{\text{GS}} + \delta\rho(\mathbf{r})$$

$$\delta\rho(\mathbf{r}) = \Re e\left\{\sum_1 \delta u_1^{(1)}\delta\rho_{0,1}(\mathbf{r}) + \frac{1}{2}\sum_{1,2}\delta u_{12}^{(2)}\delta\rho_{0,12}(\mathbf{r})\right\} \tag{14}$$

(we suppress normalization factors $\sqrt{\mathcal{N}_\mathbf{m}/\mathcal{N}_\mathbf{0}}$). The boson theory[14] guides us to simplify (14) this by transforming to modified fluctuations defined as

$$\delta\rho(\mathbf{r}) \equiv \Re e \sum_1 \delta v_1^{(1)}\delta\rho_{0,1}(\mathbf{r}) \quad . \tag{15}$$

Obviously, $\delta v^{(1)}$ implicitly sums two-pair correlations. The connection can be shown to involve the matrices of Eq. (7) via

$$\delta v_1^{(1)} = \delta u_1^{(1)} + \frac{1}{2}[M^{-1}]_{1,2}N_{2,34}\,\delta u_{43}^{(2)} \quad , \tag{16}$$

(doubly appearing indices being summed). $N_{2,34}$ involves the states $(\mathbf{p}_2\mathbf{h}_2, \mathbf{p}_3\mathbf{h}_3, \mathbf{p}_4\mathbf{h}_4)$ and thus 3-particle correlations. In the local approximation, this amounts to the (approximate) knowledge of the 3-particle ground state structure factor. The coefficient of $\delta u^{(2)}$ in (16) defines a new matrix

$$M_{1,23}^{\text{I}} \equiv \left(M^{-1}\cdot N\right)_{1,23} = [M^{-1}]_{1,4}\,N_{4,23} \quad . \tag{17}$$

Diagrammatically, $M_{1,23}^{\text{I}}$ is a proper subset of $M_{1,23} = N_{1,23}$. Similarly, certain diagrams involving 4-particle correlations are canceled from $M_{12,34}$ by introducing

---

[a]In $N_{\mathbf{m},\mathbf{m}'}$ each index stands for *all* quantum numbers; *e.g.*, in the case of a single particle-hole excitation for $\mathbf{m} = (\mathbf{p}_1, \mathbf{h}_2, \dots \mathbf{h}_N)$. Instead of spelling out the occupied states, we write $(\mathbf{p}_1, \mathbf{h}_1)$.

$$M^{\mathrm{I}}_{12,34} \equiv \left(M + M^{\mathrm{I}} \cdot M \cdot M^{\mathrm{I}}\right)_{12,34} \equiv M_{12,34} + M^{\mathrm{I}}_{12,5} M_{5,6} M^{\mathrm{I}}_{6,34} \ . \tag{18}$$

3.2. *Equations of motion*

Equations of motion (eom) follow straightforwardly from minimizing the action

$$\delta \int dt \, \frac{1}{\mathcal{N}_t} \left\langle \Psi_t \,\middle|\, \widehat{H} + \widehat{H}^{\mathrm{ext}} + \frac{\hbar \partial}{i \, \partial t} \,\middle|\, \Psi_t \right\rangle = 0 \quad ; \quad \widehat{H}^{\mathrm{ext}} = \int d^3 r \, h^{\mathrm{ext}}(\mathbf{r}, t) \, \widehat{\rho}(\mathbf{r}, t) \tag{19}$$

together with Eq. (13). Invoking the transformed fluctuation amplitude $\delta v^{(1)}$, the eom for the *two-pair* fluctuations $\delta u^{(2)}$ read

$$\begin{aligned}
\tfrac{1}{2}\left[-M^{\mathrm{I}}_{12,34} \frac{\hbar \partial}{i \, \partial t} - K_{12,34}\right] \delta u^{(2)}_{43} &\equiv \tfrac{1}{2}\left[E_{12,34}(t)\right] \delta u^{(2)}_{43} \\
&= \tfrac{1}{2} K_{1234,0} \, \delta u^{(2)\,*}_{43} + K_{12,3} \, \delta v^{(1)}_3 + K_{123,0} \, \delta v^{(1)*}_3 \ .
\end{aligned} \tag{20}$$

Here, the 4-pair coefficients are (written symbolically as well as explicitly)

$$\begin{aligned}
K_{\mathbf{m},\mathbf{m}'} &\equiv \left[H' - \left(M^{\mathrm{I}} \cdot H'\right) - \left(H' \cdot M^{\mathrm{I}}\right) + \left(H' \cdot M^{\mathrm{I}} \cdot H'\right)\right]_{\mathbf{m},\mathbf{m}'} \ , \\
K_{12,34} &\equiv H'_{12,34} - M^{\mathrm{I}}_{12,5} H'_{5,34} - H'_{12,5} M^{\mathrm{I}}_{5,34} + M^{\mathrm{I}}_{12,5} H'_{5,6} M^{\mathrm{I}}_{6,34} \ , \\
K_{1234,0} &\equiv H'_{1234,0} - M^{\mathrm{I}}_{12,5} H'_{534,0} - H'_{125,0} M^{\mathrm{I}}_{5,34} + M^{\mathrm{I}}_{12,5} H'_{56,0} M^{\mathrm{I}}_{6,34} \ ,
\end{aligned} \tag{21}$$

while the 3-pair coefficients take the simpler form

$$K_{\mathbf{m},\mathbf{m}'} \equiv \left[H' - \left(M^{\mathrm{I}} \cdot H'\right)\right]_{\mathbf{m},\mathbf{m}'} \ , \qquad \begin{aligned} K_{12,3} &\equiv H'_{12,3} - M^{\mathrm{I}}_{12,4} H'_{4,3} \ , \\ K_{123,0} &\equiv H'_{123,0} - M^{\mathrm{I}}_{12,4} H'_{43,0} \ . \end{aligned} \tag{22}$$

Formally solving Eq. (20) yields $\delta u^{(2)}$ and $\delta u^{(2)*}$ as functions of the 1-pair fluctuations. These are then inserted into the eom for the 1-pair amplitudes:

$$\begin{aligned}
\left[-M_{1,3} \frac{\hbar \partial}{i \, \partial t} - H'_{1,3}\right] \delta v^{(1)}_3 - H'_{13,0} \, \delta v^{(1)*}_3 &= 2 \int d^3 r \, \delta \rho_{1,0} \, h^{\mathrm{ext}}(\mathbf{r}, t) \\
&+ \tfrac{1}{2} K_{1,34} \, \delta u^{(2)}_{43} + \tfrac{1}{2} K_{134,0} \, \delta u^{(2)*}_{43} \ .
\end{aligned} \tag{23}$$

For practical purposes we neglect the static 4-body correlations $K_{1234,0} \approx 0$. Fourier transforming from time to frequency ω this implies for $u^{(2)}$

$$\begin{aligned}
\tfrac{1}{2} \delta u^{(2)} &= \left[E^{-1}(\omega)\right] \cdot \left\{ K \cdot \delta v^{(1)} + K \cdot \delta v^{(1)*} \right\} \ , \\
\tfrac{1}{2} \delta u^{(2)}_{12} &= \left[E^{-1}(\omega)\right]_{12,56} \left\{ K_{65,3} \, \delta v^{(1)}_3 + K_{653,0} \, \delta \delta v^{(1)*}_3 \right\} \ .
\end{aligned} \tag{24}$$

Inserting (24) on the r.h.s. of (23) gives an *effective single-particle equation*; therefore knowledge of the CBF matrices K and E^{-1} provides a 1:1 mapping onto *effective dynamic interactions*:

$$H'^{\text{eff}} \cdot \delta v^{(1)} = H' \cdot \delta v^{(1)} + \tfrac{1}{2} K \cdot \delta u^{(2)} = \left[H' + K \cdot [E^{-1}(\omega)] \cdot K \right] \cdot \delta v^{(1)}$$

$$H'^{\text{eff}}_{1,3} \delta v^{(1)}_3 = H'_{1,3} \delta v^{(1)}_3 + \tfrac{1}{2} K_{1,34} \delta u^{(2)}_{43} = \left[H'_{1,3} + K_{1,45} [E^{-1}(\omega)]_{54,67} K_{76,3} \right] \delta v^{(1)}_3 \tag{25}$$

(and the analogue for the $H'_{13,0}$ term).

4. Application of the Theory

4.1. *Approximations*

How exchange effects correct the cRPA, was recently studied[20] for bulk ^{3}He; their influence on multi-pair correlations is, presently, beyond numerical tractability within reasonable effort. We therefore decrease the number of variables by approximating $N_{\mathbf{m},\mathbf{m}'}$ and, consequently, M^{I} and K by their Fermi-sea average

$$N_{\mathbf{m},\mathbf{m}'} \;\rightarrow\; \frac{\sum_{\mathbf{h}_1 \ldots \mathbf{h}_{m'}} N_{\mathbf{m},\mathbf{m}'}}{\sum_{\mathbf{h}_1 \ldots \mathbf{h}_{m'}} n_{\overline{\mathbf{p}}_1} n_{\mathbf{h}_1} \cdots n_{\overline{\mathbf{p}}'_m} n_{\mathbf{h}'_{m'}}} \;. \tag{26}$$

Here, $n_{\mathbf{h}}$ denotes the Fermi distribution function and $n_{\overline{\mathbf{p}}} \equiv 1 - n_{\mathbf{p}}$. The kinetic energy being intrinsically non-local, we split off the diagonal parts of the Hamiltonian:

$$H'_{\mathbf{m},\mathbf{m}'} \equiv W_{\mathbf{m},\mathbf{m}'} + \frac{1}{2} \left(H'_{\mathbf{m},\mathbf{m}} + H'_{\mathbf{m}',\mathbf{m}'} \right) N_{\mathbf{m},\mathbf{m}'} \qquad (\mathbf{m} \neq \mathbf{m}') \;, \tag{27}$$

The $W_{\mathbf{m},\mathbf{m}'}$ can again be approximated locally by the procedure (26). The optimization of the ground state (3)-(4) ensures that the Fermi-sea average of $H'_{\mathbf{m},0} = 0$, relating the local approximation of $W_{\mathbf{m},\mathbf{m}'}$ uniquely to that of $N_{\mathbf{m},\mathbf{m}'}$.

Denoting the Fourier transform of $\rho(\mathbf{r})$ with $\rho_{\mathbf{q}}$, the static structure factor $S(q)$ is expressed by its free value S^0_q and CBF matrix elements via

$$S(q) = \frac{\langle \Psi_{\text{GS}} | \widehat{\delta\rho}_{\mathbf{q}} \, \widehat{\delta\rho}_{-\mathbf{q}} | \Psi_{\text{GS}} \rangle}{N \langle \Psi_{\text{GS}} | \Psi_{\text{GS}} \rangle} = S^0_q + \frac{1}{N} \sum_{\mathbf{h} \neq \mathbf{h}'} N_{\mathbf{h}+\mathbf{q}\,\mathbf{h}, \, \mathbf{h}'+\mathbf{q}\,\mathbf{h}'} \;. \tag{28}$$

This uniquely determines the local approximation of $N_{1,2}$. Similarly, $N_{1,23}$ is related to the 3-particle ground state structure factor, leading finally to

$$M^{\mathrm{I}}_{1,23} \;\rightarrow\; \frac{S^{(3)}(\mathbf{q}_1, \mathbf{q}_2, \mathbf{q}_3)}{S(q_1) \, S^0_{q_2} \, S^0_{q_3}} - \frac{S^{(3)\,0}_{\mathbf{q}_1,\mathbf{q}_2,\mathbf{q}_3}}{S^0_{q_1} \, S^0_{q_2} \, S^0_{q_3}} \tag{29}$$

Using the uniform limit approximation,[13] $S^{(3)}$ factorizes this into products of $S(q_i)$. The 4-particle $M_{12,34}$ in (18) factorizes, too (neglecting terms of $\mathcal{O}(\frac{1}{N})$). Therefore, the only input for the dynamics is the ground state $S(q)$.

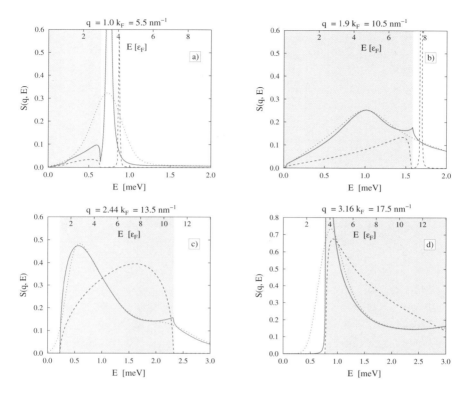

Fig. 2. (Color on line) Comparison of the dynamic structure factor of ^{3}He, as obtained from the cRPA, Eq. (2), (dashed blue lines) and from the pair fluctuation theory (full red lines). The density and wave vectors q match those of Fig. 1. The dotted curves show the theoretical results convoluted with the experimental resolution.

4.2. ^3He

In the experiment[21] a mono layer of liquid ^{3}He was adsorbed on high quality exfoliated graphite, preplated by a mono layer of solid ^{4}He. The latter has the advantage of a weaker adsorption potential than the bare graphite; in addition, it smoothes out surface defects. At temperatures well below 1K the motion of the ^{3}He fluid is two-dimensional. It thus forms an atomically thick layer of known areal density. For the case reported here this is 4.9 atoms/nm^2.

On this layer we performed inelastic neutron scattering experiments at the Institut Laue Langevin (ILL) on the time-of-flight neutron spectrometer IN6. The accessible momentum transfers range from wave vectors $q = 2.54\,\mathrm{nm}^{-1} \ldots 20.46\,\mathrm{nm}^{-1}$. The energy resolution is of the order of 0.1 meV.

For our calculations we use the $S(q)$ obtained from Fermi hypernetted chain theory[12,22] as input. The results are shown in Fig. (2), for the same wave vectors and areal density as in the ILL experiment. Clearly, there is a large disagreement between the cRPA prediction (dashed lines) and the one including dynamic pair

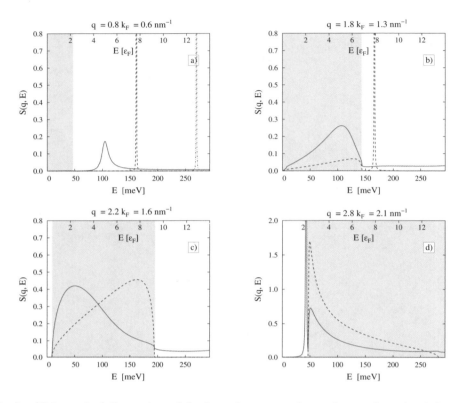

Fig. 3.   (Color on line) Comparison of the dynamic structure factor of a two-dimensional electron liquid at $r_s = 36$, in the cRPA, Eq. (2), (dashed blue lines) and from the pair fluctuation theory (full red lines). In the upper left graph also the bare RPA is seen (chained magenta line).

fluctuations (full lines), the spectra being *qualitatively* different. The dotted curves are obtained by convoluting the theoretical results with the experimental resolution of 0.1meV and 1nm^{-1}. This reproduces the main features of the experiment very well: a sharp mode above the PH continuum for q close to the Fermi momentum k_{F} (Fig. 2a), a very broad mode at intermediate q-values (Fig. 2b,c), and a resharpening of the mode at $3k_{\mathrm{F}}$ (Fig. 2d).

The main effect of dynamic pair fluctuations is to shift strength towards lower energies; this is also true in the region of large Landau damping as is clearly seen in Fig. 2c. This type of qualitative change in $S(q, E)$ cannot be described by a static effective interaction. For the given areal density of $4.9\,\mathrm{nm}^{-2}$ the phonon-roton curve does not emerge from the PH band. However, for higher densities a maximum in $S(q, E)$ *below* the PH band is obtained, inaccessible to an RPA description.

4.3. Electron layers

For two-dimensional (2D) electrons the first prediction of a collective mode traversing the PH band was made by Neilson *et al.*[23] from quantum kinetic equations. Dynamic correlations were found important for density parameters[2,3] r_s larger than 10. We here apply our theory to a 2D electron gas at $r_s = 36$, close to Wigner crystallization. In analogy to ^{3}He we study the paramagnetic case.

For a 2D electron gas on graphite this means an areal density of the order of $10^{-1}\,\mathrm{nm}^{-2}$. Other realizations are electron layers in MOSFETS and/or semiconductor hetero-structures. In a GaAlAs-GaAs-GaAlAs quantum well, with a background dielectric constant $\epsilon_b = 12$ and an effective mass of 0.067 electron masses, the effective Bohr radius is roughly 100Å; $r_s = 36$ corresponds to $n \approx 10^{-2}\,\mathrm{nm}^{-2}$.

We obtained the input $S(q)$ for our calculations from Monte Carlo results by Gori-Giorgi *et al.*[24] Figure (3) shows that the effect of the dynamic pair fluctuations is the same as in ^{3}He: At small q the plasmon is sharp, though two-pair excitations cause a finite width while the cRPA plasmon is undamped. The one-pair excitations account for a large portion of the decrease of the dispersion in comparison to the bare RPA (Fig. 3a). In Fig. 3b, at intermediate q values, we find a broad and highly Landau-damped plasmon, in marked contrast with a sharp cRPA peak (the bare RPA plasmon is at a higher energy outside the range displayed). As in ^{3}He, with a further increase of q the spectrum "leans towards the left" (Fig. 3c) and, finally, resharpens on the lower side of the PH continuum in Fig. 3d. This recollection of strength is thus independent of the inter-particle interaction, the same effect arises for the long-ranged, soft-core Coulomb potential and the short-ranged, hard-core potential of helium. A close inspection of the data shows that the plasmon, indeed, re-emerges from the continuum, a feature that is beyond a cRPA approach with any effective static interaction.

5. Summary

We showed that variational dynamic quantum many body theory based on optimizing time-dependent fluctuation amplitudes provides a powerful tool for explaining the excitations of two-dimensional ^{3}He. For a full quantitative agreement with the experimental results an inclusion of triplet fluctuations, exchange effects, or effective mass corrections may be required. The fermionic case being much more demanding than its bosonic counterpart, we concentrated our studies on the inclusion of two-pair fluctuations and could demonstrate that these give a major improvement over RPA-like approaches. The spectra obtained with pair fluctuations agree well with the scattering data on ^{3}He and qualitatively differ from any RPA result, both for helium and for electrons.

Our approach satisfies the ω^0 and ω^1 sum-rules and holds the potential for a systematic improvement. Aside from significantly changing the collective mode it also describes multi-pair damping and double-phonon / double-plasmon excitations.[25]

Acknowledgments

This work was supported by the Austrian research fund FWF (project P21264) and under the Partenariat Hubert Curien "Amadeus", as well as by the ILL neutron center and the EU FRP7 low temperature infrastructure (Grant 228464).

References

1. D. Pines, *Physics Today* **34**, 106 (Nov. 1981).
2. D. Pines and P. Nozieres, *The Theory of Quantum Liquids* (Benjamin, New York, 1966).
3. G. Giuliani and G. Vignale, *Quantum Theory of the Electron Liquid* (Cambridge University Press, Cambridge, 2005).
4. L. D. Landau, *Sov. Phys. JETP* **3**, p. 920 (1957).
5. H. Glyde, *Excitations in liquid and solid helium* (Oxford University Press, Oxford, 1994).
6. J. M. C. Chen, J. W. Clark and D. G. Sandler, *Z. Physik A* **305**, 223 (1982).
7. K. S. Singwi and M. P. Tosi, *Solid State Phys.* **36**, 177 (1981).
8. H. R. Glyde, B. Fåk, N. H. van Dijk, H. Godfrin, K. Guckelsberger and R. Scherm, *Phys. Rev. B* **61**, 1421 (2000).
9. M. M. Calbi and E. S. Hernández, *J. Low Temp. Phys.* **120**, 1 (2000).
10. J. Boronat, J. Casulleras, V. Grau, E. Krotscheck and J. Springer, *Phys. Rev. Lett.* **91**, 085302 (2003).
11. E. Feenberg, *Theory of Quantum Fluids* (Academic, New York, 1969).
12. E. Krotscheck, Theory of correlated basis functions, in *Introduction to Modern Methods of Quantum Many–Body Theory and their Applications*, eds. A. Fabrocini, S. Fantoni and E. Krotscheck, Advances in Quantum Many–Body Theory, Vol. 7 (World Scientific, Singapore, 2002) pp. 267–330.
13. C. C. Chang and C. E. Campbell, *Phys. Rev. B* **13**, 3779 (1976).
14. C. E. Campbell and E. Krotscheck, *Phys. Rev. B* **80**, 174501 (2009).
15. C. E. Campbell and E. Krotscheck, *to be published* (2010).
16. R. P. Feynman, *Phys. Rev.* **94**, 262 (1954).
17. A. Bijl, *Physica* **7**, 869 (1940).
18. E. Krotscheck and J. W. Clark, *Nucl. Phys. A* **328**, 73 (1979).
19. E. Krotscheck, *Phys. Rev. A* **26**, 3536 (1982).
20. M. Panholzer, H. M. Böhm, R. Holler and E. Krotscheck, *J. Low Temp. Phys.* **158**, 135 (2009).
21. H. Godfrin, M. Meschke, H. Lauter, H. M. Böhm, E. Krotscheck and M. Panholzer, *J. Low Temp. Phys.* **158**, 147 (2009).
22. E. Krotscheck, *J. Low Temp. Phys.* **119**, 103 (2000).
23. D. Neilson, L. Swierkowski, A. Sjölander and J. Szymanski, *Phys. Rev. B* **44**, 6291 (1991).
24. P. Gori-Giorgi, S. Moroni and G. Bachelet, *Phys. Rev. B* **70**, 115102/1 (2004).
25. H. M. Böhm, R. Holler, E. Krotscheck and M. Panholzer, *Int. J. Mod. Phys. B* **22**, 4655 (2008).

DISSECTING AND TESTING COLLECTIVE AND TOPOLOGICAL SCENARIOS FOR THE QUANTUM CRITICAL POINT

J. W. CLARK

Department of Physics & McDonnell Center for the Space Sciences, Washington University
St. Louis, Missouri 63130, U.S.A
jwc@wustl.edu

V. A. KHODEL*

Department of Physics & McDonnell Center for the Space Sciences, Washington University
St. Louis, Missouri 63130, U.S.A
vak@wuphys.wustl.edu

M. V. ZVEREV

Russian Research Centre Kurchatov Institute
Moscow 123182, Russia
zverev@pretty.mbslab.kiae.ru

Received 9 December 2009

In a number of strongly-interacting Fermi systems, the existence of a quantum critical point (QCP) is signaled by a divergent density of states and effective mass at zero temperature. Competing scenarios and corresponding mechanisms for the QCP are contrasted and analyzed. The conventional scenario invokes critical fluctuations of a collective mode in the close vicinity of a second-order phase transition and attributes divergence of the effective mass to a coincident vanishing of the quasiparticle pole strength. It is argued that this *collective scenario* is disfavored by certain experimental observations as well as theoretical inconsistencies, including violation of conservation laws applicable in the strongly interacting medium. An alternative *topological scenario* for the QCP is developed self-consistently within the general framework of Landau quasiparticle theory. In this scenario, the topology of the Fermi surface is transfigured when the quasiparticle group velocity vanishes at the QCP, yet the quasiparticle picture remains meaningful and no symmetry is broken. The topological scenario is found to explain the non-Fermi-liquid behavior observed experimentally in Yb-based heavy-fermion systems close to the QCP. This study suggests that integration of the topological scenario with the theory of second-order, symmetry-breaking quantum phase transitions will furnish a proper foundation for theoretical understanding of the extended QCP region.

Keywords: Strongly correlated fermions; quantum critical point; Fermi Liquid Theory; Landau quasiparticles; phase transitions.

---

*Permanent address: Russian Research Centre Kurchatov Institute, Moscow 123182, Russia.

1. Introduction: Quantum Phase Transitions and Non-Fermi-Liquid Behavior

The quest for a fundamental understanding of non-Fermi-liquid (NFL) behavior of Fermi systems in the vicinity of quantum phase transitions[1] persists as one of the most challenging objectives of modern condensed-matter research. In a number of heavy-fermion materials studied at very low temperatures T, e.g., Yb-based compounds[2] or the high-T_c superconductors $CeCoIn_5$ (Ref. 3) and $YBa_2Cu_3O_{6+x}$ (Ref. 4), critical temperatures $T_N(B)$ of these transitions can be driven to zero by imposition of weak magnetic fields B, creating a quantum critical point (QCP).

Divergence of the effective mass M^* as $T \to 0$ is now recognized as a pivotal experimental signature of QCP criticality. Evidence for this behavior comes from measurements of thermodynamic properties such as specific heat and magnetic susceptibility, as well as studies of de Haas-van Alfven and Shubnikov-de Haas magnetic oscillations, in diverse types of strongly correlated Fermi systems. These include heavy-fermion metals,[2,3,4] the 2D electron gas in MOSFETs,[5], and quasi-two-dimensional (2D) liquid ^{3}He,[6,7,8]

Landau's quasiparticle (qp) picture of low-T phenomena in Fermi systems,[9] epitomized in Fermi Liquid Theory (FLT), has long been a cornerstone of condensed matter physics.[10,11] The clear violations of FLT that have been revealed experimentally in the aforementioned many-fermion systems suggest that the electronic correlations in these systems become so strong that the quasiparticle concept loses validity at the QCP, at least in the version represented by standard FLT. However, we propose that although this canonical interpretation of Landau's quasiparticle pattern is no longer applicable beyond the QCP, one may still seek an explanation of NFL behavior within the *original* quasiparticle framework. Theoretical and experimental evidence will be presented to substantiate this position, in terms of a *topological scenario* for the QCP. In this view, it is the single-particle degrees of freedom, rather than critical collective fluctuations, that are responsible for the divergence of the effective mass at the QCP.

2. Competing Scenarios for the Quantum Critical Point

Working within the framework of microscopic quantum many-body theory,[12,10] let us first consider what conditions, at temperature $T = 0$, may give rise to a divergence of the effective mass (and hence the density of single-particle states) at the Fermi surface of an interacting many-fermion system. For simplicity, we restrict the analysis to a homogeneous, isotropic nonsuperfluid system. (The assumptions of homogeneity and isotropy will subsequently be lifted.) Is is assumed, in agreement with experiment for the relevant systems identified in Sec. 1, that the system obeys standard Fermi-liquid (FL) theory below the QCP, i.e, on the weakly correlated side of the phase diagram. (This is commonly regarded as the "disordered side" of a disorder-order phase diagram, although the actual situation may be more subtle.)

In the FL domain upon approach to the QCP, the effective mass M^* is inversely proportional to a product of two factors, one governed by the momentum dependence of the self-energy $\Sigma(p, \varepsilon)$ and the other by its energy dependence. Specifically,

$$\frac{M}{M^*} = z \left[1 + \frac{1}{v_F^0} \left(\frac{\partial \Sigma(p, \varepsilon)}{\partial p} \right)_0 \right], \tag{1}$$

where $v_F^0 = p_F/M$,

$$z = \left[1 - \left(\frac{\partial \Sigma(p, \varepsilon)}{\partial \varepsilon} \right)_0 \right]^{-1}, \tag{2}$$

and the derivatives are evaluated at $p = p_F$ and $\varepsilon = 0$, measuring energies from the chemical potential. Here, z represents the quasiparticle strength in single-particle (sp) states at the Fermi surface. Conventional belief, traced back over four decades to Doniach and Engelsberg,[13] has it that the divergence of M^* at the QCP is due to vanishing of the z factor in Eq. (1), rather than to the other factor, which is responsible for the renormalization of the Fermi velocity v_F^0 stemming from the momentum-dependence of the interactions between quasiparticles. We may note here that the renormalized group velocity $v_F = v(p_F) = [d\epsilon(p)/dp]_0$ is proportional to $1 + [\partial \Sigma(p, \varepsilon = 0)/\partial \epsilon_p^0]_0$.

Doniach and Engelsberg analyzed the role of critical spin fluctuations in the vicinity of ferromagnetic phase transitions based on the Ornstein-Zernike approximation, which neglects the scattering of critical fluctuations. They concluded that the z factor vanishes at the QCP, thus attributing the divergence of M^* to divergence of the derivative $[\partial \Sigma(p, \varepsilon; \rho_c)/\partial \varepsilon]_0$ at implicated second-order phase transition points ρ_c. This conclusion has become a maxim of the conventional scenario for the QCP, the essence of which is expressed as "quasiparticles get heavy and die at the quantum critical point."[14]

Thus, in the conventional or *collective scenario* for the QCP, the energy dependence plays the decisive role, causing the factor z in Eq. (1) to vanish exactly at the QCP. There is, however, another possibility, namely that the square-bracket factor on the right side of this equation vanishes instead. This is equivalent to the vanishing of the quasiparticle group velocity v_F at the QCP. In this case, it is the momentum dependence of the self-energy that is decisive. Whereas collective degrees of freedom drive the phase transition at the QCP in the conventional scenario, it is a rearrangement of the single-particle degrees that is the dominant mechanism in this alternative possibility. In this alternative view, the quantum critical point reflects a *topological phase transition*.[15] Volovik[16] has provided an authoritative classification and analysis of such topological transitions in quantum many-body systems.

3. The Original Quasiparticle Picture

To continue with the analysis, we need to recall the essential ingredients of the *original* Landau description[9] of interacting Fermi systems. It is founded on the postulate that there exists a one-to-one correspondence between the totality of real, decaying single-particle excitations of the actual Fermi system and a system of 'immortal' interacting quasiparticles. Accordingly, the number of quasiparticles (qp) equals the number of physical particles, which implies that

$$2 \int n(p) dv_{\mathbf{p}} = \rho, \tag{3}$$

where $n(p)$ is the qp momentum distribution, $dv_{\mathbf{p}}$ is an element of momentum space, and ρ is the particle density. Further, the entropy of the qp system, given by

$$S(T) = -2 \int \left[n(p, T) \ln n(p, T) + (1 - n(p, T)) \ln(1 - n(p, T)) \right] dv_{\mathbf{p}}, \tag{4}$$

is equal to the entropy of the physical system. The ground-state energy is expressible as a functional $E[n]$ of the qp distribution $n(p)$, and the qp energy $\epsilon(p)$, refereed to the chemical potential μ, is identified as

$$\epsilon(p) = \frac{\delta E[n(p)]}{\delta n(p)} - \mu. \tag{5}$$

From these relations there follows a formula of Fermi-Dirac form connecting the qp momentum distribution $n(p)$ with the qp energy $\epsilon(p)$ at temperature T:

$$n(p, T) = \left[1 + e^{\epsilon(p,T)/T} \right]^{-1}. \tag{6}$$

Galilean invariance provides a second relation between $n(p)$ and $\epsilon(p)$:

$$\frac{\partial \epsilon(\mathbf{p})}{\partial \mathbf{p}} = \mathbf{v}(\mathbf{p}) = \frac{\mathbf{p}}{M} + \int f(\mathbf{p}, \mathbf{p_1}) \frac{\partial n(\mathbf{p_1})}{\partial \mathbf{p_1}} dv_{\mathbf{p_1}}, \tag{7}$$

where the qp interaction function f is the product of z^2 and the scalar part of the scattering amplitude Γ^ω, the ω limit of the scattering amplitude Γ of two particles on the Fermi surface, with four-momentum transfer approaching zero such that $k/\omega \to 0$.

Standard Fermi Liquid Theory presupposes further that at $T = 0$ solutions of the Galilean invariance equation (7) always arrange themselves to guarantee a positive group velocity $\mathbf{v}(p) = \partial \epsilon(p)/\partial p$, which in turn implies that $n(p, T = 0)$ takes the Fermi-step form $n_F(p) = \theta(p_F - p)$. The properties of a "Fermi Liquid" in this sense turn out to be those of a Fermi *gas* of quasiparticles.

If one pursues the topological scenario for the QCP, the situation becomes interesting when the qp group velocity v_F changes sign and the standard FL presupposition fails. Setting $T = 0$ and $p = p_F$ in the Galilean relation and introducing the Legendre-harmonic Landau parameter $F_1^0 = p_F M f_1/\pi^2$, we find $v_F = (p_F/M)(1 - F_1^0/3)$. Accordingly, the inequality $v_F > 0$ is violated when F_1^0 reaches 3 at a density ρ_∞ that is identified with the QCP. It should be emphasized that this condition is quite distinct from the Pomeranchuk instability conditions.

For later reference, we rewrite the criticality condition $1 - F_1^0/3 = 0$ in terms of the k-limit of the dimensionless scattering amplitude $\nu \Gamma^k = A + B \sigma_1 \cdot \sigma_2$, where $\nu = z^2 p_F M^*/\pi^2$ is the quasiparticle density of states. (The k-limit of vanishing four-momentum transfer is such that $\omega/k \to 0$.) Noting the connection

$A_1 = F_1/(1 + F_1/3)$ between the Legendre harmonic A_1 of the scattering amplitude and the Landau parameter $F_1 = p_F M^* f_1 / \pi^2$, we have $M/M^* = 1 - A_1/3$ and the criticality condition becomes $A_1(\rho_\infty) = F_1^0(\rho_\infty) = 3$, where ρ_∞ is the density at which the effective mass diverges.

4. Fault Lines of the Conventional Collective Scenario

It will be established in this section that a QCP scenario based purely on collective fluctuations is vulnerable, both theoretically and experimentally.

Theoretical weaknesses may be exposed through several lines of argument. Consider that in the conventional scenario, the z factor vanishes at a critical density ρ_c where a collective mode collapses at a characteristic finite wave number k_c. By assumption, the quasiparticle (qp) picture of FL theory remains valid prior to the collapse. The requirement of antisymmetry of the qp amplitude vertex Γ with respect to interchange of the momenta and spins of the colliding particles[10,11] leads to the relation

$$A(\mathbf{p}_1, \mathbf{p}_2, \mathbf{k}, \omega = 0; \rho \to \rho_c) = -D(\mathbf{k}) + \frac{1}{2} D(\mathbf{p}_1 - \mathbf{p}_2 + \mathbf{k}) \tag{8}$$

in terms of the propagator

$$D(k \to k_c, \omega = 0; \rho \to \rho_c) = \frac{g}{\xi^{-2}(\rho) + (k - k_c)^2}, \tag{9}$$

with $g > 0$ and the correlation length $\xi(\rho)$ divergent at $\rho = \rho_c$. Now, with $\theta = \angle(\mathbf{p}_1, \mathbf{p}_2)$, we calculate harmonics $A_k(\rho)$ of the amplitude $A(p_1 = p_F, p_2 = p_F, \cos \theta)$ to obtain

$$A_0(\rho \to \rho_c) = g \frac{\pi}{2} \frac{k_c \xi(\rho)}{p_F^2}, \qquad A_1(\rho \to \rho_c) = g \frac{3\pi}{2} \frac{k_c \xi(\rho)}{p_F^2} \cos \theta_0. \tag{10}$$

The sign of $A_1(\rho \to \rho_c)$ coincides with that of $\cos \theta_0 = 1 - k_c^2 / 2p_F^2$. *First,* suppose that $\cos \theta_0 < 0$, i.e., $k_c > p_F \sqrt{2}$. Referring to $M/M^* = 1 - A_1/3$, we see that at the second-order phase transition, the ratio $M^*(\rho_c)/M$ must be less than unity. In this case one is forced to conclude that the critical density ρ_c and the density ρ_∞ where the effective mass diverges *cannot coincide.* The z-factor in Eq. (1) *does* indeed vanish at the density ρ_c due to the divergence of the derivative $[\partial \Sigma(p, \varepsilon)/\partial \varepsilon]_0$. However, the effective mass M^* remains *finite,* since the derivative $[\partial \Sigma(p, \varepsilon)/\partial \epsilon_p^0]_0$ also diverges at the QCP.[17]

If instead $1 > \cos \theta_0 > 0$, the conventional view again leads to an unacceptable situation. This is seen as follows. The harmonics $A_0(\rho_c)$ and $A_1(\rho_c)$ are related by $A_0(\rho_c) = A_1(\rho_c)/3 \cos \theta_0$. If $M^*(\rho_c)$ were infinite, then $A_1(\rho_c)$ would equal 3, while $A_0(\rho_c) = 1/\cos \theta_0 > 1$. However, the basic FL connection $A_0 = F_0/(1 + F_0)$ implies $A_0 \leq 1$, provided the Landau state is stable. Therefore in the conventional scenario, the QCP cannot be reached without violating the stability conditions for the Landau state.

The same reasoning applies for critical spin fluctuations with nonvanishing critical wave number.

Next consider that in the collective scenario, the standard derivation leading to divergence of the derivative $[\partial\Sigma(p,\varepsilon)/\partial\varepsilon]_0$ (and hence $z = 0$) is based on the fundamental relation

$$\frac{\partial\Sigma_{\alpha\delta}(p,\varepsilon)}{\partial\varepsilon} = -\frac{1}{2}\int U_{\alpha\delta\gamma\beta}(\mathbf{p},\varepsilon,\mathbf{l},\varepsilon_1)\frac{\partial G_{\beta\gamma}(l,\varepsilon_1)}{\partial\varepsilon_1}\frac{d\mathbf{l}d\varepsilon_1}{(2\pi)^4 i}. \tag{11}$$

The derivation retains only the pole part of the sp Green's function G and a singular part of the block U of scattering diagrams irreducible in the longitudinal particle-hole channel. A vulnerable feature of the derivation is its reliance on the simple Ornstein-Zernike (OZ) form

$$\chi(q, q_0 = 0) \approx \chi_{OZ}(q) \equiv \frac{4\pi}{q^2 + \xi^{-2}} \tag{12}$$

for the spin susceptibility. We draw attention to the fact that the OZ approximation may be tested by alternative evaluation of the derivative $[\partial\Sigma(p,\varepsilon)/\partial\epsilon]_0$ based on a set of relations of the same generic form as the fundamental identity (11), but following from different conservation laws (on momentum, spin, etc.) that may apply in the fermionic medium. In so doing,[18] inconsistent results are found for the required derivative of the self-energy; in particular, the signs and prefactors of the divergences vary. Since the nonsingular components of the vertex-blocks and self-energies neglected in the common derivations of the property $z = 0$ cannot compensate for the divergent discrepancies, this assertion becomes suspect and the status of the collective scenario for the QCP is questionable.

We turn now to confrontation of the conventional scenario with experimental measurements. It is true that this picture is supported to some extent by the presense of a logarithmic correction $\delta\gamma_{FL}(T) \propto \ln(1/T)$ to the FL Sommerfeld ratio $\gamma_{FL}(T) = \text{const.}$, observed in a number of heavy-fermion metals. If indeed $z(\rho_c)$ vanishes at the critical density ρ_c, the mass operator $\Sigma(\varepsilon, \rho_c)$ then contains a marginal term $\sim \varepsilon\ln\varepsilon$ that produces logarithmic corrections to the Sommerfeld ratio.[14] However, this implication cannot serve as unambiguous evidence for a vanishing z factor and breakdown of the quasiparticle picture at the QCP, since the same logarithmic corrections to the Sommerfeld ratio arise when spin-fluctuation contributions to the group velocity are included, even if these fluctuations are modest. Moreover, as T goes to zero, the logarithmic behavior of the experimental QCP Sommerfeld ratio fades away, being replaced by an upturn $\gamma(T \to 0) \propto T^{-\alpha}$ that is reminiscent of the critical behavior of thermodynamic quantities in the scaling theory of second-order phase transitions.[1,19,20]

A number of other experimental findings appear to contradict the conventional scenario for the QCP. For example, measurements[21] of the de Haas-van Alfven effect in the heavy-fermion metal CeRhIn$_3$ support the validity of the quasiparticle picture on both sides of the QCP. Furthermore, comprehensive experimental studies in the regime of a $T = 0$ ferromagnetic phase transition in heavy-fermion compounds

CePd$_{1-x}$Rh$_x$ have given evidence for the absence of a QCP [140] – in conflict with a basic tenet of the conventional scenario.

5. Generic Features of the Topological Scenario

With the condition $z(\rho_c) \equiv z(\rho_\infty) = 0$ of the collective scenario ruled out, the effective mass in Eq. (1) can only diverge at a density ρ_∞ where the factor in square brackets (or equivalently the group velocity v$(p = p_F)$) turns negative. In this topological scenario, the QCP is associated with a rearrangement of single-particle degrees of freedom induced by the interactions between quasiparticles. No collective parameter is involved, and no symmetry of the ground state is not violated.

In essence, a topological phase transition is associated with a change in the number of roots of the equation

$$\epsilon(p; n_F, \rho) = 0, \tag{13}$$

where, as usual, the quasiparticle energy ϵ is measured from the chemical potential μ. In standard homogeneous Fermi liquids such as liquid ^{3}He, this equation has a single root at the Fermi momentum p_F. The Fermi surface is singly connected, and the spectrum $\epsilon(p)$ is evaluated with the usual Fermi step $n_F(p) = \theta(p_F - p)$. As the system passes a bifurcation point corresponding to the topological phase transition, the number of roots of Eq. (13) ordinarily increases by two, such that the Fermi surface contains three sheets. The number of sheets grows as the control parameter (e.g., the density or coupling constant) moves further beyond the critical value. However, another more exotic possibility was uncovered about twenty years ago.[22,23,24] It can also happen that beyond the critical point, the number of sheets becomes *uncountably infinite* and in 3D the Fermi surface "swells" into a volume (or the Fermi line "spreads" into a surface, in a 2D system).

In contrast to second-order phase transitions associated with violation of Pomeranchuk stability conditions, topological transitions are induced by violation of a *necessary* condition for stability of the conventional Landau ground state, corresponding to the quasiparticle momentum distribution $n_F(p)$. The operative stability condition reads

$$\delta E_0 = \int \epsilon(p; n_F) \delta n(p) dv_{\mathbf{p}} > 0. \tag{14}$$

For stability of the Landau state, the variation of its ground-state energy E_0 must be positive for any admissible incremental variation from $n_F(p)$. Also in contrast to the Pomeranchuk stability conditions, the relation (14) is seen to involve only the sp spectrum.

System behavior near such a topological QCP can be studied quantitatively based on the Galilean relation of Landau theory, Eq. (7), together with the Fermi-Dirac relation (6) between the quasiparticle spectrum $\epsilon(p)$ and momentum distribution $n(p)$. Analytical and numerical calculations[15] proceed by self-consistent solution of these two equations for appropriate choices of the quasiparticle interaction function $f(\mathbf{p}, \mathbf{p}_1)$. Extensive results may be found in Ref. 15 over a range of temperatures for different interaction choices, including some long-range interactions with singular behavior at small momentum transfer k (see also Ref. 18). Illustrative examples of the self-consistent solutions obtained for $\epsilon(p)$ and $n(p)$ are provided in Figs. 1 and 2. The local qp interaction adopted in this case has the form

$$f(k) = -g_2 \frac{\pi}{M} \frac{1}{(k^2/4p_F^2 - 1)^2 + \beta_2^2}, \tag{15}$$

also used in microscopic simulations[26] of the 2D electron gas that show a point of inflection of the spectrum $\epsilon(p)$ at $p = p_F$ for $r_s = 7$, where $r_s = \sqrt{2}Me^2/p_F$. The temperature T_m characterizes the crossover from one Fermi surface topology to another, in a 2D analog of the 3D transition from a multisheet Fermi surface at low temperature to a swollen, finite-volume Fermi surface at higher temperature:

Proliferating set of Fermi lines $\rightarrow$ Fractal structure $\rightarrow$ Surface continuum.   (16)

(See Refs. 15, 25 for more details.)

A simple model demonstrating how NFL behavior arises in homogeneous matter was developed in Ref. 27, where we have investigated the case in which the sp spectrum has an inflection point at the QCP; specifically, the situation where $\epsilon(p) \propto (p - p_F)^3$, yielding $p - p_F \propto \epsilon^{1/3}$. The density of states in this model, given by

$$N(T, \rho_\infty) \propto \frac{1}{T} \int n(\epsilon, T)(1 - n(\epsilon, T)) \frac{1}{v(\epsilon, \rho_\infty)} d\epsilon, \tag{17}$$

diverges at $T \to 0$ as $T^{-2/3}$. This comes at the expense of a vanishing group velocity at the Fermi surface, because $1/v(\epsilon) = dp/d\epsilon \propto \epsilon^{-2/3}$. On the FL side of the QCP where the Sommerfeld ratio $\gamma(T) = C(T)/T$ differs from the density of states only by a numerical factor, one then obtains $\gamma(T) \propto T^{-2/3}$. Other topological critical indexes may be evaluated in the same manner.

6. Topological Quantum Critical Point in an Anisotropic System

Thus far homogeneous systems have been considered for simplicity. To deal with NFL behavior of heavy-fermion metals, effects of anisotropy must be included, especially in the QCP region. As a salient example, consider the 2D electron liquid in a quadratic lattice, assuming the QCP electron Fermi line to be a circle of radius approaching p_∞, with angular variable ϕ and the origin shifted to $(\pi/a, \pi/a)$. Since the normal component $v_n(\mathbf{p}) = \nabla_{\mathbf{p}} \epsilon(\mathbf{p}) \cdot \hat{\mathbf{n}}$ of the group velocity now has a well pronounced angular dependence, the *anisotropic* topological QCP is specified by the vanishing of $v_n(p, \phi, T = 0, \rho_\infty)$ at the single point $p = p_\infty, \phi = 0$. Derivation of critical exponents of thermodynamic properties for this anisotropic problem involves straightforward adaptation of the strategy employed in Ref. 27 for a homogeneous medium.

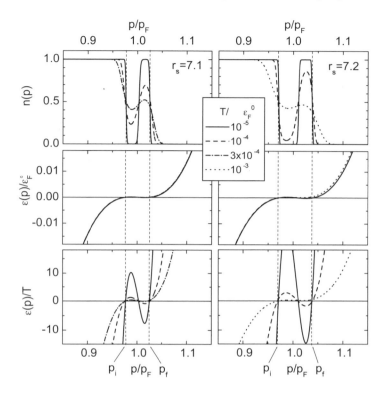

Fig. 1. Quasiparticle momentum distribution $n(p)$ (top panels), qp spectrum $\epsilon(p)$ in units ϵ_F^0 (middle panels), and ratio $\epsilon(p)/T$ (bottom panels) for the regular interaction model (15) at $r_s = 7.1$ (left column) and $r_s = 7.2$ (right column), both exceeding the QCP value $r_\infty = 7.0$. The interaction parameters are $g_2 = 0.16$ and $\beta_2 = 0.14$. Quantities are plotted as functions of p/p_F at different temperatures (in units ε_F^0) below the crossover temperature $T_m = 10^{-3}\varepsilon_F^0$.

On the FL side of and very near the QCP, where $\partial \mathrm{v}_n(p,\phi)/\partial\phi > 0$, we can take

$$\mathrm{v}_n(p,\phi; T = 0, \rho) = \mathrm{v}_n(p, \phi = 0) + a_\phi \phi^2 + a_\rho(\rho - \rho_\infty), \qquad (18)$$

with $\mathrm{v}_n(p, \phi = 0) = a_p(p - p_\infty)^2$, while

$$\mathrm{v}_n(p, \phi = 0; T) \propto T^{2/3}. \qquad (19)$$

(This is quite analogous to what was done in Ref. 27).

As a basic thermodynamic property, the density of states

$$N(T, \rho) \propto \int n(\epsilon)(1 - n(\epsilon)) \frac{d\epsilon \, d\phi}{\mathrm{v}_n(p(\epsilon), \phi; T, \rho)} \qquad (20)$$

determines the specific heat

$$C(T) = T\frac{dS}{dT} \sim T N(T) \qquad (21)$$

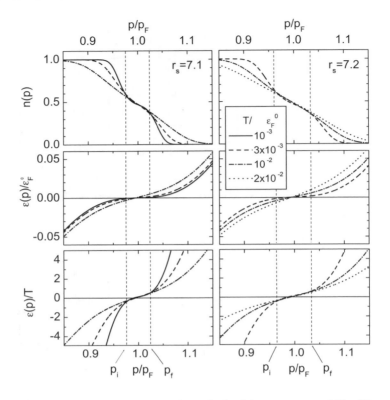

Fig. 2.   Same as Fig. 1, but at temperatures above the "melting temperature" T_m. The remarkable scaling behavior of the curves for $n(p,T)$ and $\epsilon(p,T)/T$ in this temperature regime is to be noted. For each quantity, the curves for different temperatures collapse to a single curve within the momentum range $p_i < p < p_f$. Compare with the strong excursions and deviations seen in this region in Fig. 1.

and the thermal expansion coefficient

$$\beta(T) \sim -\frac{\partial S(T,\rho)}{\partial P} \sim -T\frac{\partial N(T,\rho)}{\partial \rho}. \tag{22}$$

We now use Eq. (18) to evaluate $N(T)$, integrating over ϕ analytically.

The results for the behavior of the specific heat and thermal expansion coefficient in the QCP regime are

$$C(T,\rho_\infty) \propto \int n(\epsilon)(1-n(\epsilon))\frac{d\epsilon}{v_n^{1/2}(p(\epsilon),\phi=0;T)} \propto T^{2/3}, \tag{23}$$

$$\beta(T,\rho_\infty) \propto \int n(\epsilon)(1-n(\epsilon))\frac{v_n'(\rho_\infty)d\epsilon}{v_n^{3/2}(p(\epsilon),\phi=0;T)} \propto O(1), \tag{24}$$

where $v_n'(\rho) \equiv \partial v_n(\rho)/\partial \rho = a_\rho$. In arriving at the overall temperature behavior of these quantities, we have introduced the estimate $v_n(p,\phi=0;T) \propto T^{2/3}$ stemming from the extension of Eq. (18) to finite T and valid in the relevant interval $\epsilon \simeq T$

(see Ref. 27 for analogous details). The results (23) and (24) may be combined to determine the corresponding behavior of the Grüneisen ratio,

$$\Gamma(T, \rho_\infty) = \beta(T, \rho_\infty)/C(T, \rho_\infty) \propto T^{-2/3}, \qquad (25)$$

thus a divergent behavior for $T \to 0$ at variance with the FL result $\Gamma \propto O(1)$. Evidently the system is entering a NFL regime, which is still a disordered phase since no symmetry has been broken by the topological phase transition.

Imposition of an external magnetic field greatly enlarges the scope of challenging NFL behavior, as reflected in the *magnetic* Grüneisen ratio

$$\Gamma_m(T, B) = -\frac{\partial S(T, B)/\partial B}{C(T, B)}. \qquad (26)$$

We now analyze this key thermodynamic quantity within the topological scenario, again following the path established in Ref. 27 (see also Ref. 28). The original Fermi line is split into two, modifying field-free relations involving sp energies through the replacement of the field-free qp occupancy $n(\epsilon) = [1 + \exp(\epsilon/T)]^{-1}$ by half the sum of the field-split occupancies

$$n_\pm(\epsilon) = \{1 + \exp\left[(\epsilon \pm \mu_e B)/T\right]\}^{-1}. \qquad (27)$$

Consequently v_n^{-1} is replaced by half the sum of quantities $(\nabla_\mathbf{p}\,\epsilon(p, \phi) \cdot \hat{\mathbf{n}})^{-1}$ with $\epsilon(p, \phi)$ replaced by $\epsilon(p, \phi) \pm \mu_e B$. Analytic integration over ϕ yields half the sum of square roots of these quantities. In the limit $T \gg \mu_e B$, terms in this sum linear in $r = \mu_e B/T$ cancel each other, such that the net result is proportional to r^2, leading to a magnetic Grüneisen ratio

$$\Gamma_\mathrm{m}(T \gg \mu_e B) \propto T^{-2}. \qquad (28)$$

More specifically, the ratio $S(B, T)/C(T)$ is determined as a function of r^2 only, leading to the behavior (28) at $r \ll 1$. In the opposite limit $r \gg 1$, the density of states $N(T = 0, B)$ diverges at a critical magnetic field B_∞, where the function $v_n(p, \phi, T = 0, B_\infty)$ vanishes on one of the two Fermi lines $p^\pm(\phi)$ specified by

$$\epsilon(p^\pm, \phi) \pm \mu_e B = 0. \qquad (29)$$

It is noteworthy that the field-induced splitting which alters the Fermi-surface group velocity can be compensated, for example, by doping. This would provide a means for driving the system toward the QCP.

At $B > B_\infty$, the key ingredient $v_n(p, \phi = 0; T = 0, B) \equiv v_n(B)$ is positive and FL behavior prevails, as in the isotropic problem.[27] To evaluate the critical index specifying the divergence of the density of states $N(T = 0, B \to B_\infty) \propto v_n^{-1/2}(B)$, we calculate the qp spectrum based on Eq. (29) and insert the result into the field-present extension of Eq. (18). Thereby we obtain $v_n(B) \propto (B - B_\infty)^{2/3}$ for $B \sim B_\infty$, and hence $N(T = 0, B) \propto (B - B_\infty)^{-1/3}$. In turn, this result for the density of states leads to the critical behaviors

$$C(T \to 0, B) \sim S(T \to 0, B) \propto T(B - B_\infty)^{-1/3}, \tag{30}$$

$$\beta(T \to 0) \propto T(B - B_\infty)^{-1}, \tag{31}$$

$$\Gamma(T \to 0) \propto (B - B_\infty)^{-2/3}. \tag{32}$$

Significantly, we arrive at

$$\Gamma_m(T \to 0) = \frac{1}{3}(B - B_\infty)^{-1}. \tag{33}$$

Such a divergence was first predicted within scaling theory,[20] in which the peak of $\Gamma_{\mathrm{mag}}(T = 0, B)$ is located at B_c, the end point of the line $T_N(B)$ where $T_N(B_c) = 0$. We note that within the topological scenario, B_∞ is not expected to coincide with B_c.

7. Confrontations of Theory with Experiment

The experimental group led by Steglich has carried out extensive and precise studies of the thermodynamic properties of heavy-fermion metals, notably Yb-based compounds YbRh$_2$Si$_2$ and YbRh$_2$(Si$_{0.95}$Ge$_{0.05}$)$_2$. The large body of high-quality data now available provides tests of modern phenomenological scaling theories of second-order phase transitions in the QCP region[20,29] as well as mechanistic theories of the QCP regime based on critical collective fluctuations or a topological phase transformation. The predictions of the *topological scenario* for the critical behavior of thermodynamic properties upon approach to a quantum critical point obtained in Sec. 6 (see Eqs. (23)-(25), (28), and (30)-(33)) are in agreement with all the relevant measurements reported in Refs. 2, 30, 31, 32, 33. On the other hand, the outcome of confronting *scaling theories* with experimental results is that no single model invoking 2D or 3D fluctuations can describe the data, which require a set of critical indexes having low probability.[29]

Recent studies of peaks in the specific heat $C(T, B)$ in Yb-based compounds reveal another problem for the scaling theories, which hold that at $B = 0$ the difference $T - T_N$ is the single relevant parameter determining the structure of the fluctuation peak of the Sommerfeld ratio $C(T)/T$. Comparative analysis of corresponding experimental data in YbRh$_2$Si$_2$ and YbRh$_2$(Si$_{0.95}$Ge$_{0.05}$)$_2$ shows clearly that the structure of this peak *is not universal*. Moreover, Fig. 1 of 33 depicts quite dramatically that the fluctuation peak shrinks to naught as the QCP is approached.

The latter observation again brings into question the applicability of the spin-fluctuation *collective scenario* in the vicinity of the QCP. While the spin-fluctuation mechanism does remain applicable at *finite* $T \simeq T_N(B)$, it becomes inadequate at the QCP itself. Accordingly, the relevant critical indexes of scaling theory must be inferred anew from appropriate experimental data on the fluctuation peak located at $T_N(B)$. Importantly, the existing description of thermodynamic phenomena in the extended QCP region, including the peak at $T_N(B)$, must be revised by integrating the topological scenario with the theory of quantum phase transitions.

8. Conclusion

The conventional view of quantum critical phenomena, in which the quasiparticle weight z vanishes at points of related $T = 0$ second-order phase transitions, has been determined to be problematic on several grounds. In particular, we point to its incompatibility with a set of identities derived by gauge transformations associated with prevailing conservation laws. This failure may be traced to inadequacy of the Ornstein-Zernike approximation $\chi^{-1}(q) \approx q^2 + \xi^{-2}$ for the static correlation function $\chi(q)$ in the limit $\xi \to \infty$.

We have discussed an alternative topological scenario for the quantum critical point that places emphasis on the role of single-particle degrees of freedom in the onset of non-Fermi-liquid behavior. Developing this scenario for an appropriate lattice model, we have demonstrated that its predictions for thermodynamic properties in the vicinity of the quantum critical point are in agreement with available experimental data on heavy-fermion systems. We infer from these data that upon entering the quantum critical regime from the ordinary Fermi liquid on the disordered side, the role of single-particle degrees of freedom is paramount. The effects of critical collective fluctuations build up as the system moves further toward an associated second-order transition leading to an ordered phase.

We suggest that a satisfactory theory of the quantum critical point in systems of the class considered here will involve a unification of the two competing scenarios through an quantitative understanding of the interplay between single-particle and collective degrees of freedom.

Acknowledgments

This research was supported by the McDonnell Center for the Space Sciences, by Grants Nos. NS-3004.2008.2 and 2.1.1/4540 from the Russian Ministry of Education and Science, and by Grants Nos. 07-02-00553 and 09-02-01284 from the Russian Foundation for Basic Research.

References

1. S. Sachdev, *Quantum Phase Transitions* (Cambridge University Press, Cambridge, 1999).
2. J. Custers, P. Gegenwart, H. Wilhelm, K. Neumaier, Y. Tokiwa, O. Trovarelli, F. Steglich, C. Pepin, and P. Coleman, *Nature* **424**, 524 (2003).
3. A. Bianchi, R. Movshovich, I. Vekhter, P. G. Pagliuso, and J. L. Sarrao, *Phys. Rev. Lett.* **91**, 257001 (2003).
4. S. E. Sebastian, N. Harrison, M. M. Altarawneh, C. H. Mielke, Ruixing Liang, D. A. Bonn, W. N. Hardy, and G. G. Lonzarich, Metal-insulator quantum critical point beneath the high T_c superconducting dome. [arXiv:0907.2958].
5. S. V. Kravchenko and M. P. Sarachik, *Rep. Prog. Phys.* **67**, 1 (2004).
6. C. Bäuerle, Yu. M. Bun'kov, A. S. Chen, S. N. Fisher, and H. Godfrin, *J. Low Temp. Phys.* **110**, 333 (1998).
7. A. Casey, H. Patel, J. Nyeki, B. P. Cowan, and J. Saunders, *Phys. Rev. Lett.* **90**, 115301 (2003).

8. M. Neumann, J. Nyeki, B. P. Cowan, and J. Saunders, *Science* **317**, 1356 (2007).
9. L. D. Landau, *Zh. Eksp. Teor. Fiz.* **30**, 1058 (1956) [*Sov. Phys. JETP* **3**, 920 (1957)].
10. P. Nozières, *Theory of Interacting Fermi Systems* (W. A. Benjamin, N.Y., 1964).
11. G. Baym and C. Pethick, *Landau Fermi-Liquid Theory: Concepts and Applications* (Wiley-VCH, Weinheim, 2004).
12. A. A. Abrikosov, L. P. Gor'kov, and I. E. Dzyaloshinski, *Methods of Quantum Field Theory in Statistical Physics* (Prentice-Hall, Englewood Cliffs, NJ, 1963).
13. S. Doniach and S. Engelsberg, *Phys. Rev. Lett.* **17**, 750 (1966).
14. P. Coleman, C. Pepin, Q. Si, and R. Ramazashvili, *J. Phys.: Condens. Matter* **13**, R723 (2001).
15. V. A. Khodel, J. W. Clark, and M. V. Zverev, *Phys. Rev. B* **78**, 075120 (2008).
16. G. E. Volovik, *Springer Lecture Notes in Physics* **718**, 31 (2007).
17. V. A. Khodel, *JETP Lett.* **86**, 721 (2007).
18. V. A. Khodel, J. W. Clark, and M. V. Zverev, *JETP Lett.* **90**, 639 (2009).
19. T. Westerkamp, M. Deppe, R. Kuechler, M. Brando, C. Geibel, P. Gegenwart, A. P. Pikul, F. Steglich, *Phys. Rev. Lett.* **102**, 206404 (2009).
20. L. Zhu, M. Garst, A. Rosch, and Q. Si, *Phys. Rev. Lett.* **91**, 066404 (2003).
21. Hiroaki Shishido, Rikio Settai, Hisatomo Harima, and Yoshichika Onuki, *Proc. Phys. Soc. Japan* **74**, 1103 (2005).
22. V. A. Khodel and V. R. Shaginyan, *JETP Lett.* **51**, 553 (1990); V. A. Khodel, V. R. Shaginyan, and V. V. Khodel, *Phys. Rep.* **249**, 1 (1994).
23. G. E. Volovik, *JETP Lett.* **53**, 222 (1991).
24. P. Nozières, *J. Physique I* **2**, 443 (1992).
25. J. W. Clark, V. A. Khodel, and M. V. Zverev, *Int. J. Mod. Phys. B* **23**, 4059 (2009).
26. V. V. Borisov and M. V. Zverev, *JETP Lett.* **81**, 503 (2005).
27. J. W. Clark, V. A. Khodel, and M. V. Zverev, *Phys. Rev. B* **71**, 012401 (2005).
28. V. R. Shaginyan, *JETP Lett.* **79**, 286 (2004).
29. T. Senthil, *Phys. Rev. B* **78**, 035103 (2008).
30. P. Gegenwart, Y. Tokiwa, K. Neumaier, C. Geibel, and F. Steglich, *Physica* **B359-361**, 23 (2005).
31. R. Küchler, N. Oeschler, P. Gegenwart, T. Cichorek, K. Neumaier, O. Tegus, C. Geibel, J. A. Mydosh, F. Steglich, L. Zhu, and Q. Si, *Phys. Rev. Lett.* **91**, 066405 (2003).
32. Y. Tokiwa, T. Radu, C. Geibel, F. Steglich, and P. Gegenwart, *Phys. Rev. Lett.* **102**, 066401 (2009).
33. N. Oeschler, S. Hartmann, A. P. Pikul, C. Krellner, C. Geibel, and F. Steglich, *Physica* **B403**, 1254 (2008).

HELIUM ON NANOPATTERNED SURFACES AT FINITE TEMPERATURE

E. S. HERNANDEZ

Departamento de Física, Facultad de Ciencias Exactas y Naturales,
Universidad de Buenos Aires Buenos Aires 1428, Argentina
shernand@df.uba.ar

F. ANCILOTTO

Dipartimento di Fisica 'G. Galilei',
Università di Padova, via Marzolo 8, I-35131 Padova, Italy,

and DEMOCRITOS National Simulation Center, Trieste, Italy

M. BARRANCO, A. HERNANDO and M. PI

Departament d'Estructura i Constituents de la Matèria, Facultat de Física,

and IN² UB, Universitat de Barcelona, 08028 Barcelona, Spain

Received 23 October 2009

We investigate the wetting behavior of helium on nanostructured alkali metal surfaces, at temperatures below and slightly above the bulk superfluidity threshold. Starting from a determination of the phase diagram of helium on semiinfinite planar Cs up to 3 K, performed within finite–range, temperature–dependent density functional theory, we examine the modifications of the isotherms introduced by an infinite array of nanocavities. We compare the hysterectic loops of helium on nonwettable Cs surfaces and on wettable Na substrates in the same temperature range.

Keywords: Helium; finite temperature; finite systems.

1. Motivation

This paper is a progress report on our research about adsorption of helium on nontrivial substrates, employing a finite range density functional description. In past articles in this Workshop Series, we have discussed results obtained at zero temperature T, involving adsorption inside and outside cylinders and spheres,[1] in linear wedges [2] and in infinite polygonal pores.[3] This paper call attention on the newest results concerning adsorption of helium in finite cavities and nanopatterned surfaces at temperatures $0 \leq T \leq 3$ K, whose details can be found in a recent review paper,[4] and its outline is as follows. In Secs. 2 to 4 we classify the adsorption problems according to the dimensionality of the substrate plus fluid combination, reviewing briefly the current state of the art regarding onedimensional (1D), twodimensional (2D) and threedimensional (3D) configurations, together with the basic layout of the thermodynamics of wetting phase diagrams and of density functional theory (DFT) in Sec. 2. Moreover, in Sec. 4 we give a short overview of the situation encountered at finite $T's$ within the DFT frame. This presentation is summarized in Sec. 5.

2. Class of Onedimensional Adsorption Problems

The theoretical description of the formation of helium films on adsorbers with high symmetry like flat substrates, cylinders and spheres involves a single degree of freedom, locally perpendicular to the adsorbing surface. The equilibrium liquid–vapor interface is parallel to the liquid–solid interface and the fluid density is just a function $\rho(u)$ with u the transverse coordinate. These 1D problems are, relatively speaking, easily dealt with resorting to a variety of methods. Simple models based on thermodynamic arguments, incorporate macroscopic data such as interfacial tensions and/or the contact angle of deposited drops, along with a few interaction parameteres such as the strength of the van der Waals forces on the adsorbate atoms. These purely phenomenological descriptions are able to estimate the equilibrium film thickness and the approximate location of the wetting temperature T_W. On the way to the most elaborate many body variational and Quantum (QMC) and Path Integral (PIMC) Monte Carlo methods, several intermediate approximations are very reasonably provided by effective Hamiltonian models for the liquid–vapor interface, and by mean field descriptions based on density functional theory (DFT). The latter support a rather full account of the equation of state (EOS) of the adsorbed fluid samples, together with density profiles that accurately illustrate the layering process and the breakdown of stability of films on curved surfaces.

In order to fix ideas, in Fig. 1 we visualize a qualitative plot of adsorption isotherms of a fluid on a lyophobic surface, *i.e.*, nonwettable at temperatures below T_W, and extending up to a prewetting temperature T_{PW}. In this coexistence region, the van der Waals loops support Maxwell constructions that define two endpoints, namely a thin film coexisting with a thick one, at an equilibrium chemical potential $\mu(T)$ smaller than the bulk figure $\mu_0(T)$ at the given temperature. This plot can be compared with the textbook isotherms of a van der Waals gas, which allows us to identify the rigorous correspondence summarized in Table 1.

Adsorption	van der Waals
Thin film	Vapor
Thick film	Liquid
Bulk Liquid	Solid
Prewetting temperature T_{PW}	Critical temperature T_c
Wetting temperature T_W	Triple temperature T_t

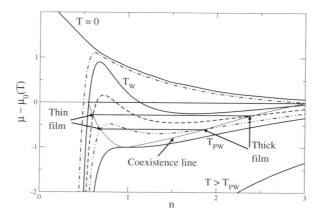

Fig. 1. Qualitative plot of the adsorption diagram of a fluid on a nonwettable surface, indicating the wetting (T_W) and prewetting (T_{PW}) temperatures, the coexisting thin and thick film on each isotherm, and the coexistence line. Temperatures increase from above to below. The variable on the vertical axis is the chemical potential of the adsorbate particles referred to the bulk figure $\mu_0(T)$ at each temperature. In the horizontal axis we plot the coverage, *i.e.*, number of particles per unit area.

In addition, for future reference we mention here that within DFT and for any dimensionality, one minimizes, with respect to the density profile ρ, a thermodynamic grandpotential $\Omega = F - \mu N$, with μ the chemical potential of the helium atoms and with N the grandcanonical ensemble average of the particle number operator

$$N = \int d^3\mathbf{r}\,\rho(\mathbf{r}) \tag{1}$$

The total free energy F is in turn expressed as a spatial integral of a density kernel $f[\rho(\mathbf{r}), T]$, designed so as to reproduce the EOS of liquid ^{4}He, and its bulk and surface properties experimentally determined for a number of temperatures. The variation $\delta\Omega/\delta\rho$ gives rise to the integrodifferential Euler-Lagrange equation

$$\left[-\frac{\hbar^2}{2m}\nabla^2 + V[\rho(\mathbf{r}), T] + V_s(\mathbf{r})\right]\sqrt{\rho(\mathbf{r})} = \mu(T)\sqrt{\rho(\mathbf{r})} \tag{2}$$

Here $V[\rho, T]$ is the effective potential arising from functional differentiation of the potential energy density appearing in $f[\rho, T]$, and $V_s(\mathbf{r})$ is the adsorber potential. With the free energy density of Ref. 5, parameterized for temperatures up to 3 K, our DFT calculations of adsorption isotherms of ^{4}He on planar Cs, by solving Eq. (2) for the density profile $\rho(z)$ and corresponding eigenvalue $\mu(T)$ for each temperature T, qualitatively reproduce the phase diagram in Fig. 1. The coexistence line thus obtained permits us, by a simple extrapolation, to extract the wetting temperature of helium on planar Cs at a value $T_W = 2.05$, in perfect agreement with reported measurements (see Ref. 4 for details) within experimental uncertainties.

3. Class of 2D Adsorption Problems

Among this class one finds a rich survey of adsorbers symmetric with respect to one degree of freedom, either translational or rotational. Power–law wedges with cross section in the (x, z) plane described by $z = |x|^\gamma$ are a useful tool to investigate the crossover from a flat surface with $\gamma = 0$ to a parallel slit at $\gamma = \infty$, through a continuous sequence of grooves ($\gamma < 1$) and cavities ($\gamma > 1$), limited by the linear wedge at $\gamma = 1$. As reported in Ref. 6, an almost phenomenological analysis of adsorption of fluids attracted to the substrate by van der Waals forces, is sufficient to exhibit the increasing importance of geometric over dominating interaction effects in the critical exponents of the filling transition. Previous studies of adsorption of macroscopic samples of either classical or quantum fluids in linear wedges, relying on thermodynamic arguments[7] and corroborated by effective Hamiltonian models involve a few variables such as the bulk liquid density ρ_0, a fixed number of particles N –or volume V– and the main assumption of a sharp liquid–vapor interface. Under such simplificatory assumptions, one finds that the interfacial line $z(x)$ is a circumference, whose radius R uniquely depends on only two variables, the contact angle $\theta(T)$ of the fluid on a flat surface and the half opening α of the wedge. While the contact angle is a signature of the specific adsorber–adsorbate combination at the selected temperature, the wedge aperture characterizes the geometry of the environment. The sequence of shape transitions upon variations of either θ or α has been recently revised in Ref. 8, together with an analysis of the stability of density and capillary waves for ^{4}He adsorbed in linear wedges. We note here that DFT calculations of EOS and density profiles of ^{4}He adsorbed in linear Cs and Na wedges at $T = 0$ have been presented in this series in Ref. 2 and in an extended version in Ref. 9.

Finally, for the sake of completeness we also point out that adsorption of ^{4}He in the interior of infinite pores with polygonal cross section has been also investigated within DFT[10] and reported in these workshops.[3] For future reference, it is worthwhile recalling that one of the most impressive outcomes of calculations carried in infinite rhombic pores is the appearance of metastable, symmetry–breaking density patterns, together with hysterectic cycles in the adsorption isotherms. This is a peculiarity of DFT calculations, already encountered in applications to classical fluids.

4. Class of 3D Problems

Given the fact that Cs flat surfaces are not wetted by ^{4}He below $T_W \approx 2$ K, we have examined the possibility of introducing irregularities on the substrate that may change the helium–cesium wetting phase diagram. For this sake, we have deviced a simple, although reliable, model of a rough surface, that makes room to DFT calculations, consisting of a periodic square array of heliophilic sites on a heliophobic surface. This is an ambitious project that includes some "regular" roughness at the nanoscale, inspired in an experimental realization[11] of a triangular lattice of parabolic cavities on a silica material exposed to organic vapor.

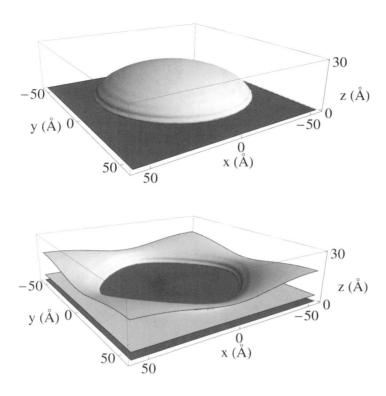

Fig. 2.   Illustrative finite configurations for helium on planar Cs at T =2.3 K. Top panel: a drop. Bottom panel: a bubble. These panels display isodensity surfaces for $\rho = 0.011$ Å^{-3}, *i.e.*, half the bulk liquid density at $T = 0$ K. The dark area represents the Cs planar surface. In the lower panel, liquid helium fills the region between the two sheets representing the $\rho = 0.011$ Å^{-3} isodensity surface. (From Ref. 4).

Our threedimensional DFT code[4] allows us to undertake the construction of finite configurations of He on planar Cs, such as bubbles or droplets. In particular, we have been able to anticipate studies of cavitation and nucleation of helium on flat surfaces.[12] Some illustrations of these finite configurations can be seen in Fig. 2, where the top and bottom panels respectively display 3D density profiles of a drop and a bubble, computed at $T = 2.3$ K.

In order to perform these calculations for our nanopatterned surface, we first construct the potential of a parabolic nanocavity of height $h = 35$ Å and outer radius $R = 25$ Å according to the procedure developed in Ref. 12, and adopt a 3D computation cell that hosts such cavity at its center, with sufficient room for the fluid to spread and wet the flat surrounding surface, and for the top vapor to reach bulk equilibrium. Periodic boundary conditions guarantee that we are describing an infinite lattice of such nanocavities drilled on a flat surface. We have computed adsorption isotherms for ^{4}He on a cesiated array and the 3D density profiles of a

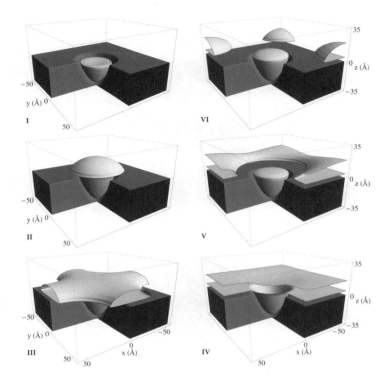

Fig. 3.   Illustrative equilibrium configurations for helium on the nanopatterned Cs substrate at $T = 2.3$ K. The corresponfing coverages are, respectively, 2.73×10^{-2} Å^{-2}; 0.105 Å^{-2}; 0.219 Å^{-2}; 0.267 Å^{-2}; 0.189 Å^{-2}; 0.108 Å^{-2}, from I to IV. The panels display isodensity surfaces for $\rho = 0.011$ Å^{-3}, *i.e.*, half the bulk liquid density at $T = 0$ K. The dark area represents the Cs planar surface. (From Ref. 4).

typical hysterectic loop are displayed in Fig. 2. For the sake of comparison, Fig. 3 shows a similar loop for an identical array of cavities on flat sodium.

As one traces the panels in Fig. 2 counterclockwise starting at the upper left, one finds the following sequence. For low coverage (panel I), the cavity is partially filled and no substantial coverage appears on the surrounding flat surface. As one proceeds along the adsorption isotherm increasing the number of adsorbed material, the fluid completely fills the cavity and spreads out forming a spherical cap on top of the hole (panel II). In panel III we observe that the fluid covers the flat surface of the computation cell almost uniformly, leaving only dry patches at the corners, which appear completely covered in panel IV. Panels V and VI illustrate the desorption path; starting from IV and decreasing the coverage we provoke dewetting, first by emptying the annular ring at the rigde of the cavity and subsequently drying the flat surface, from the center and outwards. We note that the stratification of the density due to the effect of the structured Cs surface is clearly visible at the edge of all these finite configurations.

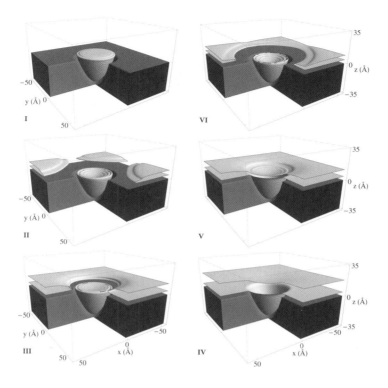

Fig. 4.   Same as in Fig. 2 for helium on a nanopatterned Na substrate at $T = 2.3$ K. The respective coverages are 3.04×10^{-2} Å^{-2}; 3.78×10^{-2} Å^{-2}; 9.20×10^{-2} Å^{-2}; 0.194 Å^{-2}; 8.93×10^{-2} Å^{-2}; 3.59×10^{-2} Å^{-2}, from I to IV. (From Ref. 4).

The equivalent sequence for the Na adsorber is interesting, due to the fact that opposite to Cs, this alkali is wetted by ^{4}He at zero temperature. Since the planar adsorbing potential is stronger than that of Cs, not only the cavity offers a deeper absorbing site; in this case, the repulsion at the rim of the pore is more intense as well. This repulsion prevents the fluid from spilling out of the cavity (cf. panels II and III) until the flat surface far from the dip is almost uniformly covered.

5. Summary

In this report, we have reviewed some aspects of the current state of the art concerning adsorption of quantum fluids on one, two and three dimensional substrates. The latest results obtained within finite temperature DFT correspond to fully 3D calculations, that involve finite configurations of helium filling parabolic nanocavities, drilled on a flat alkali surface so as to form a periodic lattice. This is the most sophisticated DFT computation performed up-to-date for this class of systems. Although it is not straightforward to derive a coexistence line from the set of isotherms, as done for a homogeneous substrate, and establish in this way possible changes in the wetting and/or prewetting temperature, the present approach gives a complete qualitative and semiquantitative description of cavity filling and emptying, as well as on wetting and drying processes on a rough surface.

Acknowledgments

This work has been performed under grants FIS2008-00421/FIS from DGI, Spain (FEDER), 2009SGR1289 from Generalitat de Catalunya and in Argentina, X099 from University of Buenos Aires and PICT 31980/05 from ANPCYT. F.A. acknowledges the support of Padova University through project CPDA077281-07.

References

1. E. S. Hernández, *Condensed Matter Theories 28*, edited by John Clark, Robert Panoff and Haochen Li, Nova Science Publishers, New York, ISBN 15-945-498-93, pp. 231-239 (2006).
2. E. S. Hernández, F. Ancilotto, M. Barranco, R. Mayol and M. Pi, *Int. J. Mod. Phys. B* **21**, 2067 (2007).
3. E. S. Hernández, A. Hernando, R. Mayol and M. Pi, *Int. J. Mod. Phys. B* **22**, 4338–4345 (2008).
4. F. Ancilotto, M. Barranco, E. S. Hernández and M. Pi, *J. Low Temp. Phys.* **157**, 174–205 (2009).
5. F. Ancilotto, F. Faccin, and F. Toigo, *Phys. Rev. B* **62**, 17035 (2000).
6. C. Rascón and A. O. Parry, *Nature* **407**, 986 (2000).
7. K. Rejmer, S. Dietrich, and M. Napiórkowski, *Phys. Rev. E* **60**, 4027 (1999).
8. E. S. Hernández, to be published.
9. E. S. Hernández, F. Ancilotto, M. Barranco, R. Mayol, and M. Pi, *Phys. Rev. B* **73**, 245406 (2006).
10. A. Hernando, E. S. Hernández, R. Mayol, and M. Pi, *Phys. Rev. B* **77**, 195431 (2008).
11. O. Gang, K. J. Alvine, M. Fukuto, P. S. Persha, C. T. Black, and B. M. Ocko, *Phys. Rev. Lett.* **95**, 217801 (2005); Erratum *ibid.* **98**, 209904 (2007).
12. F. Ancilotto, M. Barranco, E. S. Hernández, A. Hernando and M. Pi, *Phys. Rev. B* **79**, 1045141-11 (2009)
13. A. Hernando, E. S. Hernández, R. Mayol, and M. Pi, *Phys. Rev. B* **76**, 115429 (2007).

TOWARDS DFT CALCULATIONS OF METAL CLUSTERS IN QUANTUM FLUID MATRICES

S. A. CHIN

Department of Physics, Texas A&M University, College Station, TX 77845, USA
chin@physics.tamu.edu

S. JANECEK, E. KROTSCHECK* and M. LIEBRECHT

Institute for Theoretical Physics, JKU Linz, A-4040 LINZ, Austria
* eckhard.krotscheck@jku.at

Received 8 October 2009

This paper reports progress on the simulation of metallic clusters embedded in a quantum fluid matrix such as ^{4}He. In previous work we have reported progress developing a real-space density functional method. The core of the method is a diffusion algorithm that extracts the low-lying eigenfunctions of the Kohn-Sham Hamiltonian by propagating the wave functions (which are represented on a real-space grid) in imaginary time. Due to the diffusion character of the kinetic energy operator in imaginary time, algorithms developed so far are at most fourth order in the time-step. The first part of this paper discusses further progress, in particular we show that for a grid based algorithm, imaginary time propagation of any even order can be devised on the basis of multi-product splitting. The new propagation method is particularly suited for modern parallel computing environments. The second part of this paper addresses a yet unsolved problem, namely a consistent description of the interaction between helium atoms and a metallic cluster that can bridge the whole range from a single atom to a metal. Using a combination of DFT calculations to determine the response of the valence electrons, and phenomenological acounts of Pauli repulsion and short-ranged correlations that are poorly described in DFT, we show how such an interaction can be derived.

1. Introduction

With the advance of the density functional method in examining diverse solid state physics and quantum chemistry problems,[1-3] it is of growing importance to solve Schrödinger equations on 3-D meshes that can be as large as $N = 10^6$ grid points. Even a few years ago it was expensive to represent up to a few hundred wave functions on a grid of that size, but the required amount of computer memory is nowadays commonplace. It is therefore fair to say that real-space methods are the methods of the future in density functional theory: They are easy to implement, easily parallelizable, and do not suffer from the bias implicit to basis expansions. We have previously shown that fourth-order imaginary time propagation[4] provides an effective means of solving the Kohn-Sham and related equations.[5] The use of all *forward* time-step fourth-order algorithms in solving the imaginary time Schrödinger equation has since been adapted by many research groups.[6–8]

In principle, the lowest n states of the one-body Schrödinger equation

$$H\psi_j(\mathbf{r}) = E_j\psi_j(\mathbf{r}) \tag{1}$$

with Hamiltonian

$$H = -\frac{\hbar^2}{2m}\nabla^2 + V(\mathbf{r}) \equiv T + V \tag{2}$$

can be obtained by applying the evolution operator ($\epsilon = -\Delta t$)

$$\mathcal{T}(\epsilon) \equiv e^{\epsilon(T+V)} \tag{3}$$

repeatedly on the ℓ-th time step approximation $\{\psi_j^{(\ell)}(\mathbf{r})\}$ to the set of states $\{\psi_j(\mathbf{r}),\ 1 \le j \le n\}$,

$$\phi_j^{(\ell+1)} \equiv \mathcal{T}(\epsilon)\psi_j^{(\ell+1)} \tag{4}$$

and orthogonalizing the states after every step,

$$\psi_j^{(\ell+1)} \equiv \sum_i c_{ji}\phi_i^{(\ell+1)}, \quad \left(\psi_j^{(\ell+1)}\big|\psi_i^{(\ell+1)}\right) = \delta_{ij}. \tag{5}$$

The method is made practical by approximating the exact evolution operator (3) by a general product form,

$$\mathcal{T}(\epsilon) = \prod_{i=1}^{M} e^{a_i\,\epsilon T} e^{b_i\,\epsilon V}. \tag{6}$$

The simplest second order decomposition method is

$$\mathcal{T}_2(\epsilon) \equiv e^{\frac{1}{2}\epsilon V} e^{\epsilon T} e^{\frac{1}{2}\epsilon V} = \mathcal{T}(\epsilon) + O(\epsilon^3). \tag{7}$$

When this operator acts on a state $\psi_j(\mathbf{r})$, the two operators $e^{\frac{1}{2}\epsilon V}$ correspond to point-by-point multiplications and $e^{\epsilon T}$ can be evaluated by one complete (forward and backward) Fast Fourier Transform (FFT). Both are $\mathcal{O}(N)$ processes. For $\mathcal{T}_2(\epsilon)$, $|\epsilon|$ has to be small to maintain good accuracy and many iterations are therefore needed to project out the lowest n states. To derive forward, all positive time step fourth-order algorithms, Suzuki[9] and Chin[10] have shown that a correction to the potential of the form $[V,[T,V]] = (\hbar^2/m)|\nabla V|^2$ must be included in the decomposition process. These forward fourth-order algorithms can achieve similar accuracy at step sizes several orders of magnitude larger than the second-order splitting (7).[4]

2. Multi-product Expansion

If the decomposition of $\mathcal{T}(\epsilon)$ is restricted to a single product as in (6), then there is no practical means of implementing a sixth or higher order forward algorithm.[11] However, if this restriction is relaxed to a sum of products,

$$e^{\epsilon(T+V)} = \sum_k c_k \prod_i e^{a_{k,i}\epsilon T} e^{b_{k,i}\epsilon V}, \tag{8}$$

then the requirement that $\{a_{k,i}, b_{k,i}\}$ be positive means that each product can only be second order. Since $\mathcal{T}_2(\epsilon)$ is second order with positive coefficients, its powers $\mathcal{T}_2^k(\epsilon/k)$ can form a basis for such a multi-product expansion. Recent work[12] shows that such an expansion is indeed possible and takes the form

$$e^{\epsilon(T+V)} = \sum_{k=1}^{n} c_k \mathcal{T}_2^k \left(\frac{\epsilon}{k}\right) + O(\epsilon^{2n+1}) \equiv \mathcal{T}_n(\epsilon) + O(\epsilon^{2n+1}), \tag{9}$$

where the coefficients c_k are given in closed form for any n:

$$c_i = \prod_{j=1(\neq i)}^{n} \frac{k_i^2}{k_i^2 - k_j^2} \tag{10}$$

with $\{k_1, k_2, \ldots k_n\} = \{1, 2, \ldots n\}$. Since the symmetric $\mathcal{T}_2(\epsilon)$ has only odd powers in ϵ,

$$\mathcal{T}_2(\epsilon) = \exp[\epsilon(T+V) + \epsilon^3 E_3 + \epsilon^5 E_5 + \cdots], \tag{11}$$

where E_i are higher order commutators of T and V, the expansion (9) is just a systematic extrapolation which successively removes each odd order error.

Since each $\mathcal{T}_2$ requires one complete FFT, the above series of $2n$-order algorithms only requires $n(n+1)/2$ complete FFTs. Thus algorithms of order 4, 6, 8, 10 and 12 only require 3, 6, 10, 15 and 21 complete FFTs. The low order extrapolation $n = 4$ has been used previously.[13] Here, we have a systematic expansion to any even order.

Degenerate eigenvalues in general pose a notorious problem for eigenvalue solvers[2,3] that contain an orthogonalization step. In our implementation of the algorithm, we use the subspace orthogonalization method described in Ref. 5. In that method, problems only occur whenever the *highest calculated state* lies within a band of almost degenerate states.

3. Model Calculation

To demonstrate the performance of our new family of algorithms, and to compare with alternative methods, we have applied them to several representative models.

The calculations are started at a certain time step ϵ, using plane waves in an appropriate box as initial states. One iteration of the algorithm consists of propagating the states with the n-th order evolution operator $\mathcal{T}_n(\epsilon)$, and subsequently orthogonalizing the propagated states. To assess the convergence, we calculate after each iteration the variance of all states with respect to the evolution operator,

$$R_j^E = \frac{1}{|\epsilon|} \sum_k \left| \mathcal{T}_n(\epsilon)\psi_j(\mathbf{r}_k) - e^{\epsilon E_j}\psi_j(\mathbf{r}_k) \right|^2 . \tag{12}$$

Only states with $R_j^E > \gamma$, where γ is a prescribed error bound, need to be propagated and orthogonalized in the subsequent iterations. As soon as all states have converged at the time step ϵ, their variances with respect to the Hamiltonian,

$$R_j^H = \sum_k \left| H\psi_j(\mathbf{r}_k) - E_j\psi_j(\mathbf{r}_k) \right|^2 , \tag{13}$$

are calculated. If $R_j^H < \gamma$ for all states j, the iterations are terminated, otherwise the time step is reduced by a constant factor, *i.e.*, ϵ is replaced by $\alpha\,\epsilon$, and the whole process is repeated, taking the result of the previous iteration as initial values. We have empirically determined the optimal reduction factor to be $\alpha \approx 0.5$. Recently, Lehtovaara, Toivanen and Eloranta[7] have suggested that the time step size can be optimally adjusted with added efforts.

As an illustrative example we show here results for the (an-)isotropic harmonic oscillator in three dimensions,

$$V(x, y, z) = \frac{1}{2} \left(\alpha_x x^2 + \alpha_y y^2 + \alpha_z z^2 \right) . \tag{14}$$

To compare the operator factorization methods with the "Implicitly Restarted Lanczos Method" (IRLM) provided by ARPACK,[14] we have calculated the lowest $N = 120$ eigenvectors of an anisotropic harmonic oscillator with the parameters $\alpha_x = 1.0$, $\alpha_y = 1.1$ and $\alpha_z = 1.2$. Figure 1 shows the achieved average variance

$$R^H = \sqrt{\frac{\sum_j R_j^H}{\sum_j E_j^2}} \tag{15}$$

of the eigenvectors as a function of the total computation time. Results are shown for the 2nd, 4th, 6th, 8th and 12th order algorithm, for the 4th order algorithm 4A of Ref. 4, and for the Lanczos method. The computation time needed to perform the second-order propagation of *one* state, $T_{2\mathrm{nd}} = T\left[\mathcal{T}_2(\epsilon)\psi(\mathbf{r})\right]$, provides a natural time unit for the n-th order operator factorization algorithms. Since it is also comparable to the time needed for the $H\psi$ operations in the Lanczos method, we have used $T_{2\mathrm{nd}}$ as a time unit in figure 1. For the operator factorization methods, we have propagated only the lowest ("occupied") 120 states, while for the Lanczos calculation we have used 240 states, following the suggestion of the ARPACK manual that the number of "active" states should be roughly twice the number of "occupied" states.

Already the second order operator factorization is, for the same accuracy, much faster than Lanczos and does not produce variances better then $R^H \approx 10^{-3}$; higher order methods provide a significant improvement. Especially when only relatively

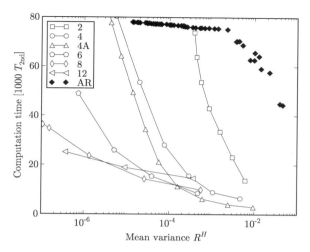

Fig. 1. (Color on line) The figure shows a comparison of the performances of the 2nd, 4th, 6th, 8th, and 12th order algorithm and the 4th order version of Ref. 4 of our imaginary time propagation method and the Lanczos method as implemented in the ARPACK[14] (labeled AR) for the anisotropic harmonic oscillator as discussed in the text.

low accuracies up to $R_j^H < 10^{-4}$ are required, the higher order operator factorization methods performed almost an order of magnitude faster than Lanczos. When higher accuracy is required (which is the case, for example, for finding equilibrium configurations of metal clusters, or for the calculation of NMR shifts) the higher order methods are somewhat more time consuming, but still outperform the Lanczos method (which did, in our tests not reach that level of accuracy) by a factor of four to five. The extra time needed for improved accuracy is largely compensated by the fact that parallelization is easy and quite efficient especially for the the high-order methods. We conclude this section by two remarks:

(1) The convergence of the operator factorization method is practically the same for the isotropic harmonic oscillator as long as the highest computed state is not in a band of degenerate states. For the isotropic case, the Lanczos method normally did not produce all states, but skipped some of the degenerate levels. We also note that our conclusions on the relative performance of the above algorithms are similar for the C_{60} model discussed above.

(2) As stated above, we envision the application of high-order algorithms in density functional calculations. Such calculations contain the solution of the Kohn–Sham equation only as a part, other parts are the update of the electron density during the iterative solution of the Kohn-Sham equations, and possibly also the annealing of the underlying ion structure. In that case, one often has reasonably good initial values for the wave functions. Then, the diffusion method needs only a few iterations, whereas good initial values provide little advantage in the Lanczos method.

4. Parallelization

The advantage of high-order propagation methods is particularly compelling on parallel computer architectures: The propagation step (4) can be parallelized efficiently without having to abandon the advantage of using FFTs by simply distributing the states ψ_j across different processors. In such an arrangement, however, the parallelization of the orthogonalization step is notoriously difficult, and we have made no specific efforts to parallelize this step. As stated above, the propagation time T_{pro}, *i.e.* the time it takes to carry out step (4) for the $2n$-order algorithm, increases as $n(n+1)/2$ whereas the orthogonalization time T_{ort} remains the same. Hence, parallelization becomes more effective since more time is spent for propagation than for orthogonalization:

The time T_{tot} for one iteration step on an ideal machine with N processors is

$$T_{\text{tot}}(N) = T_{\text{pro}}/N + T_{\text{ort}}, \tag{16}$$

and the speed up ratio for the "propagation only" and the total time step including orthogonalization for the jth order algorithm is

$$S^{(j)}_{\text{pro/tot}}(N) = \frac{T_{\text{pro/tot}}(1)}{T_{\text{pro/tot}}(N)}, \tag{17}$$

assuming the number of states is larger than the number of processors allocated for the task. The actual speed-up ratio will be less than this ideal since we have neglected communication overhead and other hardware/system specific issues.

Figure 2 shows the speed up ratio for the C_{60} model calculation in the case of the 2nd, 6th, and 12th order algorithms on a 128 Itanium[15] processor Altix[16] machine for up to twelve threads. We show the two speed-up ratios $S^{(j)}_{\text{pro}}(N)$ and $S^{(j)}_{\text{tot}}(N)$. Evidently, the speed-up of the propagation step alone is a reasonably linear function of the number of threads. The performance improves significantly with the order of the algorithm because the increase computational effort for propagation can be distributed while the cost of communication remained the same. The 12th order algorithm can reach about 80 percent of the optimal performance.

The impressive convergence of our high–order algorithms has a computing cost: One propagation step of the $2n$-th order algorithm is equivalent to $n(n+1)/2$ propagation steps of the second order algorithm. This cost is compensated by two effects: The first is shown in the figures: The much faster convergence as a function of time step implies that fewer iterations are needed to complete the calculation. The second advantage is less obvious; since orthogonalization is carried out *after* the propagation step (9), the relatively costly number of orthogonalization steps is dramatically reduced. We note, however, that the subspace orthogonalization requires mostly matrix-matrix products. These operations are beginning to be implemented on modern GPUs which promise a speedup of this performance bottlenck by more than an order of magnitude.

Thus, high order algorithms provide a number advantages: they profit from good initial values, they have faster convergence at larger time steps, and they are more

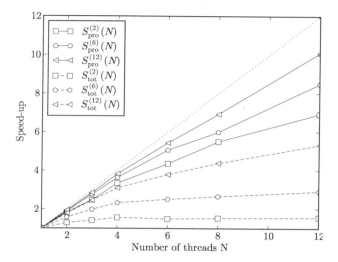

Fig. 2.   (Color on line) The total speed-up time factor S_{tot} (solid lines) and the propagation only (without orthogonalization) time speed-up factor S_{pro}, as a function of parallel threads, for the 2nd, 6th, and 12th order algorithm (filled squares, circles, and triangles, respectively). Also shown is the "ideal" speed-up factor (dotted line).

adaptive to parallel computing environments. We have made no particular effort to parallelize the orthogonalization step so far. The actual speed-up is therefore limited by that step; while the 12th order algorithm can still attain a more than five-fold speed-up, it is hardly worth parallelizing the second order algorithm. The specific speed-up factor for higher order algorithms also depends on the number n of needed eigenstates. In general, the time for propagation is essentially proportional to n, whereas the time for orthogonalization goes as n^2. Our family of algorithms gives the user the possibility to adapt the algorithm to the type of hardware used for the computation. In the single processor mode, we have determined that the 6th order algorithm performed the best. On a quad-core CPU, the 12th order method already gives a speedup of a factor of three without using a GPU.

5. Cluster-Helium Interactions

A second component of our project is the calculation of interactions between helium atoms and metal clusters. To describe the most complicated part of the physics, let us take magnesium as an example metal. Presently there are two types of theories that can provide effective helium-metal interactions. One looks at the interaction between *individual* Mg atoms and helium atoms.[17] In that case, the nuclear charge is completely screened, and the interaction is the usual van der Waals force that falls off as r^{-6} for large distances. The opposite situation is a metal or a metal surface. There, the conduction electrons are delocalized. The simplest approximation is a model where the conduction electrons are smeared out and the magnesium cores are point charges. *Neither* of these models is completely appropriate for our situation because, even in a conductor, the ion cores are partly screened. The problem has been dealt with, for the half space, by Zaremba and Kohn[18,19] and, later, by Chizmeshya, Cole and Zaremba.[20] Since, in particular for Mg, we are confronted with the situation that the cluster can undergo a transition from an insulator to a conductor, we must design a theory that is accurate in *both* situations. An effective potential describing the interaction between a single He atom and the metal cluster should therefore have the following features:

- It should be agree with the bare He-Mg interaction for a single atom[17];
- It should also be close to bare e-He interaction for a single electron[21];
- It should include the response of the conduction electrons in the cluster. This has the immediate consequence that the potential seen by the He atom is generally not the sum of pair interactions.
- Finally, it should, for large clusters, be close to the Mg-half-space - He interaction.[20]

Our model of an effective interaction between the cluster and the helium atom adopts the following picture: The electrons in the helium atom are strongly bound, the atom acts, at large distances, like a polarizable sphere. The *valence electrons* of the metal atoms are, on the other hand, relatively weakly bound; these might turn into the conduction electrons of the cluster. The *core-electrons* are again strongly bound, their interaction with a He atom can be described by a static Van der Waals force. Thus, the only dynamic effect is the response of the valence (or conduction) electrons to the induced field of the helium atoms. The induced dipole moment $\mathbf{p}$ of a polarizable sphere (the helium atom) in an external field $\mathbf{E}$ is

$$\mathbf{p} = \alpha \mathbf{E}, \qquad E_{\text{ind}} = \frac{\alpha}{2} |\mathbf{E}|^2, \tag{18}$$

where α is the polarizability, and E_{ind} the energy of that dipole in the field. The electrostatic potential seen by the helium atom is

$$U[n_e](\mathbf{r}, \{\mathbf{R}_i\}) = e \left[\sum_{i=1}^{N} Z_i \frac{1}{|\mathbf{r} - \mathbf{R}_i|} - \int d^3r' \, n_e(\mathbf{r}') \frac{1}{|\mathbf{r} - \mathbf{r}'|} \right], \tag{19}$$

where Z_i are the ion core charges, $\mathbf{R}_i$ the positions of the ions, and $n_e(\mathbf{r})$ is the electron density in the cluster. To be specific, we assume here that only the valence electrons contribute to the electron density $n_e(\mathbf{r})$. The induced energy E_{ind} is added to the energy functional, this causes an additional term in the Kohn-Sham equations. Hence, all electron orbitals and the electron density depend on the position of the helium atom. A second component of the interaction is the *Pauli repulsion*. This is due the fact that the valence or conduction electrons cannot penetrate the helium atom. The effect has been modeled by a repulsive potential that acts between the helium atom and those electrons that are close to the atom.[21] We have chosen the form

$$E_{\text{Pauli}}(\mathbf{r}) = B \int d^3r' n_e(\mathbf{r}') e^{-\frac{1}{2}\left(\frac{|\mathbf{r}-\mathbf{r}'|}{\sigma}\right)^2} \tag{20}$$

where B and σ are adjustable parameters.

The third component of the interaction is due to the polarization. of the core electrons of the Mg. Such interactions are, for example, obtained by Coupled Cluster calculations,[17] but are outside of the regime of validity of density functional theory. We have fitted this effect by a static interaction of the form[22]

$$V_{\text{core}}(\mathbf{r}) = \sum_i \left[\frac{C_6}{|\mathbf{r} - \mathbf{R}_i|^6 + r_0^6} + \frac{C_8}{|\mathbf{r} - \mathbf{R}_i|^8 + r_0^8} \right].$$ (21)

The asymptotic form of the dispersion terms is damped, at small distances, by the parameter r_0.

To determine the parameters of the potential we have proceeded as follows: The coefficients C_6, C_8, and r_0 are responsible for those long-ranged parts of the interaction that are not determined by the response of the valence electrons. We have therefore chosen these to fit the long-range part of the bare He-Mg interaction for a single atom. The parameters B and σ models to the spatial spread of the electron-density of one Helium atom and is responsible for the short-ranged repulsion between an electron and the He atom. We have used $3\sigma = r_{\text{VdW}} = 1.22$ Å, where r_{VdW} is the Van der Waals radius of Helium, B was then adjusted to reproduce the hard-core of the He-Mg interaction. We have also tried the exponential form of the bare e-He interaction.[21] From the numerical point of view, this functional is more cumbersome, because it requires a very fine discretization is necessary. More importantly, it also leads to a core-radius of the Mg-He interaction that is at least 1 Å too large; it was therefore abandoned.

As stated above, we should also compare with the interaction between a helium atom and a metallic half space. Substrate potentials are often $3 - 9$ potentials characterized by their *range* C_3 and their *well depth* D. They have the analytic form

$$U_{\text{sub}}(z) = \left[\frac{4C_3^3}{27D^2} \right] \frac{1}{z^9} - \frac{C_3}{z^3}.$$ (22)

Such a potential can be be obtained by averaging Lennard-Jones potentials

$$V_{\text{LJ}}(r) = 4\epsilon \left[\left(\frac{\sigma}{r} \right)^{12} - \left(\frac{\sigma}{r} \right)^6 \right].$$ (23)

between metal atoms via

$$U_{\text{sub}}(z) = \rho_{\text{metal}} \int_{z' < 0} V_{LJ}(r - r'),,$$ (24)

the relationships between the coefficients of the substrate potential $U_{\text{sub}}(z)$ and the Lennard-Jones potential $V_{\text{LJ}}(r)$ are then

$$C_3 = \frac{2\pi}{3} \rho_{\text{metal}} \epsilon \sigma^6, \qquad \frac{4C_3^3}{27D^2} = \frac{4\pi}{45} \rho_{\text{metal}} \epsilon \sigma^{12}.$$ (25)

Figure 3 shows the results of two different calculations of the energy curve between a Mg atom and a He atom. In the first calculation (blue curve labeled with $E[\rho_0]$), we have first carried out a DFT calculation for the valence electrons, and

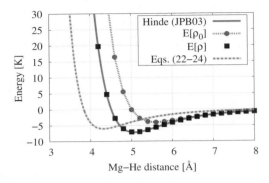

Fig. 3. (Color on line) The figure shows the effective interaction between a single helium atom and a Mg atom. The blue line depicts the result when the electrons are not allowed to relax, whereas the squares show the calculated interaction allowing the relaxation of the valence electrons. Also shown is the Hinde-potential [17] (red line) and the Lennard-Jones potential obtained from the half-space potential as described in the text (pink).

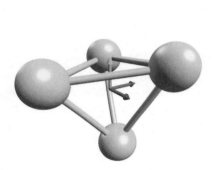 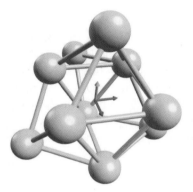

Fig. 4. (Color on line) The figure shows the ground state configuration of the Mg_4 and a Mg_{10} cluster used in our calculation. These configurations have been obtained by annealing LDA calculations for the ground state. The three arrows in the center of the cluster indicate the x (blue), y (red) and z (green) direction.

have then kept the resulting electron density $\rho_0(\mathbf{r})$ fixed as the He atom approaches the atom. The effect of the polarization is clearly seen in the figure. The figure also shows a "Lennard-Jones" potential (23) that would be obtained from the half-space potential (22) by identifying the parameters through Eq. (25). Evidently, a "helium-magnesium" potential obtained from the half-space this way is very different from the Hinde potential. This simply shows that the physics at work is very different: For the metallic half-space, the interaction is mostly due to the polarization of the conduction electrons which are much weaker bound than in a single Mg atom.

The interaction between a single Mg atom and a He atom provided the means to adjust the phenomenological components of our procedure. Turning now to larger clusters, we have used the same procedure for calculating the interaction between a single He atom and a Mg_4 and a Mg_{10} cluster. The structures were obtained by minimizing the LDA energy of the cluster with simulated annealing,[23] they are shown in Fig. 4.

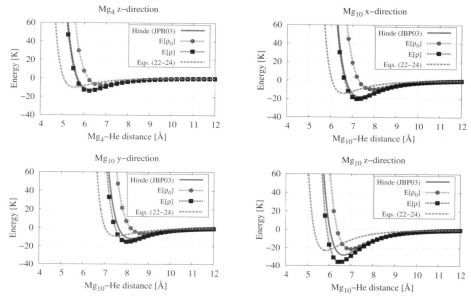

Fig. 5. (Color on line). The potentials seen by a single He atom in the vicinity of a Mg_4 cluster (upper left pane) and a Mg_{10} cluster, approaching it from the x, (upper right pane), y (lower left pane), and z direction (lower right pane), as a function of the distance between the center of mass of the cluster and the He atom. The red curve shows the potential seen by the He atom when Hinde-potentials are just added up, and the ones obtained by adding up the Lennard-Jones interactions constructed from the half space are shown in pink. The blue and black curves with markers show the energy curves obtained by keeping the electrons rigid (blue) or letting them respond to the presence of the He atom.

Evidently, the energy depends now on the direction in which the He atom approaches the cluster. In each case, we show four results: For comparison we show the potential seen by the He atom when Hinde-potentials are just added up (red curves) and the ones obtained by adding up the Lennard-Jones interactions constructed from the half space (green curves). The blue and black curves with markers show the energy curves obtained by keeping the electrons rigid (blue) or letting them respond to the presence of the He atom.

The message of these calculations is clear: While we can understand the small Mg_4 cluster still as a superposition of single atoms interacting with Mg through by pairwise potentials, this is no longer true for the Mg_{10} cluster. First, the energy surface is quite anisotropic. Only in the z direction, the superposition of the single-atom potentials gives a reasonable representation to our best potential. In all cases, the energy curve constructed by the procedure describe above starts to approach the potential that one might expect from a Mg sphere. The agreement with the Lennard-Jones interaction constructed from the half space is generally poor. The reason for that is that evidently we have not reached the particle number where the cluster is conducting.

Acknowledgments

This work was supported, in part, by the Austrian Science Fund FWF under grants No. P18134 and P21924. We also would like to thank E. Hérnandez and R. Zillich for useful discussions and correspondence.

References

1. T. L. Beck, *Rev. Mod. Phys.* **72**, 1041 (2000).
2. T. Torsti, T. Eirola, J. Enkovaara, T. Hakala, P. Havu, V. Havu, T. Höynälänmaa, J. Ignatius, M. Lyly, I. Makkonen, T. T. Rantala, J. Ruokolainen, K. Ruotsalainen, E. Räsänen, H. Saarikoski and M. J. Puska, *Physica Status Solidi* **B 243**, 1016 (2006).
3. N. R. Wijesekera, G. Feng and T. L. Beck, *Phys. Rev.* **B 75**, 115101 (2007).
4. J. Auer, E. Krotscheck and S. A. Chin, *J. Chem. Phys.* **115**, 6841 (2001).
5. M. Aichinger and E. Krotscheck, *Computational Materials Sciences* **34**, 188 (2005).
6. L. Brualla, K. Sakkos, J. Boronat and J. Casulleras, *J. Chem. Phys.* **121**, 636 (2004).
7. L. Lehtovaara, J. Toivanen and J. Eloranta, *J. of Comput. Phys.* **221**, 148 (2006).
8. G. B. Ren and J. M. Rorison, *Phys. Rev.* **B 77**, 245318/1 (2008).
9. M. Suzuki, New scheme of hybrid exponential product formulas with applications to quantum Monte Carlo simulations, in *Computer Simulation Studies in Condensed Matter Physics*, eds. D. P. Landau, K. K. Mon and H.-B. Schüttler (Springer, Berlin, 1996) pp. 1-6.
10. S. A. Chin, *Phys. Lett.* **A 226**, 344 (1997).
11. S. A. Chin, *Phys. Rev.* **E 71**, 016703/1 (2005).
12. S. A. Chin, Multi-product splitting and Runge-Kutta-Nyström integrators (2008), arXiv:0809.0914.
13. K. E. Schmidt and M. A. Lee, *Phys. Rev.* **E 51**, 5495 (1995).
14. http://www.caam.rice.edu/software/ARPACK/.
15. Itanium is a registerd trademark of Intel Corp.
16. Altix is a registered trademark of Silicon Graphics Inc.
17. R. J. Hinde, *J. Phys. B (At. Mol. Opt. Phys.)* **36**, 3119 (2003).
18. E. Zaremba and W. Kohn, *Phys. Rev.* **B 13**, 2270 (1976).
19. E. Zaremba and W. Kohn, *Phys. Rev.* **B 15**, 1769 (1977).
20. A. Chizmeshya, M. W. Cole and E. Zaremba, *J. Low Temp. Phys.* **110**, 677 (1998).
21. J. Eloranta and V. A. Apkarian, *J. Chem. Phys.* **117**, p. 10139 (2002).
22. F. O. Kannemann and A. D. Becke, *Journal of Chemical Theory and Computation* **5**, 719 (2009).
23. R. Wahl, DFT calculations of small sodium and magnesium clusters, Master's thesis, JKU (2006).

ACOUSTIC BAND GAP FORMATION IN METAMATERIALS

D. P. ELFORD, L. CHALMERS, F. KUSMARTSEV* and G. M. SWALLOWE

*Department of Physics, Loughborough University,
Leicestershire, LE11 3TU, UK*
*F.Kusmartsev@lboro.ac.uk

Received 10 February 2010

We present several new classes of metamaterials and/or locally resonant sonic crystal that are comprised of complex resonators. The proposed systems consist of multiple resonating inclusion that correspond to different excitation frequencies. This causes the formation of multiple overlapped resonance band gaps. We demonstrate theoretically and experimentally that the individual band gaps achieved, span a far greater range ($\approx 2kHz$) than previously reported cases. The position and width of the band gap is independent of the crystal's lattice constant and forms in the low frequency regime significantly below the conventional Bragg band gap. The broad envelope of individual resonance band gaps is attractive for sound proofing applications and furthermore the devices can be tailored to attenuate lower or higher frequency ranges, i.e., from seismic to ultrasonic.

Keywords: Acoustic metamaterials; sonic crystals; resonance band gaps.

1. Introduction

Acoustic wave propagation in sonic crystals (SC's) has been studied since the late 1980's.[1,2] Such structures forbid the transmission of sound in a band gap when its wavelength is comparable to the periodic spacing of the scattering units.[3-9] As such, Bragg's law is satisfied and a band gap is formed (Bragg band gap - BBG). This condition imposes a restriction in designing conventional SC's as the lattice spacing of the array is intrinsically linked to the attenuated frequency range. Subsequently a new class of SC was proposed, known as a locally resonant sonic material (LRSM),[10] which form resonance band gaps (RBG) that occur at frequencies independent of the structures lattice constant. A further extension upon this concept utilised several resonators to form multiple RBG's that can be overlapped.[11]

We present here a theory and experiment of acoustic metamaterials and/or locally resonant sonic crystal (we call such systems as LRSC), which are comprised of two classes of complex resonators, being analogous to Helmholtz and quarter-wave resonators respectively. Also detailed is a third LRSC system with resonators that have a geometry similar to Nautilus shells, that combine aspects from the first two designs. All systems have been found to exhibit multiple broad band gaps, in the lower frequency range below Bragg gap formation. We show that the position and width of the RBG's can be tuned by adjusting the resonator properties, such that, multiple band gaps can be combined to form broad regions of frequency attenuation.

This can be exploited to create band gaps of any desired width. The acoustic resonators described here operate with a low quality (Q) factor and form RBG's much
wider than previously reported.[11]

To investigate the RBG formation experimentally in the first LRSC system, C-
shaped resonators were constructed by machining a single 4mm wide longitudinal
slot along the length of a steel tube, with external diameter 13mm, internal diameter
9.7mm and length 300mm. The LRSC was constructed from an array of such resonators, positioned in a 10×10 square configuration with lattice constant 22mm.
The length of the scatterers was sufficiently longer than the relevant wavelength
range being tested ensuring that the environment can be assumed to be quasi-two-
dimensional. To measure attenuation levels a 150W Realistic Nova 800 speaker with
a maximum frequency response of 20 kHz was driven by a Bruel & Kjaer beat frequency oscillator (type 1022), which was in turn driven by a Bruel & Kjaer level
recorder (type 2305). This allowed a six-second rising tone sample from 1-10 kHz to
be generated. The signal was transmitted through the LRSC systems and recorded
using a dual Behringer C-2 studio condenser microphone setup. Control recordings
were obtained by performing the experiment without a crystal present. A frequency
spectrum was obtained by performing a Fourier analysis of the recording and sound
pressure levels were calculated from the relative amplitudes of the control and LRSC
system being tested. The frequency spectra were then compared to locate regions of
attenuation. All acoustic experiments were performed inside an anechoic chamber.

2. Numerical Simulations of the Sound Propagation through the Resonance Media

In order to interpret the experimental results extensive finite element (FE) simulations were performed modelling acoustic wave propagation and scattering through
LRSC's comprised of C-shaped resonators. A time harmonic analysis of the acoustic
wave propagation was performed using the FE method (see, for details in Ref.[12]).
For a time harmonic pressure wave excitation of the form $p = p_0(x, y)e^{i\omega t}$, the wave
equation reduces to the Helmholtz equation, which is used to describe the acoustic
wave propagation:

$$\nabla \cdot \left(-\frac{1}{\rho_0} \nabla p_0 \right) - \frac{\omega^2 p_0}{\rho_0 c^2} = 0 \qquad (1)$$

Where $p_0(x, y, z)$ is the pressure, depending on x and y spatial dimensions, c is the
speed of sound, ρ_0 is the density and $\omega = 2\pi f$ is the angular frequency. We apply the
impedance matching boundary conditions, where the c-resonators are made from
steel tubes. By solving this equation inside the air domain, which is inside steel
tubes with slots we obtain the resulting pressure field. The resonating units in the
LRSC system consist of steel tubes that are modelled as fluid inclusions with a high
stiffness and mass embedded in air. Therefore, the high contrast between the steel
and air parameters gives rise to a high reflection coefficient, even though a small
pressure field may be present inside the resonators walls.

The transmitted acoustic frequency spectrum obtained from the FE simulation for the C-shaped resonator LRSC shows the appearance of two distinct band gaps as seen in Fig. 1(A). It is possible to distinguish between the BBG and RBG mechanisms by comparison with a spectrum obtained from a conventional SC of the same size but with non-resonating inclusions. This structure forms a BBG but does not demonstrate any local resonance characteristics and hence, no RBG is formed. The BBG is located at 8.15 kHz, as predicted by Bragg's law, and spans from 6.9–9.4 kHz. This is in good agreement with the experimentally obtained band gap shown in Fig. 1(B), which forms from 6.8–9.8 kHz, centered at 8.3 kHz. The simulation shows the C-shaped cavities acting analogously to Helmholtz resonators that have a maximum response at their excitation frequency, of 4.7 kHz and form a RBG centered at this frequency spanning from 3.9–5.5 kHz. The experimentally obtained spectrum confirms the appearance of this gap from 3.8–6.1 kHz. Attenuation levels achieved in the experimental results reach a maximum at 16dB and 20dB for the BBG and RBG respectively. This is a suitably attractive level of attenuation for soundproofing applications, as a 9dB decrease would remove 80% of unwanted noise.

Each C-shaped resonator within the LRSC acts as an individual Helmholtz resonator, with a broad neck, containing a fixed volume of air. The broad neck induces the formation of broad band excitations in the resonators. When a pressure variation, in the form of a sound wave, interacts with the air in the neck, the pressure of the air inside increases. As the external force is removed, the pressure equalizes and forces air back through the neck. Due to the inertia of the air in the neck, a region of low pressure is created in the cavity, which in turn causes air to be drawn back in. The air then continues to oscillate at frequency ω_0, which is dependent upon the characteristic dimensions of the resonators cavity and neck, but is shape and material independent. When an incoming wave of angular frequency ω is incident upon the resonating unit with excitation frequency ω_0 the linear response function is proportional to $1/(\omega_0^2 - \omega^2)$ causing exponential attenuation of the wave.[10] At frequencies below Bragg's condition and away from resonance, the LRSC systems present a homogeneous structure to the sound waves. The resonant frequency f_0 of the C-shaped resonating units may be tuned using an empirically obtained equation derived from Helmholtz's equation for resonators.

$$f_0 = \frac{c}{\pi^{5/4}} \frac{\sigma^{1/4}}{S^{1/2}} \qquad (2)$$

Where σ is the slot width and $S = \pi r^2$ for our two-dimensional scenario. We can assume from this equation that when the slot width is increased for a fixed diameter resonator, the frequencies should increase by a factor of $\sigma^{1/4}$ and for a constant slot width, the frequencies should decrease inversely proportionally to the diameter.

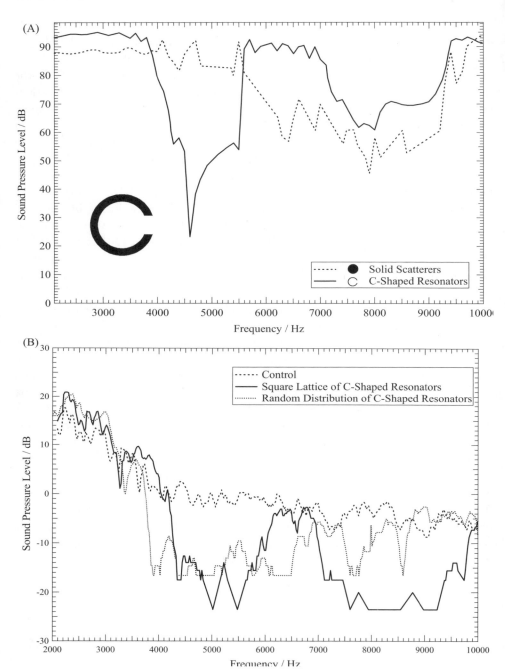

Fig. 1.   **(A)** Frequency spectrum obtained by FE simulation of a LRSC consisting of C-shaped resonator units (solid line). Included is a frequency spectrum obtained for a conventional sonic crystal comprised of solid scattering units for comparison (dashed line). **(B)** Experimentally obtained frequency spectrum of a LRSC consisting of C-shaped resonators as described in Fig. 1(A) (solid line). Included on the plot are spectra for a control sample (dashed line) and for a LRSC with a random distribution of C-shaped resonators (dotted line).

3. Experiments and Comparison with the Theory

To validate the existence of the RBG mechanism, experiments were conducted upon a random distribution of C-shaped resonators, see Fig. 1(B). Due to the lack of periodicity in this structure, the array is disordered and BBG formation cannot occur. However, the RBG is still observed due to its independence from the arrays periodicity. Each C-shaped resonator acts independently and accounts for a small amount of absorption, that when summed for the whole system, gives appreciable levels of sound attenuation forming a RBG for this disordered system. Similar measurements were performed using the same Bruel & Kjaer system as detailed above. The experimentally obtained frequency spectrum shows the existence of an RBG from 3.5–6.5 kHz.

In order to enlarge the wavelength range in which the LRSC is active, a multiple RBG system is simulated using several C-shaped resonators of varying dimensions in a mixed array configuration. In such an arrangement, several RBG's are formed, and by careful selection of the cavity dimensions, these individual gaps can be overlapped to form a combined broader gap. The frequency spectrum, see Fig. 2(A), shows RBG's for 11mm and 14mm diameter resonators, from 5.5–7.5 kHz with 65dB of attenuation at its peak and at 4.8–6.0 kHz with 60dB of attenuation. A frequency spectrum obtained from a mixed LRSC with alternating layers shows the formation of a combined band gap system enveloping the two individual resonance peaks, 5.0–7.5 kHz with similar levels of attenuation. In effect, the width of the band gap is doubled using two carefully selected resonator sizes. Furthermore, the RBG can be widened with more than two resonator sizes in a single array, thus allowing band gaps of any width to be created.

4. Acoustic Metamaterials made of the Matryoshka Resonators

We propose an alternative design to this basic mixed system that is comprised of an arrangement of concentric resonators coined the Matryoshka (Russian doll) configuration, see insert on Fig. 2(B). The inclusion of multiple sized resonating units gives rise to the formation of many RBG's with substantial levels of attenuation, and the concentric distribution gives an overall reduction in array size. The frequency spectrum for a Matryoshka array comprised of 11mm and 14mm diameter resonators forms two individual RBG's from 5.6–6.5 kHz and 3.5–5.1 kHz respectively, see Fig. 2(B). These RBG's can again be overlapped, by careful selection of the resonator dimensions, allowing LRSC systems to be created with very broad frequency ranges. The compactness of the Matryoshka configuration allows an acoustic barrier to be constructed from a single layer of resonators, whilst still offering reasonable levels of attenuation.

An alternative LRSC system is proposed that utilises a second class of acoustic resonator inclusions and is again able to form an enlarged RBG by overlapping resonances. The inclusions are quarter wave resonators that have a fundamental excitation frequency f_0, which is active when the length of the resonator is a quarter

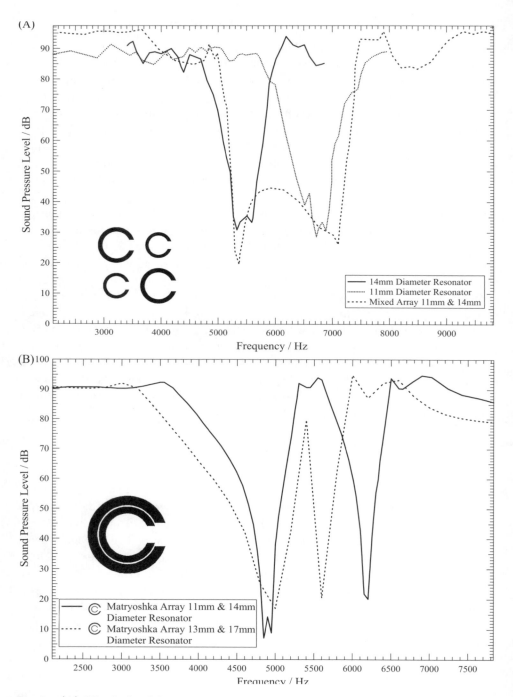

Fig. 2. **(A)** FE calculated frequency spectrum for a LRSC comprised of two different diameter C-shaped resonators, 14mm and 11mm, (dashed line) indicating the presence of a RBG envelope. Included on the plot are the frequency spectra for LRSC's comprised of each of the individual sized resonators for comparison. **(B)** Frequency spectrum for a LRSC with Matryoshka (Russian doll) inclusions, comprised of 11mm and 14mm C-shaped resonators.

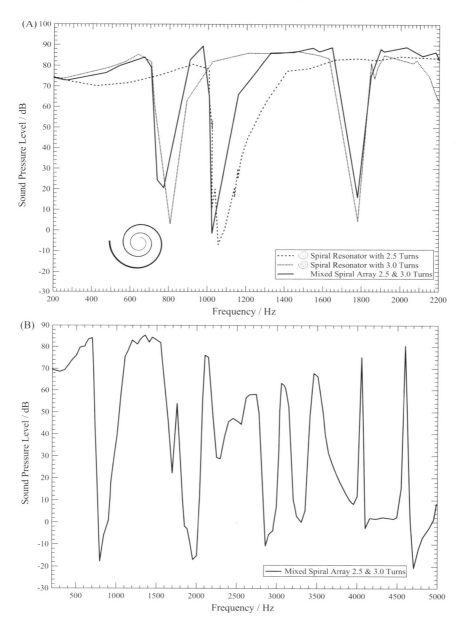

Fig. 3. **(A)** Frequency spectra obtained from FE simulations for a LRSC comprised of spiral resonators with 2.5 turns (dashed line), 3.0 turns (dotted line) and a mixed configuration (solid line). The spectra are shown over the low frequency regime 0.2–2.2 kHz to detail the overlapping mechanism. **(B)** An extended frequency spectrum ranging from 0–5 kHz for the LRSC with the mixed spiral inclusions which, shows an envelope of multiple RBG's corresponding to the fundamental and harmonic excitation frequencies of each resonating unit.

of a wavelength long. Harmonics of this fundamental frequency also exist at integer multiples of f_0. As an acoustic wave is incident upon the resonator, part of the wave enters the cavity and is reflected back. In the time the acoustic wave takes to travel down the resonator and back to the opening, the acoustic wave outside of the resonator has shifted half a wavelength, and the two waves interfere destructively causing attenuation. As the length of the resonator is intrinsically linked to the excitation frequency, the characteristic length of such resonators needed to attenuate low frequency noise is approximately 0.5m for 150 Hz. This is an undesirable scale for sound proofing applications and therefore the length of the resonator can be coiled into a more compact spiral design, reducing the resonator size to mm scale. The spiral resonator schematic is inset on Fig. 3(A). This geometry allows the formation of quarter wave resonating units that are active at wavelengths significantly larger than their size.

FE simulations have been performed on a two-dimensional LRSC comprised of such spiral resonators. The first unit used was a Bernoulli type spiral with external radius 0.0128m and decay per 90° of 86% with 3.0 turns. This spiral has a characteristic path length of 0.16m and a corresponding fundamental frequency $f_0 = 0.74$ kHz as given by quarter wave resonator theory.[13] When included in a LRSC system a fundamental RBG is observed from 0.68–0.9 kHz with 60dB of attenuation, see Fig. 3(A). A higher order harmonic also exists at double the fundamental frequency at 1.72 kHz with similar levels of attenuation. A second spiral with external radius 0.0103m and the same decay, but with only 2.5 turns is also included. This resonator unit has a shorter path length of 0.11m, and hence a higher fundamental excitation frequency $f_0 = 1.05$ kHz. The frequency spectrum demonstrates this with the appearance of a RBG from 0.96–1.38 kHz with attenuation peaking at 70dB. A LRSC comprised of a mixed array of the 2.5 and 3.0 turn Bernoulli spiral resonators was then simulated and demonstrates combined resonances as with the mixed array of C-shaped resonators. Similarly, the band gap envelope covers the individual resonances of each unit. The overlapping of the fundamental RBG's also causes combined RBG formation in the harmonic bands. This premise has been exploited to form a LRSC with multiple overlapping fundamental and harmonic RBG's, see Fig. 3(B).

5. Media Made of the 'Seashell' Resonators

A final LRSC system was modelled that again exhibits a large number of broad local resonances. It is comprised from a third class of acoustic resonator unit, coined the 'seashell' resonator. The design of this unit incorporates aspects of each of the previous geometries, namely the inclusion of multiple sized resonating cavities from the Matryoshka resonator and the space saving feature of the spiral resonator. The appearance of this system resembles that of one found in nature, namely the Nautilus shell, see Fig. 4(A). It has a spiral profile with logarithmic progression and is divided into several separate chambers, each section of which is connected by a small

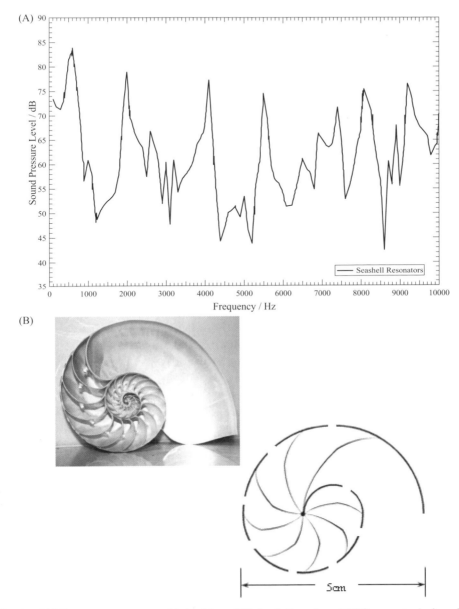

Fig. 4. (**A**) Frequency spectrum obtained from FE simulations for a LRSC comprised of seashell resonators with eight chambers. (**B**) Nature's example of a broad band resonator, in the form of a Nautilus shell, and a schematic of the similar seashell resonators used in the FE investigation.

opening. This system is analogous to our seashell resonators, which are a collection of individual resonating cavities, with different excitation frequencies. Therefore, both examples should have a similar acoustic response and form a large number of broad resonances spanning over a large frequency range. For a Nautilus shell, the amplification of background noise over a broad range of frequencies produces pink noise. This is in fact what is commonly mistaken to be sound of the sea when a shell is placed next to one's ear.

FE simulations have been performed for a LRSC comprised of seashell resonators with eight resonant cavities, and its frequency response is presented in Fig. 4(B). This system demonstrates multiple RBG's, corresponding to each of the cavities dimensions, that span a broad range of frequencies from 10 Hz–20 kHz. The excitation frequencies of the cavities lie very close to each other and thus, due to the large width of each RBG, an enlarged band gap envelope is formed with an average level of attenuation of approximately 30dB. Again, a band gap of any desired width can be created by including more cavities in the seashell resonator.

6. Conclusion

In summary, LRSC's constructed from the proposed complex resonators exhibit broad regions of acoustic attenuation, which arise due to the overlapping of several close resonance band gaps. This is most easily achieved for the C-shaped and spiral resonating system by using a simple mixed configuration of multiple sized resonators. The proposed Matryoshka and seashell LRSC systems do not require this arrangement as each unit exhibits multiple resonances itself and thus, is independent of lattice spacing. This reduces the overall size of the LRSC and allows reasonable levels of attenuation to be achieved from a single layer of resonators. Furthermore, the attenuated frequency range can be tailored by adjusting the dimensions, and hence excitation frequencies, of the resonating cavities. This can be exploited to form band gaps over lower and higher frequency regimes, i.e., seismic to ultrasonic.

Acknowledgments

The work was supported by the ESF network-program AQDJJ.

References

1. Lakhtakia, A., Varadan, V.V. and Varadan, V.K., *J. Acoust. Soc. Am.* **83**, 1267 (1988).
2. Economou, E.N. and Zdetsis, A., *Phys. Rev. B.* **40**, 1334 (1989).
3. Kushwaha, M.S., Halevi, P., Dobrzynski, L. and Djafari-Rouhani, B., *Phys. Rev. Lett.* **71**, 2022 (1993).
4. Chen, Y.Y. and Ye, Z., *Phys. Rev. Lett.* **87**, 184301 (2001).
5. Chen, Y.Y. and Ye, Z., *Phys. Rev. E.* **64**, 036616 (2001).
6. Larabi, H., Pennec, Y., Djafari-Rouhani, B. and Vasseur, J.O., *Journal of Physics: Conference Series.* **92**, 012112 (2007).
7. Larabi, H., et al, *Phys. Rev. E.* **75**, 066601 (2007).
8. Vasseur, J.O., et al., *Phys. Rev. E.* **65**, 056608 (2002).
9. Sanchez-Perez, J.V., et al, *Phys. Rev. Lett.* **80**, 5325 (1998).
10. Liu, Z., et al., *Science.* **289**, 1734 (2000).

11. Ho, K.M., Cheng, C.K., Yang et al, P., *Appl. Phys. Lett.* **83**, 5566 (2003).
12. *COMSOL Multiphysics v3.4a.* (2007).
13. Kinsler, L.E., Frey, A.R., Coppens, A.B. and Sanders, J.V., *Fundamentals of Acoustics, Fourth Edition,* **Wiley & Sons**, p274 (2000).

DISSIPATIVE PROCESSES IN LOW DENSITY STRONGLY INTERACTING 2D ELECTRON SYSTEMS

DAVID NEILSON

Dipartimento di Fisica, Università di Camerino, I-62032 Camerino, Italy

NEST CNR-INFM, I-56126 Pisa, Italy

david.neilson@unicam.it

Received 1 December 2009

A glassy phase in disordered two dimensional (2D) electron systems may exist at low temperatures for electron densities lying intermediate between the Fermi liquid and Wigner crystal limits. The glassy phase is generated by the combined effects of disorder and the strong electron-electron correlations arising from the repulsive Coulomb interactions. Our approach here is motivated by the observation that at low electron densities the electron pair correlation function, as numerically determined for a non-disordered 2D system from Monte Carlo simulations, is very similar to the pair correlation function for a 2D classical system of hard discs. This suggests that theoretical approaches to 2D classical systems of hard discs may be of use in studying the disordered, low density electron problem. We use this picture to study its dynamics on the electron-liquid side of a glass transition. At long times the major relaxation process in the electron-liquid will be a rearrangement of increasingly large groups of the discs, rather than the movement of the discs separately. Such systems have been studied numerically and they display all the characteristics of glassy behaviour. There is a slowing down of the dynamics and a limiting value of the retarded spatial correlations. Motivated by the success of mode-coupling theories for hard spheres and discs in reproducing experimental results in classical fluids, we use the Mori formalism within a mode-coupling theory to obtain semi-quantitative insight into the role of electron correlations as they affect the time response of the weakly disordered 2D electron system at low densities.

1. Introduction

At lower electron densities in two–dimensional electron systems at semiconductor heterojunction interfaces, the correlations between electrons arising from exchange and interaction effects are strong and the standard linear response approach, the RPA, breaks down. There are two categories of corrections to the RPA.

The first type of correction is a static effect. Since each electron repels all other electrons from its surroundings, the neighbourhood surrounding each electron contains a reduced density of electrons which is less than the average electron density for the sample as a whole. This reduces the average strength of the repulsion between electrons since electrons spend more time further apart than the overall average electron density would suggest.

The second effect resulting from corrections to the RPA is dynamic in nature and this leads to various dissipative phenomenon, including damping of the RPA plasmon. However there are also much more dramatic effects and these form the centrepiece of our subsequent discussions.

The static exchange-correlation hole surrounding each electron, inside which the density of electrons is less than the average density, includes the effect of the depletion of density arising from exchange corrections to linear response and the effect of the depletion from the repulsive Coulomb interaction acting between the electrons.[1] For densities $r_s \gtrsim 5$ (r_s is the average spacing between electrons in units of the effective Bohr radius), there is a region around each electron where there is essentially a total exclusion of other electrons. By density $r_s \sim 10$, the radius of this exclusion region has become comparable to the average spacing between electrons.[2]

For the static exchange correlation hole, the depleted surroundings around each electron rigidly follows as the electron propagates through the system. The static exchange correlation hole instantly follows each deviation in the electron path as it scatters off other electrons or off defects. Thus the electron always remains at the centre of its static exchange-correlation hole.

Since the time scale for electron relaxation is given by the inverse of the plasmon frequency $\omega_{pl}^{-1} = \sqrt{m/ne^2}$, the static exchange correlation hole model is a good approximation at high electron densities n. However as the electron density is lowered, the reaction of the exchange-correlation hole lags the changes in motion of its associated electron. The electron can no longer always remain at the centre of its hole, and dissipative processes become important.

2. Dynamic Corrections to Linear Response

The exact third–moment sum rule is given by

$$\int_0^\infty d\omega\, \omega^3 S(\mathbf{q}, \omega) =$$

$$nq^2 \left[\frac{q^4}{8} + q^2 \langle E_{kin} \rangle + \frac{\omega_{pl}^2}{2} \left[1 - \frac{1}{N} \sum_k \frac{\mathbf{k} \cdot \mathbf{q}}{kq} [S(k) - S(|\mathbf{q} - \mathbf{k}|)] \right] \right] . \quad (1)$$

$S(\mathbf{q}, \omega)$ and $S(\mathbf{q})$ are the dynamic and static structure factors, respectively, and $\langle E_{kin} \rangle$ is the expectation value for the kinetic energy in the interacting system. Because of the ω^3 weighting in the integral, the third–moment sum rule is sensitive to the relative transient movement over short–time scales between the exchange-correlation hole and the electron at the centre of the exchange-correlation hole. The sum rule has been exploited to determine the frequency dependence of a local field factor $G(\mathbf{q}, \omega)$.[3]

One can systematically investigate the dynamics using a conserving perturbative expansion of the most important corrections to the RPA dynamic Response Function. Dynamics are expressed as a sequence of independent binary collisions.[4] The results can be expressed in terms of modifications to the linear response density-density response function $\chi(\mathbf{q}, \omega)$,

$$\chi(\mathbf{q}, \omega) = \frac{\chi^0(\mathbf{q}, \omega)}{1 - V_{coul}(\mathbf{q})\chi^0(\mathbf{q}, \omega)} \rightarrow \frac{\tilde{\chi}(\mathbf{q}, \omega)}{1 - V_{coul}(\mathbf{q})G(\mathbf{q}, \omega)\tilde{\chi}(\mathbf{q}, \omega)} . \tag{2}$$

$\chi^0(\mathbf{q}, \omega)$ is the Lindhard response function for the non-interacting system, $V_{coul}(\mathbf{q})$ is the bare Coulomb interaction. In Eq. 2, the RPA expression for $\chi(\mathbf{q}, \omega)$ is modified by a dynamic local field factor $G(\mathbf{q}, \omega)$ and $\tilde{\chi}(\mathbf{q}, \omega)$. The $\tilde{\chi}(\mathbf{q}, \omega)$ contains self-energy correction insertions to $\chi^0(\mathbf{q}, \omega)$. The simultaneous modifications $G(\mathbf{q}, \omega)$ and $\tilde{\chi}(\mathbf{q}, \omega)$ ensure that the conservation laws are maintained.

3. Dynamics of Density Fluctuations

Retaining only independent binary collisions[4] omits effects that become increasing important at low electron densities. The expansion of the region of total density exclusion in the exchange-correlation hole surrounding each electron has an analogy in the development of a virtual "hard core" in the electron-electron interaction potential. At low densities the area of the "hard cores" are comparable to the average area occupied by each electron. For a classical 2D system of hard discs it is known that a hard disc can be trapped for long periods of time in a cage temporarily formed by the hard discs surrounding it. It is useful to ask if there is not an analogous effect for electrons with their large virtual "hard cores". To adequately study this effect we must allow each electron to couple to the fluid of electrons surrounding it. The surrounding fluid acts as a whole on the electron. Sequential binary collisions cannot represent this.

By coupling the electron to density fluctuations we can emulate the forces associated with the dynamic density profile surrounding each electron. Density fluctuations involve the collective motion of large numbers of particles, and their time evolution is significantly affected by conservation requirements. When the time evolution becomes much slower than the single–electron scattering time, then the density excitations do not have time to adjust to changes in the electron motion. In this case the density fluctuations exert a significant force on the propagating electron, thus affecting the subsequent motion of the propagating electron.

We want to calculate the behaviour of the density fluctuation $\rho_q(t)$. When q is small compared with the inverse of the system size L, we know that

$$\rho_q(t) = \rho_q e^{i\omega_{pl} t} . \tag{3}$$

However when q is large, $\rho_q(t)$ will have structure on a scale of the inverse of the average electron spacing.

When we include dynamic effects, there will be randomly fluctuating forces $f(t)$ acting on the $\rho_q(t)$. These fluctuating forces have a damping effect, causing $\rho_q(t)$ to lose amplitude and energy. As the average electron density is decreased, correlations between electrons become stronger and the short-distance structure is more pronounced. If the dissipation effects become strong enough they can change the state of the system.[5]

The time correlation function for density fluctuation is the ensemble average

$$F(\mathbf{q}, t) = \frac{1}{N} <\rho_{-\mathbf{q}}(0)\rho_{\mathbf{q}}(t)> \tag{4}$$

$$= \frac{1}{N} \sum_{ij} < e^{-i\mathbf{q}\cdot\mathbf{r}_i(0)} e^{-i\mathbf{q}\cdot\mathbf{r}_j(t)} > , \tag{5}$$

with initial value $F(\mathbf{q}, 0) = S(\mathbf{q})$, the static Structure Factor.

For a normal electron liquid we expect

$$F(\mathbf{q}, t) = F(\mathbf{q}, 0)e^{-t/\tau} , \tag{6}$$

with the decay time τ of order of the time scale for electron-electron collisions. Thus for a normal electron liquid the probability of finding an electron at distance $\mathbf{r}$ from the origin at time $t > 0$, when there had been an electron at the origin at $t=0$, goes to unity on a time scale τ,

$$\langle \rho(\mathbf{r}, t) | \rho(0, 0) \rangle \to 1 . \tag{7}$$

4. Dynamics in the Classical System

Let us first consider the $\rho_q(t)$ as a classical function of the phase space variables. Its time dependence is obtained from the equation of motion

$$\frac{d\rho_q}{dt} = \{\rho_q(t), H\} = \frac{1}{m} \sum_i \left(p_i \cdot \frac{\partial}{\partial r_i}\right) \rho(t) = i\mathcal{L}\rho_q(t) , \tag{8}$$

where the $\{,\}$ are Poisson brackets, the sum is over all electron coordinates and $\mathcal{L}$ is the Liouvillian operator. We introduce a scalar product between any two variables A and B

$$(A, B) = < A^\star B > \tag{9}$$

and a projection operator $\mathcal{P}$ which for any variable A projects A on to the variable ρ:

$$\mathcal{P}A = (\rho, A)(\rho, \rho)^{-1}\rho . \tag{10}$$

Geometrically, the effect of $\mathcal{P}$ is to extract the ρ-variable part contained in A.

For small q, the operator ρ_q is a hydrodynamic variable that varies slowly in time, and its equation of motion in this case is simply the continuity equation,

$$\frac{d\rho}{dt} = \nabla \cdot \mathbf{j} . \tag{11}$$

However we want to also determine the equation of motion for $\rho_q(t)$ when q is not small and fast fluctuations in the density can be important. The formal solution of Eq. 8 is $\rho_q(t) = e^{i\mathcal{L}t}\rho_q$, so

$$\frac{d\rho_q(t)}{dt} = i\mathcal{L}e^{i\mathcal{L}t}\rho_q \tag{12}$$

$$= e^{i\mathcal{L}t}[\mathcal{P} + (1 - \mathcal{P})]i\mathcal{L}\rho_q \tag{13}$$

$$= i\Omega \cdot \rho_q(t) + e^{i\mathcal{L}t}(1 - \mathcal{P})i\mathcal{L}\rho_q \tag{14}$$

where

$$i\Omega = (\rho_q, i\mathcal{L}\rho_q).(\rho_q, \rho_q)^{-1} . \tag{15}$$

The first term in Eq. 14 is the ρ_q component of the time-development of $\rho_q(t)$,

$$i\Omega\dot{\rho}_q(t) = i(\rho_q(t), \mathcal{L}\rho_q).(\rho_q, \rho_q)^{-1}\rho_q(t) = i\mathcal{P}\mathcal{L}\rho_q = i\mathcal{P}(d\rho_q/dt) . \tag{16}$$

The second term in Eq. 14, $e^{i\mathcal{L}t}(1 - \mathcal{P})i\mathcal{L}\rho_q$, represents that part of $\rho_q(t)$ which has been forced out of its original slowly varying variable space. This term is generated by random forces $f(t)$ that vary rapidly in space and time. $f(t)$ is caused by random interactions of the density fluctuation with the electrons surrounding it. $f(t)$ acts dissipatively on $\rho_q(t)$. We define

$$f(t) = e^{i(1-\mathcal{P})\mathcal{L}t}i(1 - \mathcal{P})i\mathcal{L}\rho_q . \tag{17}$$

By introducing an auxiliary function

$$\mathcal{O}(t) = 1 - e^{-i\mathcal{L}t}e^{i(1-\mathcal{P})\mathcal{L}t} \tag{18}$$

we can show that

$$e^{i\mathcal{L}t}(1 - \mathcal{P})i\mathcal{L}\rho_q = \int_0^t d\tau e^{i\mathcal{L}(t-\tau)}\mathcal{P}i\mathcal{L}f(\tau) + f(t) \tag{19}$$

as follows:

$$\int_0^t d\tau e^{i\mathcal{L}(t-\tau)}\mathcal{P}i\mathcal{L}f(\tau) = \int_0^t d\tau e^{i\mathcal{L}(t-\tau)}(\rho_q, \rho_q)^{-1}(\rho_q, \rho_q)\mathcal{P}i\mathcal{L}f(\tau)$$

$$= \int_0^t d\tau \rho_q(t - \tau)(\rho_q, \rho_q)^{-1}i(\mathcal{L}f(\tau), \rho_q)$$

$$= \int_0^t d\tau \rho_q(t - \tau)(\rho_q, \rho_q)^{-1}(f(\tau), i\mathcal{L}\rho_q)$$

$$= \int_0^t d\tau \rho_q(t - \tau)(\rho_q, \rho_q)^{-1}(f(\tau), i(1 - \mathcal{P})\mathcal{L}\rho_q)$$

$$= \int_0^t d\tau \rho_q(t - \tau)(\rho_q, \rho_q)^{-1}(f(\tau), f(t{=}0)) \tag{20}$$

$$\frac{d\mathcal{O}(t)}{dt} = (-i\mathcal{L} + i(1 - \mathcal{P})\mathcal{L})(1 - \mathcal{O}(t))$$

$$= (-i\mathcal{P})\mathcal{L})e^{-i\mathcal{L}t}e^{i(1-\mathcal{P})\mathcal{L}t} \tag{21}$$

$$\mathcal{O}(t) = \int dt(-i\mathcal{P})\mathcal{L})e^{-i\mathcal{L}t}e^{i(1-\mathcal{P})\mathcal{L}t} \tag{22}$$

$$e^{i\mathcal{L}t} = e^{i\mathcal{L}t}\mathcal{O}(t) + e^{i(1-\mathcal{P})\mathcal{L}t} \tag{23}$$

$$e^{i\mathcal{L}t}(1 - \mathcal{P})i\mathcal{L}\rho_q = \left[e^{i\mathcal{L}t}\mathcal{O}(t) + e^{i(1-\mathcal{P})\mathcal{L}t}\right](1 - \mathcal{P})i\mathcal{L}\rho_q . \tag{24}$$

Hence

$$e^{i\mathcal{L}t}(1-\mathcal{P})i\mathcal{L}\rho_q = \left[e^{i\mathcal{L}t} \int dt(-i\mathcal{P})\mathcal{L})e^{-i\mathcal{L}t}e^{i(1-\mathcal{P})\mathcal{L}t} + e^{i(1-\mathcal{P})\mathcal{L}t} \right] (1-\mathcal{P})i\mathcal{L}\rho_q , \quad (25)$$

from which Eq. 19 follows.

Using Eq. 19 we can write Eq. 14 as

$$\frac{d\rho_q}{dt} = i\Omega.\rho_q(t) + \int_0^t d\tau e^{i\mathcal{L}(t-\tau)}\mathcal{P})i\mathcal{L}f(\tau) + f(t) . \quad (26)$$

It is important to note that $d\rho_q(t)/dt$ is affected not only by the instantaneous $f(t)$ at time t, but also by the past *history* of $f(t)$. The time over which the system remembers its earlier states in Eq. 26 is $e^{i\mathcal{L}(t-\tau)}\mathcal{P}i\mathcal{L}$.

By introducing the memory function $K(t)$ defined as

$$K(t) = (\rho_q, \rho_q)^{-1}(f(t), f(t=0)) , \quad (27)$$

we can write Eq. 26 as

$$\frac{d\rho_q}{dt} = i\Omega.\rho_q(t) + \int_0^t d\tau \rho_q(t-\tau).K(\tau) + f(t) . \quad (28)$$

5. Dynamic Effects in Quantum Systems

The Hamiltonian for our quantum system is,

$$H = \sum_k \epsilon_k a_k^\dagger a_k + \frac{1}{2}\sum_q v(q)\rho_q\rho_{-q} + \sum_q u(q)\rho_{-q} . \quad (29)$$

$\epsilon_k = \hbar^2 k^2/(2m)$, $v(q)$ is the Coulomb interaction between electrons and $u(q)$ is the electron-defect potential

$$u(q) = w(q) \sum_{imp} \exp(-\,i\mathbf{q}\cdot\mathbf{R}_{imp}) . \quad (30)$$

The summation is over the position of all defects at $\mathbf{R}_{imp}$ and $w(q)$ is the electron-defect interaction.

While the extension of the Mori formalism[6] to quantum fluids[7] has suffered from controversy regarding the treatment of quantum interference effects, this is not an issue for the phenomena in which we are interested here since, as we have seen, the strong exchange–correlation hole makes the classical density fluctuations a dominating effect at low electron densities. We study their effect on the propagation of electrons and so we continue to work in the basis set formed by the density fluctuations.

Introducing the normalised density fluctuations $N_q(t) = \rho_q(t)/\sqrt{\chi(q)}$, where $\chi(q) = \langle \rho_q(0)|\rho_q(0)\rangle$ is the static density-density response function, the Kubo-relaxation function for $N_q(t)$ is defined as

$$\Phi(q,t) \equiv \langle N_q(t)|N_0(0)\rangle . \quad (31)$$

Its equation of motion is

$$i\frac{d\Phi(q,t)}{dt} = \langle [N_{\mathrm{q}}(t), N_0(0)] \rangle \ . \tag{32}$$

Therefore

$$i\int_0^\infty dt\, e^{izt}\frac{d\Phi(q,t)}{dt} = ie^{izt}\Phi(q,t)\Big|_0^\infty + \int_0^\infty dt\, ze^{izt}\Phi(q,t)$$

$$= -i(1 - z\Phi(q,z))$$

$$= \int_0^\infty dt\, e^{izt}\langle [N_{\mathrm{q}}(t), N_0(0)]\rangle$$

$$= -i\chi(q,z) \tag{33}$$

$$\Phi(q,z) = \frac{1 - \chi(q,z)}{z} \ . \tag{34}$$

The quantum equation of motion for $N_{\mathrm{q}}(t)$ is given by the commutator of $N_{\mathrm{q}}(t)$ with the Hamiltonian,

$$-i\frac{dN_{\mathrm{q}}(t)}{dt} = [H, N_{\mathrm{q}}(t)] = \mathcal{L}N_{\mathrm{q}}(t) \tag{35}$$

$$N_{\mathrm{q}}(t) = e^{i\mathcal{L}t}N_{\mathrm{q}}(0) \tag{36}$$

$$\Phi(q,z) = i\int_0^\infty dt\, e^{izt}\langle N_{\mathrm{q}}(t)|N_0(0)\rangle \tag{37}$$

$$= i\int_0^\infty dt\, e^{izt}\langle N_{\mathrm{q}}(0)|e^{-i\mathcal{L}t}|N_0(0)\rangle \tag{38}$$

$$= \langle N_{\mathrm{q}}(0)\Big|\frac{1}{\mathcal{L}-z}\Big|N_0(0)\rangle \ . \tag{39}$$

Let us consider the propagation of a density fluctuation forward in time from $N_{\mathrm{q}}(0)$ to $N_{\mathrm{q}}(t)$. One mode for this propagation is via the slow variable channel of hydrodynamic variables in the subspace characterised by the operator

$$\mathcal{P} = |N_q)\,(N_q| \tag{40}$$

In this mode the density fluctuation ρ can only be redistributed in space via a current flow $\mathbf{j}$.

The remaining propagation occurs in the complementary subspace characterised by the operator $\bar{\mathcal{P}} = 1 - \mathcal{P}$. We can represent this propagation through alternating channels (see Fig. 1) as self-energy corrections to the propagation through the slow channel (Fig. 2),

$$\Sigma(z) = \mathcal{P}\mathcal{L}\bar{\mathcal{P}}\frac{1}{z - \bar{\mathcal{P}}\mathcal{L}\bar{\mathcal{P}}}\bar{\mathcal{P}}\mathcal{L}\mathcal{P} \ . \tag{41}$$

The full propagation of the density fluctuation represented by Fig. 1 is given by the Dyson series

$$\mathcal{P}\frac{1}{z-\mathcal{L}}\mathcal{P} = \frac{1}{z - \mathcal{P}\mathcal{L}\mathcal{P}}\left(\frac{1}{1 - \Sigma(z)(z - \mathcal{P}\mathcal{L}\mathcal{P})^{-1}}\right) \tag{42}$$

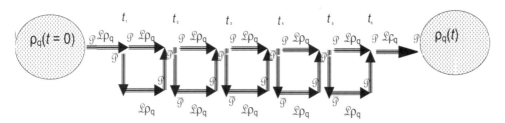

Fig. 1. *Projection operator $\mathcal{P}$ and its complement $\bar{\mathcal{P}}$ repeatedly split the propagation of the density fluctuation $\rho_q(t)$ into the slow-variable and fast-variable channels. $\mathcal{L}$ is the Liovillian time development operator*

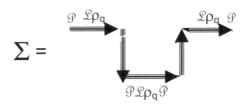

Fig. 2. *Propagating process described by Eq. 41*

which can be rearranged to give

$$\mathcal{P}\frac{1}{z-\mathcal{L}}\mathcal{P} = \frac{1}{z - \mathcal{P}\mathcal{L}\mathcal{P} - \Sigma(z)} \tag{43}$$

$$\mathcal{P}\frac{1}{z-\mathcal{L}}\mathcal{P}\left[z - \mathcal{P}\mathcal{L}\mathcal{P} - \Sigma(z)\right] = 1 \ . \tag{44}$$

Hence

$$\mathcal{P} = \left[\mathcal{P}\mathcal{L}\mathcal{P} - z - \mathcal{P}\mathcal{L}\bar{\mathcal{P}}\frac{1}{\bar{\mathcal{P}}\mathcal{L}\bar{\mathcal{P}} - z}\bar{\mathcal{P}}\mathcal{L}\mathcal{P}\right]\mathcal{P}\frac{1}{\mathcal{L}-z}\mathcal{P} \tag{45}$$

Taking a scalar product of Eq. 45 with the density fluctuation $\left|N_q\right)$ we obtain,

$$1 = \left(N_q\middle|N_q\right) = \left(N_q\middle|\mathcal{P}\middle|N_q\right) =$$

$$= \left[\left(N_q\middle|\mathcal{P}\mathcal{L}\mathcal{P}\middle|N_q\right) - \left(N_q\middle|z\middle|N_q\right) - \left(N_q\middle|\mathcal{P}\mathcal{L}\bar{\mathcal{P}}\frac{1}{\bar{\mathcal{P}}\mathcal{L}\bar{\mathcal{P}} - z}\middle|\bar{\mathcal{P}}\mathcal{L}\mathcal{P}\middle|N_q\right)\right] \times$$

$$\times \left(N_q\middle|\mathcal{P}\middle|N_q\right)\left(N_q\middle|\frac{1}{\mathcal{L}-z}N_q\right)\left(N_q\middle|N_q\right) \tag{46}$$

$$= \left[\left(N_q\middle|\mathcal{L}\middle|N_q\right) - z - \left(N_q\middle|\mathcal{L}\bar{\mathcal{P}}\frac{1}{\bar{\mathcal{P}}\mathcal{L}\bar{\mathcal{P}} - z}\middle|\bar{\mathcal{P}}\mathcal{L}\middle|N_q\right)\right]\left(N_q\middle|\frac{1}{\mathcal{L}-z}\middle|N_q\right) \tag{47}$$

$$= \left[\left(N_q\middle|\mathcal{L}\middle|N_q\right) - z - \left(N_q\middle|\mathcal{L}\bar{\mathcal{P}}\frac{1}{\bar{\mathcal{P}}\mathcal{L}\bar{\mathcal{P}} - z}\bar{\mathcal{P}}\mathcal{L}\middle|N_q\right)\right]\Phi(q,z) \ . \tag{48}$$

Solving Eq. 48 for $\Phi(q, z)$,

$$\Phi(q, z) = -\frac{1}{z + \left(\mathcal{L}N_q|(\bar{\mathcal{P}}\mathcal{L}\bar{\mathcal{P}} - z)^{-1}|\mathcal{L}N_q\right)} \tag{49}$$

$$= -\frac{1}{z + K(q, z)} , \tag{50}$$

where we have defined the current relaxation function as

$$K(q, z) = \left(\mathcal{L}N_q\left|\frac{1}{\tilde{\mathcal{L}} - z}\right|\mathcal{L}N_q\right). \tag{51}$$

In deriving Eq. 50 we have used time reversal symmetry to write $(N_q|\mathcal{L}|N_q) = 0$, so

$$\bar{\mathcal{P}}\mathcal{L}|N_q) = \{\mathcal{L}|N_q) - \mathcal{P}\mathcal{L}|N_q)\} = \{\mathcal{L}|N_q) - |N_q)(N_q|\mathcal{L}|N_q)\} = |\mathcal{L}N_q) . \tag{52}$$

The continuity equation, Eq. 11, ensures the slow varying hydrodynamic normalised current density is $\mathcal{L}N_q$.

Now we introduce a projection operator $\mathcal{J}$ which projects dynamical variables into the subspace spanned by current density fluctuations

$$\mathcal{J} = |\mathcal{L}N_q)\frac{1}{(\mathcal{L}N_q|\mathcal{L}N_q)}(\mathcal{L}N_q| . \tag{53}$$

We can write an identity for $\mathcal{J}$ analogous to Eq. 45:

$$\mathcal{J} = \left[\mathcal{J}\tilde{\mathcal{L}}\mathcal{J} - z - \mathcal{J}\tilde{\mathcal{L}}\bar{\mathcal{J}}\frac{1}{\bar{\mathcal{J}}\tilde{\mathcal{L}}\bar{\mathcal{J}} - z}\bar{\mathcal{J}}\tilde{\mathcal{L}}\mathcal{J}\right]\mathcal{J}\frac{1}{\tilde{\mathcal{L}} - z}\mathcal{J} . \tag{54}$$

We define the complementary projection operator $\bar{\mathcal{J}} = 1 - \mathcal{J}$ and $\tilde{\mathcal{L}} \equiv \bar{\mathcal{P}}\mathcal{L}\bar{\mathcal{P}}$.

Taking scalar product of Eq. 54 with the current fluctuation $|\mathcal{L}N_q)$,

$$1 = \left(\mathcal{L}N_q\Big|\mathcal{L}N_q\right) = \left(\mathcal{L}N_q\Big|\mathcal{J}\Big|\mathcal{L}N_q\right) =$$

$$\left(\mathcal{L}N_q\Big|\left[\mathcal{J}\tilde{\mathcal{L}}\mathcal{J} - z - \mathcal{J}\tilde{\mathcal{L}}\bar{\mathcal{J}}\frac{1}{\bar{\mathcal{J}}\tilde{\mathcal{L}}\bar{\mathcal{J}} - z}\bar{\mathcal{J}}\tilde{\mathcal{L}}\mathcal{J}\right] \times\right.$$

$$\left.\left\{|\mathcal{L}N_q)\frac{1}{(\mathcal{L}N_q|\mathcal{J}\mathcal{L}N_q)}(\mathcal{L}N_q|\right\}\frac{1}{\tilde{\mathcal{L}} - z}\Big|\mathcal{L}N_q\right) . \tag{55}$$

Now $\mathcal{J}|\mathcal{L}N_q) = |\mathcal{L}N_q)$ and

$$(\mathcal{L}N_q|\bar{\mathcal{P}}\mathcal{L}\bar{\mathcal{P}}|\mathcal{L}N_q) = \left\{(\bar{\mathcal{P}}\mathcal{L}N_q|\mathcal{L}^2N_q) - (\bar{\mathcal{P}}\mathcal{L}N_q|N_q)(N_q|\mathcal{L}N_q)\right\} = 0 . \tag{56}$$

$(\bar{\mathcal{P}}\mathcal{L}N_q|\mathcal{L}^2N_q) = 0$ and $(N_q|\mathcal{L}N_q) = 0$ both vanish from time inversion symmetry. Thus

$$1 = \left[-z - \left(\mathcal{L}N_q\Big|\tilde{\mathcal{L}}\bar{\mathcal{J}}\frac{1}{\bar{\mathcal{J}}\tilde{\mathcal{L}}\bar{\mathcal{J}} - z}\bar{\mathcal{J}}\tilde{\mathcal{L}}\Big|\mathcal{L}N_q\right)\right]\left\{\frac{1}{(\mathcal{L}N_q|\mathcal{J}\mathcal{L}N_q)}(\mathcal{L}N_q|\right\}\frac{1}{\tilde{\mathcal{L}} - z}\Big|\mathcal{L}N_q\right) \tag{57}$$

Recalling the definition of $K(q,z)$ (Eq. 51), and solving Eq. 57 for $K(q,z)$ we obtain:

$$K(q,z) = -\frac{\left(\mathcal{L}N_q \middle| \mathcal{L}N_q\right)}{z + \left((\mathcal{L}N_q|\mathcal{L}N_q)\right)^{-1} \left(\mathcal{L}N_q \tilde{\mathcal{L}} \bar{\mathcal{J}} \middle| (\bar{\mathcal{J}} \tilde{\mathcal{L}} \bar{\mathcal{J}} - z)^{-1} \middle| \bar{\mathcal{J}} \tilde{\mathcal{L}} \mathcal{L}N_q\right)} \ . \tag{58}$$

Now

$$\left(\mathcal{L}N_q \middle| \mathcal{L}N_q\right) = \chi^{-1}(q)\left(\mathcal{L}\rho_q \middle| \mathcal{L}\rho_q\right) \equiv \chi^{-1}(q)\langle [\rho_q,[\rho_q,H]]\rangle \ , \tag{59}$$

and

$$[\rho_q, H] = \left[\sum_i e^{-i\mathbf{q}\cdot\mathbf{r}_i}, \sum_j \frac{p_j^2}{2m}\right] \tag{60}$$

$$= \frac{1}{2}\sum_i \frac{\mathbf{q}\cdot\mathbf{p}_i}{m} e^{-i\mathbf{q}\cdot\mathbf{r}_i} + e^{-i\mathbf{q}\cdot\mathbf{r}_i}\frac{q\cdot p_i}{m} \tag{61}$$

$$= \mathbf{q}\cdot\mathbf{J}_q \ . \tag{62}$$

Then

$$[\rho_q, [\rho_q, H]] = [\sum_i e^{-iq\cdot r_i}, \frac{1}{2}\sum_j \frac{\mathbf{q}\cdot\mathbf{p}_j}{m} e^{-iq\cdot r_j} + e^{-iq\cdot r_j}\frac{\mathbf{q}\cdot\mathbf{p}_j}{m}] \tag{63}$$

$$= \frac{Nq^2}{m} \ , \tag{64}$$

so

$$\left(\mathcal{L}N_q \middle| \mathcal{L}N_q\right) = \chi^{-1}(q)(q^2/m) = \Omega_q \ . \tag{65}$$

where $\Omega_q = q^2/\left(m\chi(q)\right)$. Equation 65 is a version of the f-sum rule.

Now $\mathcal{P}|\mathcal{L}N_q) = 0$ and

$$\left|\bar{\mathcal{J}}\tilde{\mathcal{L}}\mathcal{L}N_q\right) = \left\{1 - \Omega_q^{-1}|\mathcal{L}N_q)(\mathcal{L}N_q|\right\}\left|\bar{\mathcal{P}}\mathcal{L}\bar{\mathcal{P}}\mathcal{L}N_q\right) \tag{66}$$

$$= \left\{1 - \Omega_q^{-1}|\mathcal{L}N_q)(\mathcal{L}N_q|\right\}\bar{\mathcal{P}}\mathcal{L}^2 N_q) = |\bar{\mathcal{P}}\mathcal{L}^2 N_q) \tag{67}$$

since by symmetry $\left(\mathcal{L}N_q\middle|\bar{\mathcal{P}}\mathcal{L}^2 N_q\right) = 0$. Therefore Eq. 58 becomes

$$K(q,z) = -\frac{\Omega_q}{z + (m/q^2)\left(\bar{\mathcal{P}}\mathcal{L}^2\rho_q\middle|(\bar{\mathcal{J}}\tilde{\mathcal{L}}\bar{\mathcal{J}} - z)^{-1}\middle|\bar{\mathcal{P}}\mathcal{L}^2\rho_q\right)} \tag{68}$$

$$= -\frac{\Omega_q}{z + M(q,z)} \ , \tag{69}$$

where

$$M(q,z) = \frac{m}{q^2}\left(\bar{\mathcal{P}}\mathcal{L}^2\rho_q\middle|\frac{1}{\bar{\mathcal{J}}\tilde{\mathcal{L}}\bar{\mathcal{J}} - z}\middle|\bar{\mathcal{P}}\mathcal{L}^2\rho_q\right) \tag{70}$$

is the force-force correlation function. Using Eq. 50, we thus obtain for the Kubo relaxation function the following expression:

$$\Phi(q,z) = -\frac{1}{z - \Omega_q(z + M(q,z))^{-1}} \ . \tag{71}$$

To determine Eq. 70 we first evaluate $\bar{\mathcal{P}}\mathcal{L}^2\rho_q$. For our Hamiltonian the equation of motion gives

$$\bar{\mathcal{P}}\mathcal{L}^2\rho_q = \bar{\mathcal{P}}\Big[H,[H,\rho_q]\Big] \tag{72}$$

$$= \bar{\mathcal{P}}\Big[H,\frac{1}{2}\sum_i \frac{q\cdot p_i}{m}e^{-iq\cdot r_i} + e^{-iq\cdot r_i}\frac{q\cdot p_i}{m}\Big] \tag{73}$$

$$= \bar{\mathcal{P}}\Big[H,\mathbf{q}\cdot\mathbf{J}_q\Big] . \tag{74}$$

Since $\mathbf{J}_q$ depends on both r_i and p_i, we get contributions to the commutator in Eq. 74 from both the kinetic energy and interaction terms of the Hamiltonian. However the $\bar{\mathcal{P}}$ operator projects out the contribution from the kinetic energy.

In Eq. 74, the electron-electron interaction term from the Hamiltonian gives:

$$\Big[\sum_k v(k)\rho_k\rho_{-k}, \frac{1}{2}\sum_m \frac{\mathbf{q}\cdot\mathbf{p}_m}{m}e^{-iq\cdot r_m} + e^{-iq\cdot r_m}\frac{\mathbf{q}\cdot\mathbf{p}_m}{m}\Big]$$

$$= \Big[\sum_k v(k)\sum_i e^{ikr_i}\sum_j e^{-ikr_j}, \frac{1}{2}\sum_m \frac{\mathbf{q}\cdot\mathbf{p}_m}{m}e^{-iq\cdot r_m} + e^{-iq\cdot r_m}\frac{\mathbf{q}\cdot\mathbf{p}_m}{m}\Big] \tag{75}$$

$$= \sum_k v(k)\sum_i e^{ikr_i}\sum_j e^{-ikr_j}\frac{1}{2}\Big\{\frac{\mathbf{q}\cdot\mathbf{k}}{m}e^{-iq\cdot r_j} + e^{-iq\cdot r_j}\frac{\mathbf{q}\cdot\mathbf{k}}{m}\Big\} \tag{76}$$

$$= \sum_k v(k)\rho_k\rho_{q-k}\frac{\mathbf{q}\cdot\mathbf{k}}{m} , \tag{77}$$

while the electron-impurity interaction term gives:

$$\Big[\sum_k u(k)\rho_{-k}, \frac{1}{2}\sum_m \frac{\mathbf{q}\cdot\mathbf{p}_m}{m}e^{-iq\cdot r_m} + e^{-iq\cdot r_m}\frac{q\cdot\mathbf{q}\cdot\mathbf{p}_m}{m}\Big]$$

$$= \Big[\sum_k u(k)\sum_i e^{ikr_i}, \frac{1}{2}\sum_m \frac{\mathbf{q}\cdot\mathbf{p}_m}{m}e^{-iq\cdot r_m} + e^{-iq\cdot r_m}\frac{\mathbf{q}\cdot\mathbf{p}_m}{m}\Big] \tag{78}$$

$$= \sum_k u(k)\sum_i e^{ikr_i}, \frac{1}{2}\Big\{\frac{\mathbf{q}\cdot\mathbf{k}}{m}e^{-iq\cdot r_i} + e^{-iq\cdot r_i}\frac{\mathbf{q}\cdot\mathbf{k}}{m}\Big\} \tag{79}$$

$$= \sum_k u(k)\rho_{q-k}\frac{\mathbf{q}\cdot\mathbf{k}}{m} . \tag{80}$$

We conclude that Eq. 74 leads to

$$\bar{\mathcal{P}}\mathcal{L}^2\rho_q = \Big\{\frac{1}{m}\sum_k v(k)(\mathbf{q}\cdot\mathbf{k})\rho_k\rho_{\mathbf{q}-\mathbf{k}} + \frac{1}{m}\sum_k u(k)(\mathbf{q}\cdot\mathbf{k})\rho_{\mathbf{q}-\mathbf{k}}\Big\} . \tag{81}$$

Thus we can write Eq. 70 as

$$M(q,z) = \frac{1}{mq^2}\Bigg(\Big[\sum_k v(k)(\mathbf{q.k})\rho_k\rho_{\mathbf{q}-\mathbf{k}} + \sum_k u(k)(\mathbf{q.k})\rho_{\mathbf{q}-\mathbf{k}}\Big]\Big|\frac{1}{\bar{\mathcal{J}}\tilde{\mathcal{L}}\bar{\mathcal{J}} - z}\Big|$$

$$\Big[\sum_k v(k)(\mathbf{q.k})\rho_k\rho_{\mathbf{q}-\mathbf{k}} + \sum_k u(k)(\mathbf{q.k})\rho_{\mathbf{q}-\mathbf{k}}\Big]\Bigg) . \tag{82}$$

For convenience let us split Eq. 82 for $M(q, z)$ into three terms,

$$M(q, z) = M_{\text{el}-\text{el}}(q, z) + M_{\text{imp}-\text{imp}}(q, z) + M_{\text{imp}-\text{el}}(q, z) + M^{\star}_{\text{imp}-\text{el}}(q, z) , \quad (83)$$

where

$$M_{\text{el}-\text{el}}(q, z) = \frac{1}{mq^2} \left(\sum_k v(k)(\mathbf{q}.\mathbf{k})\rho_k \rho_{\mathbf{q}-\mathbf{k}} \left| \frac{1}{\bar{\mathcal{J}}\tilde{\mathcal{L}}\bar{\mathcal{J}} - z} \right| \sum_k v(k)(\mathbf{q}.\mathbf{k})\rho_k \rho_{\mathbf{q}-\mathbf{k}} \right)$$

$$M_{\text{imp}-\text{imp}}(q, z) = \frac{1}{mq^2} \left(\sum_k u(k)(\mathbf{q}.\mathbf{k})\rho_{\mathbf{q}-\mathbf{k}} \left| \frac{1}{\bar{\mathcal{J}}\tilde{\mathcal{L}}\bar{\mathcal{J}} - z} \right| \sum_k u(k)(\mathbf{q}.\mathbf{k})\rho_{\mathbf{q}-\mathbf{k}} \right)$$

$$M_{\text{imp}-\text{el}}(q, z) = \frac{1}{mq^2} \left(\sum_k u(k)(\mathbf{q}.\mathbf{k})\rho_{\mathbf{q}-\mathbf{k}} \left| \frac{1}{\bar{\mathcal{J}}\tilde{\mathcal{L}}\bar{\mathcal{J}} - z} \right| \sum_k v(k)(\mathbf{q}.\mathbf{k})\rho_k \rho_{\mathbf{q}-\mathbf{k}} \right) \quad (84)$$

The $M_{\text{imp}-\text{imp}}(q, z)$ in Eq. 84 is a two-point density relaxation function

$$M_{\text{imp}-\text{imp}}(q, z) = \frac{1}{mq^2} \sum_{kk'} u(k)(\mathbf{q}.\mathbf{k}) \left(\rho_{\mathbf{q}-\mathbf{k}} \left| \frac{1}{\bar{\mathcal{J}}\tilde{\mathcal{L}}\bar{\mathcal{J}} - z} \right| \rho_{\mathbf{q}-\mathbf{k}'} \right) u(k')(\mathbf{q}.\mathbf{k}') \quad (85)$$

To close the system of equations we simplify $\bar{\mathcal{J}}\tilde{\mathcal{L}}\bar{\mathcal{J}} = \mathcal{L}$, and obtain

$$M_{\text{imp}-\text{imp}}(q, z) = \frac{1}{mq^2} \sum_{kk'} \left\langle (u(k)(\mathbf{q}.\mathbf{k})u(k')(\mathbf{q}.\mathbf{k}')) \right\rangle \left(\rho_{\mathbf{q}-\mathbf{k}}(t) \middle| \rho_{\mathbf{q}-\mathbf{k}'}(0) \right) \quad (86)$$

$$= \sum_k \left[u^2(k) S_{\text{imp}}(k)(\mathbf{q}.\mathbf{k})^2 \right] \times \left[\chi(|\mathbf{q} - \mathbf{k}|)\Phi(|\mathbf{q} - \mathbf{k}|, t) \right] . \quad (87)$$

$S_{\text{imp}}(k)$ is the impurity structure factor. In the absence of correlations between the randomly distributed impurities, we can set $S_{\text{imp}}(k) = 1$.

The contribution to $M(q, z)$ coming from electron-electron scattering $M_{\text{el}-\text{el}}$ in Eq. 84 is a four-point relaxation function:

$$M_{\text{el}-\text{el}}(q, z) = \frac{1}{mq^2} \sum_{kk'} v(k)(\mathbf{q}.\mathbf{k}) \left(\rho_k \rho_{\mathbf{q}-\mathbf{k}} \left| \frac{1}{\mathcal{L} - z} \right| \rho_{k'} \rho_{\mathbf{q}-\mathbf{k}'} \right) v(k')(\mathbf{q}.\mathbf{k}') . (88)$$

We reduce the four-point function to two two-point-functions

$$\left(\rho_k \rho_{\mathbf{q}-\mathbf{k}} \bar{\mathcal{P}} \left| \frac{1}{\mathcal{L} - z} \right| \bar{\mathcal{P}} \rho_{k'} \rho_{\mathbf{q}-\mathbf{k}'} \right)$$

$$= \left[\langle \rho_k(t)\rho_{k'}(0) \rangle \langle \rho_{\mathbf{q}-\mathbf{k}}(t)\rho_{\mathbf{q}-\mathbf{k}'}(0) \rangle + \langle \rho_k(t)\rho_{\mathbf{q}-\mathbf{k}'}(0) \rangle \langle \rho_{\mathbf{q}-\mathbf{k}}(t)\rho_{k'}(0) \rangle \right] , \quad (89)$$

to obtain the approximate expression

$$M_{el-el}(q,t) = \frac{1}{mq^2} \sum_{kk'} v(k)(\mathbf{q}.\mathbf{k}) \Big[\langle \rho_k(t)\rho_{k'}(0)\rangle\langle \rho_{\mathbf{q}-\mathbf{k}}(t)\rho_{\mathbf{q}-\mathbf{k}'}(0)\rangle$$

$$+\langle \rho_k(t)\rho_{\mathbf{q}-\mathbf{k}'}(0)\rangle\langle \rho_{\mathbf{q}-\mathbf{k}}(t)\rho_{k'}(0)\rangle \Big] v(k')(\mathbf{q}.\mathbf{k}') \tag{90}$$

$$= \frac{1}{mq^2} \sum_k v(k)(\mathbf{q}.\mathbf{k}) \Big[\Phi(k,t)\chi(k)\Phi(q-k,t)\chi(q-k)v(k)(\mathbf{q}.\mathbf{k})$$

$$+\Phi(k,t)\chi(k)\Phi(q-k,t)\chi(q-k)v(k-q)(\mathbf{q}.(\mathbf{k}-\mathbf{q})) \tag{91}$$

$$= \frac{1}{2mq^2} \sum_k \Big[v(k)(\mathbf{q}\cdot\mathbf{k}) + v(\mathbf{q}-\mathbf{k})(\mathbf{q}\cdot(\mathbf{q}-\mathbf{k})) \Big]^2 \chi(q')\Phi(k,t)\chi(|\mathbf{q}-\mathbf{k}|)\Phi(|\mathbf{q}-\mathbf{k}|,t) \tag{92}$$

To obtain the last expression, Eq. 92, we have added a second equivalent term on the right hand side with the summation index $\mathbf{k} \to \mathbf{q} - \mathbf{k}$ interchanged, and then divided by two.

Since we focus on phenomenon where the dominant effect comes from strong electron-electron correlations and there are only weak electron-impurity interactions we neglect $M_{imp-el}(q,z)$ in Eq. 84.

Taking the limit $t \to \infty$ ($z \to 0$) in Eq. 83, we define the infinite relaxation time memory function as

$$M^{(\infty)}(q) \equiv \lim_{t\to\infty} M(q,t) = \lim_{z\to0}\{zM_{el-el}(q,z) + zM_{imp-imp}(q,z)\} . \tag{93}$$

6. Order Parameter for Quantum Glass

We introduce an order parameter for a structurally arrested state

$$f(q) = \lim_{t\to\infty} \Phi(q,t) = -\lim_{z\to0} z\Phi(q,z) . \tag{94}$$

The motivation for this choice for the order parameter is that the retarded pair correlation function at infinite time is then given by

$$\lim_{t\to\infty} g(r,t) = 1 + \int dq f(q)e^{i\mathbf{q}\dot{\mathbf{r}}} . \tag{95}$$

Thus when the order parameter $f(q) = 0$, the limit $\lim_{t\to\infty} g(r,t)$ is identically one, implying that over infinite time the system retains no memory of its earlier states.

Combining Eqs. 71, 83, 93 and 94, we obtain

$$f(q) = \frac{1}{1 + \Omega_q/M^{(\infty)}(q)} , \tag{96}$$

with

$$M^{(\infty)}(q) = \{M_{el-el}^{(\infty)}(q) + M_{imp-imp}^{(\infty)}(q)\} \tag{97}$$

$$M_{el-el}^{(\infty)}(q) = \frac{1}{2mq^2} \sum_{q'} \Big[v(q')(\mathbf{q}\cdot\mathbf{q}') + v(\mathbf{q}-\mathbf{q}')(\mathbf{q}\cdot(\mathbf{q}-\mathbf{q}')) \Big]^2 \times$$

$$\times \chi(q')f(q')\chi(|\mathbf{q}-\mathbf{q}'|)f(|\mathbf{q}-\mathbf{q}'|) \tag{98}$$

$$M^{(\infty)}_{\text{imp}-\text{imp}}(q) = \frac{n_i}{mq^2} \sum_{q'} \left[u^2(q')(\mathbf{q} \cdot \mathbf{q}')^2 \right] \chi(|\mathbf{q} - \mathbf{q}'|) f(|\mathbf{q} - \mathbf{q}'|). \qquad (99)$$

Since $M^{(\infty)}(q)$ in Eq. 97 depends on $f(q)$, Eqs. 96-99 form a closed set of non-linear equations for $f(q)$. These can be numerically solved by iteration and the convergence is rapid.

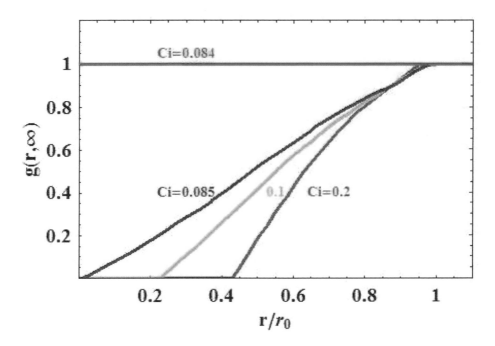

Fig. 3. *The pair correlation function at infinite time* $\lim_{t\to\infty} g(r,t)$ *at electron density corresponding to* $r_s = 10$. $c_i = n_i/n$ *is relative impurity concentration.* n_i *is impurity density.* r_0 *is average electron spacing. At* $r_s = 10$, *when impurity concentration* $c_i < 0.084$, *the limit* $\lim_{t\to\infty} g(r,t) = 1$. *This implies no infinite time memory in the system of its past states. However, for* $c_i > 0.084$, $\lim_{t\to\infty} g(r,t) < 1$ *so that the system retains a memory of its earlier states for an infinite length of time. This memory effect can block the system from relaxing to its true ground state.*

7. Typical Results

The aim of this paper has been to provide an introduction to this powerful formalism. However in this concluding section we briefly provide some representative results from calculations using the formalism (see, for example, Ref. 8).

Referring to Eqs. 98 and 99, at higher charge carrier densities the $\chi(q)$ (and equivalently, the pair correlation function $g(r)$) correspond to weak correlations. For weak disorder this makes the contributions of $\chi(q)$ to $M^{(\infty)}_{\text{el}-\text{el}}(q)$ and $M^{(\infty)}_{\text{imp}-\text{imp}}(q)$

small, and as a result, there are no non-trivial solutions. The only self-consistent solution is $f(q) = 0$. This means the $\rho(q,t)$ decay to zero for $t \to \infty$, the system loses all memory of its earlier states and the phase is liquid.

However as we lower the electron density there is a critical electron density at which the contributions of $\chi(q)$ to $M_{el-el}^{(\infty)}(q)$ and $M_{imp-imp}^{(\infty)}(q)$ become sufficiently strong that there suddenly appear in the coupled equations non-trivial solutions $f(q) > 0$. For the maximum critical electron density $r_s \sim 10$, the impurity concentration at which this jump occurs is $C_i = n_i/n \sim 0.085$, where n_i is the density of the randomly distributed impurities. At $C_i = n_i/n \sim 0.085$, the order parameter $f(q)$ discontinuously jumps from zero, making the retarded pair correlation function $g(r, t = \infty)$ drop from being identically equal to unity, discontinuously down to values $g(r, t = \infty) < 1$ (see Fig. 3).

As the electron density is decreased below $r_s = 10$, this discontinuous drop in $g(r, t = \infty)$ occurs at lower impurity concentrations C_i. This can be understood in the following terms. At lower electron densities the electron correlations are known to be stronger and the virtual "hard-cores" surrounding each electron larger. A larger virtual "hard-core" makes it easier to trap electrons in a cage of surrounding electrons.

In conclusion, at low electron densities we know that each electron is surrounded by a large area in which the density from the other electrons is essentially zero. These zero-density areas act dissipatively on density fluctuations and can produce complete structural arrest, a non-equilibrium glass state. We can speculate that such a state may even block experimental access to the ground state Wigner crystal in these systems.

References

1. K.S. Singwi and M.P. Tosi, in *Solid State Physics*, edited by H. Ehrenreich, F. Seitz and D. Turnbull (Academic, New York, 1981), Vol. **36**, p. 177
2. B. Tanatar and D.M. Ceperley, Phys. Rev. B **39**, 5005 (1989)
3. B. Goodman and A. Sjölander, Phys. Rev. B **8**, 200 (1973); N. Iwamoto, E. Krotschek, and D. Pines, Phys. Rev. B **29**, 3936 (1984)
4. F. Green, D. Neilson and J. Szymański, Phys. Rev. B **31**, 2779 (1985); *ibid.* 2796 (1985); *ibid.* 5837 (1985); F. Green, D. Neilson, D. Pines and J. Szymański, Phys. Rev. B **35**, 133 (1987)
5. U. Bengtzelius, W. Götze W and A. Sjölander, J. Phys. C: Solid State Phys. **17** 5915 (1984); W. Kob, 2002 Lecture Notes for Les Houches 2002 Summer SchoolSession LXXVII: Slow Relaxations and Nonequilibrium Dynamics in Condensed Matter p. 47 [cond-mat/0212344]; David R Reichman and Patrick Charbonneau, J. Stat. Mech. P05013 (2005); R. Zwanzig, in *Lectures in Theoretical Physics*, ed. by W.E. Brittin, B.W. Downs and J. Downs, Vol. III (Interscience, New York, 1961)
6. H. Mori, Prog. Theor. Phys. **33**, 423 (1965)
7. C.D. Boley and J.H. Smith, Phys. Rev. A **12**, 661 (1975); Phys. Rev. B **17**, 4260 (1978); O.T. Valls, G.F. Mazenko and H. Gould, Phys. Rev. B **18**, 263 (1978); Phys. Rev. B **25**, 1663 (1982); W. Götze J. Phys. C **12**, 1279 (1979); Phil. Mag. **43**, 219 (1981)
8. J.S. Thakur and D. Neilson, Phys. Rev. B **54**, 7674 (1996)

DYNAMICAL SPATIALLY RESOLVED RESPONSE FUNCTION OF FINITE 1-D NANO PLASMAS

THOMAS RAITZA

Institut für Physik, Universität Rostock,
18055 Rostock, Germany
thomas.raitza@uni-rostock.de

HEIDI REINHOLZ

Institut für Theoretische Physik, Johannes-Kepler-Universität Linz,
4040 Linz, Austria,
Institut für Physik, Universität Rostock,
18055 Rostock, Germany
heidi.reinholz@jku.at

GERD RÖPKE

Institut für Physik, Universität Rostock,
18055 Rostock, Germany
gerd.roepke@uni-rostock.de

Received 3 December 2009

The dynamical response of one dimensional chains containing 55 till 309 atoms is investigated using a *restricted molecular dynamics* simulation scheme. The total momentum correlation function of an electron cloud shows resonances that are related to different collective excitation modes of the nano plasma. Spatially resolved cross correlation functions are calculated to deduce the spatial structure and strength of these resonance modes. The dependence of the corresponding resonance frequencies on temperature, density and chain size is investigated. The width of the resonances is analyzed in terms of a mode dependent collision frequency.

Keywords: Nano plasma; dynamical response; correlation functions; linear chain; dynamical collision frequency; restricted MD simulations; dispersion; excitation modes.

1. Introduction

The optical response of homogeneous bulk plasmas with electron density n_e and inverse temperature $\beta = (k_B T_e)^{-1}$ is described by the response function $\chi(\vec{k}, \omega)$ which is related to the dielectric function $\epsilon^{-1}(\vec{k}, \omega) = 1 + \chi(\vec{k}, \omega)/(\epsilon_0 k^2)$, the polarization function $\Pi(\vec{k}, \omega) = \chi(\vec{k}, \omega)\epsilon(\vec{k}, \omega)$, and the dynamical structure factor $S(\vec{k}, \omega) = \frac{\hbar}{n_e e^2} \frac{1}{e^{-\beta\hbar\omega} - 1} \mathrm{Im}\chi(\vec{k}, \omega)$. These quantities are of fundamental interest in describing the collective behavior of the system as well as the response to external fields, in particular emission, absorption and scattering of light. Within linear response theory, transport coefficients such as the electrical conductivity are related to these quantities that can be expressed in terms of equilibrium correlation functions. As a well known example, the Kubo formula relates the dc conductivity to the momentum auto-correlation function (ACF).

A correlation function describes the influence of the system at $\vec{r}', t'$ on the action at position $\vec{r}$ and time t. In thermal equilibrium, only the time difference $t - t'$ is of relevance. As a consequence, after Laplace transformation, the response function $\chi(\vec{r}t, \vec{r}'t')$ depends on a single frequency ω. Similarly, in a homogeneous medium, only the difference $\vec{r} - \vec{r}'$ is relevant. Therefore, after a further Fourier transformation, the wave-vector and frequency dependent response function $\chi(\vec{k}, \omega)$ is obtained. Within linear response theory, the response function

$$\chi(\vec{k}, \omega) = -i\beta\Omega_0 \frac{e^2}{m_e^2} \frac{k^2}{\omega} \langle \vec{P}_k; \vec{P}_{-k} \rangle_\omega, \tag{1}$$

with normalization volume Ω_0 is expressed via the ACF of the non-local momentum density $P_k = \Omega_0^{-1} \sum_p \vec{p} a^\dagger_{p-k/2} a_{p+k/2}$ and can be evaluated using quantum statistical approaches such as Green function theory, see Ref. 1. In the long wavelength limit ($k \to 0$), the conductivity has been calculated. The corresponding dynamical collision frequency $\nu(\omega)$ follows from the generalized Drude formula, see Ref. 2,

$$\lim_{k \to 0} \chi(\vec{k}, \omega) = \frac{\varepsilon_0 k^2 \omega_{pl}^2}{\left(\omega^2 - \omega_{pl}^2\right) + i\omega\nu(\omega)} \tag{2}$$

with the plasmon frequency $\omega_{pl} = \left[n_e e^2/(m_e\epsilon_0)\right]^{1/2}$. A sharp peak arises at the plasmon frequency ω_{pl}. For finite wavelengths, the resonance is shifted and can be approximated by the so called Gross-Bohm plasmon dispersion, see Ref. 4, 5, $\omega(k) \approx \omega_{pl} + 3k^2/\kappa^2 + \cdots$ with the Debye screening length $\kappa^{-1} = [n_e e^2/(\epsilon_0 k_B T_e)]^{-1/2}$. This relation has recently been revisited in Ref. 6. It has been found that the dynamical collision frequency, see Ref. 6 and Ref. 7, is crucial for the general behavior of warm dense matter. In the two-component plasma, a phonon mode can arise in addition to the plasmon excitations.

In the classical case, correlation functions can be calculated according to

$$\langle \vec{P}_k; \vec{P}_{-k} \rangle_\omega = \int \frac{d(\vec{r} - \vec{r}')}{\Omega_0} \int_0^\infty d(t - t') \langle \vec{P}(\vec{r}, t) \cdot \vec{P}(\vec{r}', t') \rangle e^{i\vec{k} \cdot (\vec{r} - \vec{r}') - i\omega(t - t')} \tag{3}$$

with the time-dependent local momentum density

$$\vec{P}(\vec{r}, t) = \sum_l^N \delta(\vec{r}_l(t) - \vec{r}) \vec{p}_l(t). \tag{4}$$

using molecular dynamics (MD) simulations. This has been extensively done including the long-wavelength limit $k = 0$. Exemplarily we refer to Ref. 3.

The thermodynamic state of a homogeneous one-component plasma in thermodynamic equilibrium is characterized by the nonideality parameter $\Gamma = e^2 (4\pi n_e/3)^{1/3}(4\pi\varepsilon_0 k_B T_e)^{-1}$ and the degeneracy parameter $\Theta = 2m_e k_B T_e \hbar^{-2}$

$(3\pi^2 n_{\rm e})^{-2/3}$. Applying classical MD simulations techniques, the results are valid only for plasmas, which are non-degenerate ($\Theta \geq 1$). Arbitrary values for the plasma parameter Γ can be treated.

The response function $\chi(\vec{k}, \omega)$ and the related dynamical structure factor $S(\vec{k}, \omega)$ as well as the optical properties have been intensively investigated for electron-ion bulk systems, see Ref. 6 and Ref. 8, and will not be further reported here.

A new situation arises if finite systems are considered. Nano plasmas can nowadays be easily produced in laser irradiated clusters, and new physical phenomena have come into focus experimentally as well as theoretically. Interactions between laser fields of $10^{13} - 10^{16}$ W cm^{-2} and clusters have been investigated over the last few years, see Refs. 9–17. After laser interaction, extremely large absorption rates of nearly 100% were found, see Ref. 18, as well as X-ray radiation, see Refs. 19–25. In pump-probe experiments (Ref. 26), a nano plasma was generated with a first pulse in order to probe it with a second pulse. Quantum and semi-classical methods, see Refs. 27–29, respectively, were used to describe cluster excitation via laser fields. Collisional absorption processes in nano plasmas have been the subject of theoretical investigations in Ref. 30. For the question of optical properties, which are of relevance in such systems, a numerical method via a MD simulations scheme will be discussed in this paper.

Concepts that have been developed for bulk systems near thermodynamic equilibrium have to be altered for applications to finite systems, e.g. clusters. In particular, we are interested to study the dynamical structure factor or the response function for such finite nano plasmas In order to bridge from finite systems to bulk plasmas, we investigate size effects, e.g. in the dynamical collision frequency. First results in this direction have been obtained in Refs. 31–34.

Sec. 2 explains in short the application of the *restricted MDsimulations* scheme for nano plasmas. Based on earlier calculations, local thermal equilibrium can be assumed for the calculation of single-time and two-time properties of the electronic subsystem. Exemplarily, the correlation function of a three dimensional cluster is presented in Sec. 3 which leads to a motivation for the then following analysis of a one dimensional problem. Cross-correlations are discussed in Sec. 4, and analyzed in terms of collective mode excitations in Sec. 5. A further feature is the dispersion relation of the extracted modes, see Sec. 6. Finally, the collisional damping of the modes is estimated using a Drude like expression in order to relate the simulations to a dynamical collision frequency. This is done in Sec. 7 before finishing with some concluding remarks.

2. MD Simulations of Finite Plasmas

The investigation of finite systems has been done using molecular dynamics (MD) simulations based on a method by Suraud *et al.* (Ref. 35). The main difference to bulk MD simulations is the absence of periodic boundary conditions. We want to describe a two-component system of singly charged ions and electrons using an error function pseudo potential for the interaction of particles i, j,

$$V_{\mathrm{erf}}(r_{ij}) = \frac{Z_i Z_j e^2}{4\pi\varepsilon_0 r_{ij}} \mathrm{erf}\left(\frac{r_{ij}}{\lambda}\right),\tag{5}$$

where Z_i is the charge of the ith particle. The Coulomb interaction is modified at short distances, assuming a Gaussian electron distribution, which takes quantum effects into account. Considering a sodium like system, the potential parameter $\lambda = 0.318$ nm was chosen in order to reproduce an ionization energy of $I_{\mathrm{P}} = V_{\mathrm{ei}}(r \to 0) = -5.1$ eV, typically for solid sodium at room temperature.

The velocity Verlet algorithm (Ref. 36) was applied to solve the classical equations of motion for electrons numerically. This method takes into account the conservation of the total energy of the whole finite system, as long as there is no external potential. To follow the fast electron dynamics, time steps of 0.01 fs were taken to calculate the time evolution.

For three dimensional clusters, we considered icosahedral configurations of 55, 147, and 309 ions, see Ref. 34 as initial condition. A homogeneous, spherical distribution of the ions results. The number of electrons is equal to the number of ions. The electrons have been positioned nearby the ions with only small, randomly distributed deviations from the ion positions.

Considering the single-time properties immediately after the electron heating, we obtain (Ref. 34) that local thermodynamic equilibrium is established. In particular, at each time step, the momentum distribution of electrons is well described by the Maxwell distribution, and the density distribution agrees with the Boltzmann distribution with respect to a self-consistent mean field potential. This justifies the application of a *restricted MD simulations* scheme (Ref. 34). The ions are kept fixed acting as external trap potential. Starting from an initial state, the many-particle trajectory $\{\vec{r}_l(t), \vec{p}_l(t)\}$ is calculated, solving the classical equations of motion of the electrons, from which all further physical properties of the cluster are determined. Now, a temporal change of the plasma parameters is not present. A long-time run can be performed in order to replace the ensemble average by a temporal average. This has been successfully done for the single-time properties such as the momentum distribution and the density profile, see Ref. 34 and will now be applied to the two-time correlation functions.

Starting from a homogeneous ion distribution at a fixed ion density, a thermostat was used to heat the electrons. At every time step, a heating rate was calculated via comparison of the electron temperature T_{e}, calculated via kinetic energy, and the intended final temperature T_{aim}. Hot electrons are emitted so that the cluster becomes ionized. Evaluating the trajectories of electrons, sufficient time of 200 fs has to be taken before a stationary ionization degree is established and thermal equilibrium is achieved, characterized by stationary distributions in position and momentum space.

In the one dimensional case, chains of fixed ions with equal distance to each other were taken as trap potentials. Lengths of 55, 147, 200, 250 and 309 ions were investigated. As well as in the 3D case, the electrons are distributed homogeneously on top of the ions. Using the thermostat, the electron temperature was increased. The net charge of the chain was taken to be the same as for the analogous 3D systems. Temperature and ion density were also chosen in the same parameter range as for the spherical clusters.

3. Two-time Correlation Functions

Using the trajectories obtained from the *restricted MD simulations* scheme, we calculate the spatially resolved momentum ACF spectrum

$$K(\vec{r}, \vec{r}', \omega) = \int dt \, K(\vec{r}, \vec{r}', t) e^{-i\omega t}, \tag{6}$$

of the spatially resolved momentum ACF $K(\vec{r}, \vec{r}', t) = \langle \vec{P}(\vec{r}, t) \cdot \vec{P}(\vec{r}', 0) \rangle$ with the local momentum density (4). Technically, we discretize the ACF with respect to cells around a position $\vec{r}_i$ with a finite volume ΔV_i and consider

$$K_{ij}(t) = \langle \vec{P}_i(t) \cdot \vec{P}_j(0) \rangle, \tag{7}$$

with the local momentum density in the cell ΔV_i

$$\vec{P}_i(t) = \frac{1}{\Delta V_i} \sum_l^N \delta_i(\vec{r}_l(t)) \vec{p}_l(t), \tag{8}$$

where $\delta_i(\vec{r}) = 1$ if $\vec{r}$ is found in the cell ΔV_i, and $\delta_i(\vec{r}) = 0$ otherwise. The spatially resolved response function $K_{ij}(t)$ and their Laplace transform $K_{ij}(\omega)$ are matrices with respect to the cells i, j which are identical to the response function $K(\vec{r}, \vec{r}', t)$ and its Laplace transform $K(\vec{r}, \vec{r}', \omega)$, Eq. (6), in the limit $\Delta V_i \to 0$. Because of time inversion symmetry, $K(\vec{r}, \vec{r}', t) = K(\vec{r}, \vec{r}', -t) = K(\vec{r}', \vec{r}, t)$ is real and symmetric, so that $K_{ij}(\omega) = K_{ji}(\omega) = K_{ij}^*(-\omega)$.

Assuming thermal equilibrium, no time instant is singled out in the trajectory obtained from MD simulations. Therefore, we calculate the ensemble average as

$$K_{ij}(t) = \frac{1}{N_\tau} \sum_{a=1}^{N_\tau} \vec{P}_i(a\tau + t) \cdot \vec{P}_j(a\tau), \tag{9}$$

and, correspondingly, for the ACF

$$K(t) = \langle \vec{P}_e(t) \cdot \vec{P}_e(0) \rangle = \frac{1}{N_\tau} \sum_{a=1}^{N_\tau} \vec{P}_e(a\tau + t) \vec{P}_e(a\tau), \tag{10}$$

of the total momentum $\vec{P}_e(t) = \sum_l^N \vec{p}_l(t)$ of the finite system, for a sufficient number of time steps $N_\tau > 10^5$.

Summarizing over all cells or directly from Eq. (10), we obtain the frequency spectrum of the total momentum ACF

$$K(\omega) = \sum_{i,j} \Delta V_i \Delta V_j K_{ij}(\omega) = \int dt K(t) e^{-i\omega t}. \tag{11}$$

In Fig. 1, the frequency spectrum $K(\omega)$ of the total momentum ACF in a cluster of 309 atoms with ionic background density $n_i = 5.2 \cdot 10^{27}$ m^{-3} and electron temperature $T_e = 1.96$ eV is shown. These parameters are identical to those obtained directly after laser irradiation of the cluster. We obtain more than one resonance frequency. Calculations of the momentum ACF or related auto-correlation function

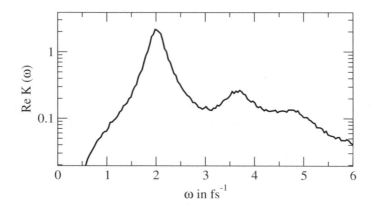

Fig. 1.   Total momentum ACF spectrum calculated via *restricted MD simulations* for a cluster of 309 atoms at electron temperature $T_e = 1.96$ eV and electron density $n_e = 5.2 \cdot 10^{27}$ m^{-3}.

such as dipole or total momentum ACF for finite clusters have also been reported by other authors, see Ref. 28 and 37. Results were usually presented on a linear scale and therefore only a pronounced main peak was observed and interpreted as the Mie resonance of the nano plasma. Additional peaks were also seen, but they are small and have not be discussed in these former works.

Before further analyzing the 3D resonance spectrum and discussing the dependence on the ionic background density and temperature in a future work, we will now consider a simpler model system and calculate the type of collective electron motions for different resonance frequencies.

4. Evaluation of the One-dimensional Bi-local Response Function

The response of an excited cluster in one dimension (a linear chain) was used to discuss the observed multiple resonance structure in more detail. To start with, a chain of 55 ions was taken as background configuration with $n_i = 35$ nm^{-1} and a thermalized electron gas consisting of 46 particles. This leads to a homogeneous ionic background potential of the same spatial extension as was investigated for 3D clusters, and an electron temperature $T_e = 2$ eV. In order to resolve the spatial excitations, the chain was divided into 10 spatial regions. In each cell $\{x_i, \Delta x\}$ the local electron momentum density, Eq.(8),

$$P_i(t) = P(x_i, t) = \frac{1}{\Delta x} \sum_l^N \delta_i(x_l(t)) p_l(t) \tag{12}$$

was determined and cross-correlations $K_{ij}(\omega)$ were calculated according to Eq.(9). For $j = i$, auto-correlation of the local momentum $P(x_i, t)$ results.

In Fig. 2, the real part of the cross-correlation $K_{ij}(\omega)$ for cell $i = 1$ (outer region of the chain) and $i = 5$ (central region of the chain) with all cells $(j = 1, \ldots, 10)$ is shown. Several resonances are obtained. At the lowest resonance frequency

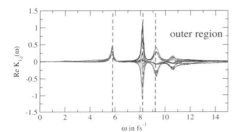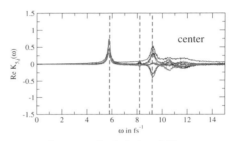

Fig. 2.   Real part of momentum cross-correlation functions for an excited chain of 55 atoms at $n_i = 35$ nm^{-1} and $T_e = 2$ eV. Left: cross-correlation of cell 1 (left outermost region) with all other cells. Right: cross-correlation of cell 5 (left central region) with all other cells. The correlations with the 5 cells on the left are shown in black. Correlations with the 5 cells on the right are shown in red. The first three resonance frequencies are marked with blue vertical dashed lines.

$\omega_1 = 5.8$ fs^{-1}, all cross correlations contribute in the same manner. In contrast, the spatial contributions to the resonance peak at $\omega = 8.2$ fs^{-1} cancel out for the outer cells. For the center of the chain (cell $i = 5$), the second resonance is barely noticeable. The next resonance is observed at $\omega_2 = 9.2$ fs^{-1} with contributions from all cross-correlations in all cells. They partially cancel, but a remaining final contribution remains in the total ACF $K(\omega) = (\Delta x)^2 \sum_{i,j} K_{ij}(\omega)$, Eq.(11), which is shown in Fig. 3. Note, that the resonance structure calculated for the linear chain looks similar to the 3D case, Fig. 1. We find one major resonance peak with two smaller satellite peaks at higher frequencies. The resonance frequencies are also seen in the spatially resolved $K_{ij}(\omega)$. Additional resonances in $K_{ij}(\omega)$, e.g. at $\omega_3 = 8.2$ fs^{-1}, see Fig. 2, are not observed in the total momentum ACF $K(\omega)$, Fig. 3.

5. Analysis of the Excitation Modes

For a more detailed analysis of the frequencies marked in Figs. 2 and 3, we divide the chain in 50 cells which leads to a 50x50 matrix $\hat{K}(x, x', \omega)$. The correlations should be symmetric with respect to interchanging x and x' as well as to space inversion $(x \to -x)$ if the center of the chain is set to $x = 0$. Therefore, we calculate averaged matrix elements $K(x, x', \omega) = [\hat{K}(x, x', \omega) + \hat{K}(x', x, \omega) + \hat{K}(-x, -x', \omega) +$

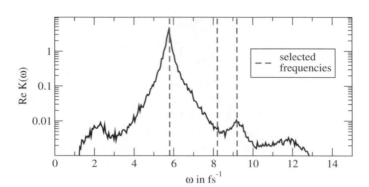

Fig. 3.   Real part of the total momentum ACF $K(\omega)$ for an excited chain of 55 atoms at $n_i = 35$ nm^{-1} and $T_e = 2$ eV. The investigated frequencies are marked with blue vertical dashed lines.

$\hat{K}(-x', -x, \omega)]/4$. For a chosen frequency ω, the eigenvalue problem:

$$\sum_{x'} \operatorname{Re} K(x, x', \omega) \Psi_\mu(x')' = K_\mu(\omega) \Psi_\mu(x). \tag{13}$$

for the matrix $K(x, x', \omega)$ is solved, leading to $N_\mu = 50$ eigenvalues $K_\mu(\omega)$ and corresponding eigenvectors $\Psi_\mu(x)$. The eigenvectors are orthonormal, $\sum_x \Psi_\mu^+(x)\Psi_{\mu'}(x) = \delta_{\mu,\mu'}$. They characterize the local spatial modes. The eigenvalue $K_\mu(\omega)$ gives the strength of the mode at the considered frequency. After summation over all cells, we obtain the total momentum ACF

$$\operatorname{Re} K(\omega) = \sum_{x,x'} \sum_\mu \Psi_\mu^+(x) K_\mu(\omega) \Psi_\mu(x')(\Delta x)^2. \tag{14}$$

Here, we restrict ourselves to the real part of the momentum ACF. The calculations for the imaginary part have been done along the same line and lead to the same resonance spectrum, which is due to the Krames-Kronig relation.

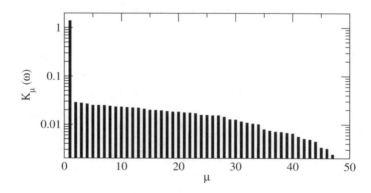

Fig. 4.   Eigenvalues for the first resonance frequency $\omega_1 = 5.8$ fs^{-1} for an excited chain of 55 atoms at $n_i = 35$ nm^{-1} and $T_e = 2$ eV.

We now choose the strongest resonance frequency at $\omega_1 = 5.8$ fs^{-1} and solve Eq. (13). The distribution of eigenvalues, shown in Fig. 4, is presented in the order of strength $K_\mu(\omega)$. We find a leading mode $\Psi_1(\omega)$ which is almost 2 orders of magnitude stronger than any other mode. Similar results are observed at all resonance frequencies, we find a pronounced occupation of just one excitation mode. Contrary, for frequencies where no resonance is observed, the strength of all eigenmodes are in the same order of magnitude. This indicates that there is a single leading mode for each of the resonance frequencies.

The type of collective motion of the electrons at the eigen frequencies can be further illustrated. On the left side of Fig. 5, the eigenvectors $\Psi_\mu(x)$ of the leading modes at the three resonance frequencies marked in Fig. 2 and 3 are shown. The first mode $\Psi_1(x)$ at $\omega_1 = 5.8$ fs^{-1} manifests as a pure dipole mode because a collective motion of all electrons in one direction is seen. This mode is deformed at the outer regions of the linear chain.

The second mode $\Psi_2(x)$ at the resonance frequency $\omega_2 = 8.2$ fs^{-1} is a breathing mode. The motion is antisymmetric with respect to the center of the linear chain. The electrons on the left hand side of the center move opposite to the electrons on the right hand side of the center. Consequently, the total momentum of this type of collective electron motion vanishes, and the mode is not visible in the total momentum correlation function, see Fig. 3.

The third mode $\Psi_3(x)$ at $\omega_3 = 9.2$ fs^{-1}, again, is symmetric with respect to the center of the linear chain. But in contrast to the first mode $\Psi_1(x)$, this one has two nodes. The surface electrons are moving collectively in the opposite direction compared to the central electrons. The total momentum does not vanish at the resonance frequency and can be seen in the spectrum of the total momentum correlation function, Fig. 3. A clear signal is found at the resonant frequency of the third mode $\Psi_3(x)$.

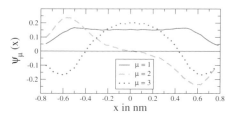 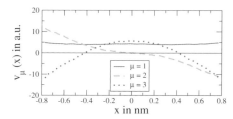

Fig. 5.   Eigenmodes for an excited chain of 55 atoms at $n_i = 35$ nm^{-1} and $T_e = 2$ eV. Left: Eigenvectors $\Psi_\mu(x)$ after solving the eigenwert problem Eq.(13). Right: The velocity mode of electrons due to Eq.(15).

According to Eq.(13), Eq.(14) and Eq.(12), the eigenvectors

$$\Psi_\mu(x) \sim P_\mu(x) = n_e(x)\, m_e v_\mu(x) \qquad (15)$$

represent the spatial contributions $P_\mu(x)$ to the local stationary total momentum densities of the modes and can be expressed in terms of the local electron density $n_e(x)$ and the electron velocity $v_\mu(x)$. On the right of Fig. 5, the mean velocity per electron $v_\mu(x)$ is shown for the same frequencies as on the left. The first mode $v_1(x)$ appears as a homogeneous motion of electrons with the same momentum per electron at $\omega_1 = 5.8$ fs^{-1}. From the second eigenvector follows $v_2(x)$, which shows fast electrons at the edges of the chain and a node of non-moving electrons in the center. The electrons positioned at opposite edges of the chain are moving in opposite directions. Considering $v_3(x)$ for the third mode, the electrons at the edge are moving in the same direction. The central electrons are oscillating opposite to the edge electrons.

We are also interested in the eigenvalues $K_\mu(\omega)$ of the discussed modes in the off-resonant frequency region. As pointed out before, in non-resonant regions, the strength of all eigenmodes is of the same order and can't be distinguished with respect to their strength. Therefore, we decompose the eigenmodes with respect to plane waves using the Fourier transformation

$$\Psi_\mu(k_\mu, \omega) = \int dx \, e^{ik_\mu x} \Psi_\mu(x, \omega). \tag{16}$$

Each eigenmode μ is related to the corresponding wavenumber $k_\mu = 2\pi(\mu - 1)/L$, where L ist the length of the linear chain. In this way, the eigenmodes can be followed over the whole frequency range. Subsequently, we are able to seperate independent

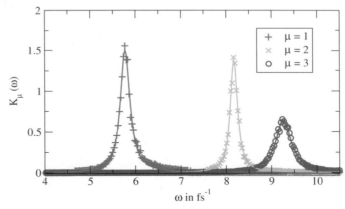

Fig. 6.   Frequency dependent eigenmodes $K_\mu(\omega)$ of the cross correlation matrix $K(x, x', \omega)$ for an excited chain of 55 atoms at $n_i = 35$ nm^{-1} and $T_e = 2$ eV.

frequency spectra of the eigenmodes $K_\mu(\omega)$. In Fig. 6, the first three eigenmodes are shown. For each eigenmode, a single resonance is observed. The resonance of the first mode is the strongest. The third mode is weaker than the first two ones, as expected from the total momentum correlation spectrum, see Fig. 3. However, the resonance peak is broader. The width can be used to deduct a damping rate ν_μ of the eigenmode μ. For this, a Lorentz fit

$$K_\mu(\omega) = K_0 \, \mathrm{Re} \left[\frac{\omega_\mu^2 \omega}{\nu_\mu \omega - \mathrm{i} \left(\omega^2 - \omega_\mu^2 \right)} \right] \tag{17}$$

has been applied which is the behaviour known from the plasmon resonance in bulk, see Eq.(2). The results of the fits are listed in Tab. 1 where also the corrsponging wavenumbers are given. The resonance frequencies ω_μ for odd μ coincide with the resonances obtained from the total momentum ACF spectrum, shown in Fig. 3. The collision frequency increases with increasing wavenumber. The fitted resosonance profiles are also shown in Fig. 6.

Table 1. Resonance frequency ω_μ, corresponding wavenumber k_μ and collision frequency ν_μ for first three eigenmodes μ of a linear chain with 55 ions and 46 electrons at $T_e = 2$ eV and $n_i = 35$ nm^{-1}.

mode μ	ω_μ in fs^{-1}	k_μ in nm^{-1}	ν_μ in fs^{-1}
1	5.77	0	0.253
2	8.17	1.62	0.19
3	9.26	3.70	0.419

6. Dispersion Relations of Collective Modes in a Chain

We shall now discuss the size dependence of the resonance modes considering the cross correlation functions for linear chains of different lengths. For each length, taking the number of ions $N_i = 55, 200, 250, 309$ with the same ion density $n_i = 35$ nm^{-1} and temperature $T_e = 2$ eV, we were able to extract up to 11 independent modes with their resonant frequency and a corresponding wave number. In Fig. 7, the systematic behaviour of the resonance frequencies in dependence on the wavenumber is shown. In the limiting case of an infinite chain, we expect a continuous function $\omega(k)$, the dispersion relation.

The resonance frequency for at least the fundamental mode ($k_1 = 0$) of linear collective oscillations of electrons in the chain can be obtained applying a harmonic approximation for the potential energy of the electron cloud shifted by a small amount x against the equilibrium position. The resonance frequency is given by

$$m\omega^2 = \left. \frac{\partial^2 U(x)}{\partial x^2} \right|_{x=0}, \tag{18}$$

where $U(x) = U_0 + \Delta U(x)$ is the potential energy of an electron at position x and $m = N_e m_e$ is the total mass of N_e electrons in the cloud. Additionally to the potential U_0 at the equilibrium position, the change of potential

$$\Delta U(x) = \int n_e(x' - x) V_{\mathrm{ion}}(x') \mathrm{d}x' \tag{19}$$

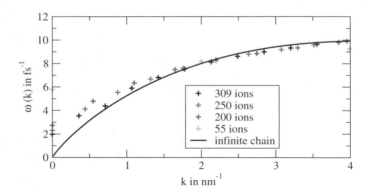

Fig. 7. Resonance frequencies ω in dependence on wavenumber k for excited chains of various lengths at $n_i = 35$ nm^{-1} and $T_e = 2$ eV.

due to the shift of electrons can is expressed in terms of the external ion potential $V_{\rm ion}(x)$ and the electron density $n_e(x)$. The latter is taken from the simulation data. The external ion potential $V_{\rm ion}(x) = \sum_i V_{\rm erf}(x - x_i)$ at a position x is due to the electron-ion interactions, Eq.(5), with the ions at positions x_i. The resonance frequencies of the first mode for all considered chain lengths are presented in Tab. 2. Good agreement is found for the chain of 55 atoms. Strong deviations from the resonance frequency are observed for the long chains. Obviously, the harmonic approximation of the mean ion potential is not justified.

Table 2.   Resonance frequencies of the first mode ω_1 calculated via MD simulations in comparison with ω_r from calculations assuming an oscillation of a rigid electron cloud in an external ion potential at $T_e = 2$ eV and $n_i = 35$ nm^{-1}.

$N_{\rm ion}$	ω_1 in fs^{-1}	ω_r in fs^{-1}
55	5.77	5.84
200	2.73	3.21
250	2.30	2.87
309	1.97	2.59

Alternatively, we use hydrodynamics in order to take into account the inhomogeneous current density profile $j(x,t)$ of the electron motion. We consider the one-dimensional hydrodynamical equation of motion for an electron in a mean field potential $U(x,t)$

$$\frac{\partial j(x,t)}{\partial t} = -\frac{\partial j(x,t)v(x,t)}{\partial x} - \frac{1}{m_e}\frac{\partial p(x,t)}{\partial x} - \frac{n_e(x,t)}{m_e}\frac{\partial U(x,t)}{\partial x}, \qquad (20)$$

where $p(x,t) = n_e(x,t)k_B T_e$ is the pressure of the ideal electron gas. This equation

Table 3.   Resonance frequency ω_μ, corresponding wavenumber k_μ, and collision frequency ν_μ of different modes μ for linear chains of 200 ions and 164 electrons, as well as 250 ions and 205 electrons, both at $T_e = 2$ eV and $n_i = 35$ nm^{-1}.

mode μ	$N_i = 200$			$N_i = 250$		
	ω_μ in fs^{-1}	k_μ in nm^{-1}	ν_μ in fs^{-1}	ω_μ in fs^{-1}	k_μ in nm^{-1}	$\nu_m u$ in fs^{-1}
1	2.73	0	0.0253	2.30	0	0.00314
2	4.79	0.55	0.0434	4.12	0.44	0.0335
3	6.34	1.10	0.0542	5.54	0.88	0.0418
4	7.49	1.65	0.0491	6.68	1.32	0.0524
5	8.32	2.20	0.103	7.59	1.76	0.0388
6	8.86	2.75	0.174	8.29	2.20	0.0524
7	9.34	3.30	0.257	8.81	2.64	0.155
8	9.81	3.85	0.652	9.18	3.08	0.239
9	10.17	4.4	0.574	9.59	3.52	0.379
10	10.38	4.95	0.541	9.90	3.96	0.555
11	10.50	5.50	0.844	10.16	4.40	0.511

is linearized using the following ansatz

$$j(x,t) = \delta j(x)e^{i\omega t}, \tag{21}$$

$$v(x,t) = \delta v(x)e^{i\omega t}, \tag{22}$$

$$n_e(x,t) = n_0(x) + \delta n(x)e^{i\omega t}, \tag{23}$$

$$U(x,t) = U_0(x) + \delta U(x)e^{i\omega t}, \tag{24}$$

assuming density profile $n_0(x)$ and mean field potential $U_0(x)$ of an electron cloud in thermodynamic equilibrium. Restricting ourselves to linear terms with respect to perturbations of the equilibrium, the hydrodynamical equation reads

$$i\omega\delta j(x) = -\frac{k_B T_e}{m_e}\frac{\partial n_0(x)}{\partial x} - \frac{k_B T_e}{m_e}\frac{\partial \delta n(x)}{\partial x} - \frac{n_0(x)}{m_e}\frac{\partial U_0(x)}{\partial x} \tag{25}$$
$$-\frac{\delta n(x)}{m_e}\frac{\partial U_0(x)}{\partial x} - \frac{n_0(x)}{m_e}\frac{\partial \delta U(x)}{\partial x}.$$

In case of thermodynamic equilibrium, we can use the Boltzmann distribution

$$\frac{k_B T_e}{n_0(x)}\frac{\partial n_0(x)}{\partial x} = -\frac{\partial U_0(x)}{\partial x}. \tag{26}$$

The contribution $\delta U(x)$ to the mean field results from the deviation of electrons due to the oscillation

$$\delta U(x) = -\frac{e^2}{4\pi\varepsilon_0}\int dx'\delta n(x')\frac{\mathrm{erf}\left(\frac{x-x'}{\lambda}\right)}{x-x'}. \tag{27}$$

With this, Eq.(25) simplifies to

$$i\omega\delta j(x) = -\frac{k_B T_e}{m_e}\frac{\partial \delta n(x)}{\partial x} - \frac{\delta n(x)}{m_e}\frac{\partial U_0(x)}{\partial x} \tag{28}$$
$$-\frac{e^2}{4\pi\varepsilon_0}\frac{n_0(x)}{m_e}\frac{\partial}{\partial x}\int dx'\delta n(x')\frac{\mathrm{erf}\left(\frac{x-x'}{\lambda}\right)}{x-x'}.$$

Note that the first derivative of the equilibrium mean field $\partial U_0(x)/\partial x = F(x)$ is the mean force. We have calculated the mean force from the MD simulations as described in Ref. 34. Furthermore, using the continuity equation, the change of the current density $\delta j(x)$ is related to the change of the particle density via

$$\mathrm{i}\omega\delta n(x) = -\frac{\partial\,\delta j(x)}{\partial\,x} \tag{29}$$

On the other hand, see Eq.(15), the current density is related to the eigenvectors $\delta j(x) = A\Psi(x)$. Subsequently, the frequency can be calculated using the following equation

$$\omega^2\Psi(x) = -\frac{k_\mathrm{B}T_\mathrm{e}}{m_\mathrm{e}}\frac{\partial^2\,\Psi(x)}{\partial\,x^2} - \frac{1}{m_\mathrm{e}}\frac{\partial\,\Psi(x)}{\partial\,x}F(x) \tag{30}$$

$$-\frac{e^2}{4\pi\varepsilon_0}\frac{n_0(x)}{m_\mathrm{e}}\frac{\partial^2}{\partial\,x^2}\int \mathrm{d}x'\Psi(x')\frac{\mathrm{erf}\left(\frac{x-x'}{\lambda}\right)}{x-x'}.$$

The first term on the right hand side consitutes a contribution due to the ideal gas pressure of the electrons. The second term is due to the mean force on the electrons in equilibrium. The third term results from the deviation of electrons from their equilibrium position. Inserting the eigenvectors $\Psi_\mu(x)$, obtained from the simulations for the various chains considered before, the corresponding resonance frequencies ω_μ are reproduced with an accuracy better than 5 %. The main contribution is due to the third term on the right hand side of Eq.(30).

Table 4.   Resonance frequency ω_μ, corresponding wavenumber k_μ, and collision frequency ν_μ for different modes μ of a linear chain of 309 ions and 250 electrons at $T_\mathrm{e} = 2$ eV and $n_\mathrm{i} = 35$ nm^{-1}.

mode μ	ω_μ in fs^{-1}	k_μ in nm^{-1}	ν_μ in fs^{-1}
1	1.97	0	0.00654
2	3.54	0.36	0.00744
3	4.99	0.71	0.00842
4	5.90	1.07	0.0498
5	6.81	1.42	0.0401
6	7.54	1.78	0.0471
7	8.14	2.13	0.0815
8	8.62	2.49	0.0860
9	9.00	2.84	0.178
10	9.33	3.20	0.250
11	9.65	3.56	0.480

For further investigation of the size dependence, the dispersion relation of an infinte chain is calculated assuming a plane wave $\Psi(x) = A\mathrm{e}^{\mathrm{i}kx}$. From Eq.(30), we have

$$\omega^2\mathrm{e}^{\mathrm{i}kx} = -\frac{k_\mathrm{B}T_\mathrm{e}}{m_\mathrm{e}}\frac{\partial^2}{\partial\,x^2}\mathrm{e}^{\mathrm{i}kx} - \frac{e^2}{4\pi\varepsilon_0}\frac{n_0(x)}{m_\mathrm{e}}\frac{\partial^2}{\partial\,x^2}\int \mathrm{d}x'\mathrm{e}^{\mathrm{i}kx'}\frac{\mathrm{erf}\left(\frac{x-x'}{\lambda}\right)}{x-x'}. \tag{31}$$

The second term vanishes in the case of an infinite chain. Performing the Fourier transformation of the error function potential, we obtain the dispersion relation

$$\omega(k) = \frac{k_{\mathrm{B}} T_{\mathrm{e}}}{m_{\mathrm{e}}} k^2 + \sqrt{\frac{e^2}{4\pi\varepsilon_0} \frac{n_{\mathrm{e}}}{m_{\mathrm{e}}} \Gamma\left(0, \frac{(k\lambda)^2}{4}\right)} \, k \qquad (32)$$

with $n_{\mathrm{e}} = n_{\mathrm{i}}$ the ionic background density due to the neutrality of the infinite chain, and $\Gamma(x, y) = \int_y^\infty t^{x-1} \mathrm{e}^{-t} \mathrm{d}t$. For our plasma parameters, see Fig. 7, the result for an infinite chain is shown for comparison. The first term of Eq.(30) is negligible for all observed wavenumbers k. We find good agreement for large wavenumbers. For small wavenumbers, the wavelength of the mode is comparable to the chain length. Subsequently, the differences between finite chains and an infinite chain differs considerably.

7. Size Effects of Mode Damping and Dynamical Collision Frequency

As already discussed for the chain of 55 ions, for every eigenmode μ, a resonance frequency ω_μ, a corresponding wavenumber k_μ, and a resonance width ν_μ can be found. Additionally, chains of 200 and 250 ions have been inverstigated, see Tab. 3, as well as a chain of 309 ions, see Tab. 4.

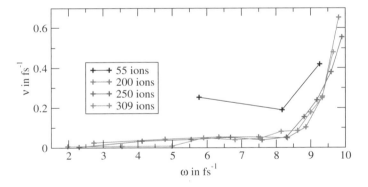

Fig. 8.   Width of the resonance modes ν for different chain sizes.

The width of the resonance mode ν_μ can be interpreted as damping and quantified via a collision frequency. The collision frequencies for different resonance frequencies and different chain lengths are shown in Fig. 8. For the small chain with 55 ions, an enhanced damping compared to the longer chains of more than 100 ions is found. The collective motion of the electrons occurs in the center of the chain. However, for the chain of 55 ions, the ionic trap potential is fairly weak. The electrons are moving in the Coulombic part of the potential at the edge of the ionic trap. As a result, the damping of the resonance frequency is mainly caused by an anharmonic potential. For the longer chains (200, 250 and 309 ions), the electrons are moving solely in the harmonic central part of the ionic trap potential. In general, we observe an increase of the damping rate with increasing wavenumber.

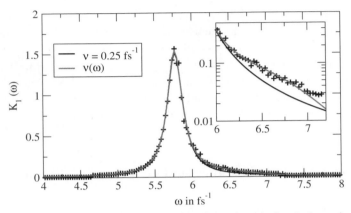

Fig. 9.   Result of the generalized Lorentz fit to the calculated eigenvalues.

An even better fit to the resonance peaks can be obtained if a generalized Lorentz form for the correlation function

$$K(\omega) = K_0 \operatorname{Re}\left[\frac{\omega_{\mathrm{R}}^2 \omega}{\nu(\omega)\omega - \mathrm{i}\left(\omega^2 - \omega_{\mathrm{R}}^2\right)}\right],\tag{33}$$

with a dynamical collision frequency, is used. Exemplarily, we investigated the first resonance mode for a chain of 55 atoms at the ion density $n_{\mathrm{i}} = 35$ nm^{-1} and electron temperature $T_{\mathrm{e}} = 2$ eV. For the frequency dependence of the collision frequeny we assume

$$\nu(\omega) = Ae^{-\frac{\omega^2}{B^2}} + Ce^{-\frac{(\omega-D)^2}{E^2}}.\tag{34}$$

We determined $A = 0.333$ fs^{-1}, $B = 9.63$ fs^{-1}, $C = 0.354$ fs^{-1}, $D = 6.71$ fs^{-1}, and $E = 0.534$ fs^{-1} by fitting to the MD simulations. The comparison is shown in Fig. 9. The generalized Lorentz form fits the calculated eigenvalues better than if assuming a constant collision frequency, see Tab. 1. This can be extended to all of the observed resonance modes. leading to a dynamical collision frequency $\nu(\omega, k)$ which depends on wavenumber as well.

8. Conclusion

Investigations of the total momentum ACF of spherical clusters using *restricted MD simulations*, show several resonance frequencies, that can be attributed to collective excitation modes. One-dimensional *restricted MD simulations* of electron motion in highly excited chains of atoms have been investigated in order to study this phenomenon in a simple model system. The total momentum ACF of the one dimensional system shows the same general behavior as for the spherical clusters.

A more detailed investigation of the collective electron motion is possible if the spatially resolved ACF is investigated, dividing the chain into cells. Solving the eigenvalue problem for the spatially resolved ACF matrix, different excitation modes with corresponding resonance frequencies are found. In order to characterize the different modes, a Fourier analysis of the mode amplitudes has been performed. The largest component in k space was selected to relate the mode excitation to exitations in bulk plasmas. To compare with the plasmon dispersion relation of an infinite system (bulk limit), a large number of modes (about 10) was analyzed.

A hydrodynamic approach was used to determine the plasmon dispersion relation for the infinite system. Good agreement with the resonance frequencies of the higher modes from MD simulations was obtained, in particular for long chains. The fundamental modes are more sensitive to the chain extension. Using the formfactor of the eigenmodes (eigenvector) from simulations, also the resonance frequencies of the fundamental excitations are reproduced from hydrodynamic considerations.

From the separated eigenmodes, the collision frequency was calculated as well. We found an increasing collision frequency for increasing wavenumber. First attempts have been performed to introduce a dynamical collision frequency for each individual excitation mode.

Acknowledgment

We would like to acknowledge financial support of the SFB 652 which is funded by the DFG, as well as support from the conference organizers of the CMT33 meeting in Quito/Ecuador. We particularly thank Igor Morozov and Eric Suraud for fruitful discussions.

References

1. H. Reinholz; *Ann. Phys. Fr.* **30**, N° 4 - 5 (2006).
2. H. Reinholz, R. Redmer, G. Röpke, and A. Wierling; *Phys. Rev. E* **62**, 5648 (2000).
3. I. Morozov, H. Reinholz, G. Röpke, A. Wierling, and G. Zwicknagel; *Phys. Rev. E* **71**, 066408 (2005).
4. D. Bohm, and E. P. Gross; *Phys. Rev.* **75**, 1864 (1949).
5. W.-D. Kraeft, D. Kremp, W. Ebeling, and G. Röpke; *Quantum Statistics of Charged Particle Systems*, Akademie-Verlag, Berlin (1986).
6. R. Thiele, T. Bornath, C. Fortmann, A. Höll, R. Redmer, H. Reinholz, G. Röpke, A. Wierling, S. H. Glenzer, and G. Gregori; *Phys. Rev. E* **78**, 026411 (2008).
7. H. Reinholz, I. Morozov, G. Röpke, and T. Millat; *Phys. Rev. E* **69**, 066412 (2004).
8. C. Fortmann; *Phys. Rev. E* **79**, 016404 (2009).
9. A. McPherson, K. Boyer, and C. K. Rhodes; *J. Phys. B* **27**, L637 (1994).
10. T. Ditmire, J. W. G. Tisch, E. Springate, M. B. Mason, N. Hay, R. A. Smith, J. Marangos, and M. H. R. Hutchinson; *Nature* **386**, 54 (1997).
11. M. Lezius, S. Dobosz, D. Normand, and M. Schmidt; *Phys. Rev. Lett.* **80**, 261 (1998).
12. L. Köller, M. Schumacher, J. Köhn, S. Teuber, J. Tiggesbäumker, and K.-H. Meiwes-Broer; *Phys. Rev. Lett.* **82**, 3783 (1999).

13. R. Schlipper, R. Kusche, B. v. Issendorff, and H. Haberland; *Phys. Rev. Lett.* **80**, 1194 (1998).
14. V.P. Krainov, and M.B. Smirnov; *Physics Uspekhi* **43**, 901 (2000).
15. P.-G. Reinhard, and E. Suraud; *Introduction to Cluster Dynamics*, Wiley, New York, 2003.
16. U. Saalmann, Ch. Siedschlag, and J. M. Rost; *J. Phys. B* **39**, R39 (2006).
17. T. Döppner, T. Diederich, A. Przystawik, N. X. Truong, T. Fennel, J. Tiggesbäumker, and K. - H. Meiwes-Broer; *Phys. Chem.* **9**, 4639 (2007).
18. T. Ditmire, R. A. Smith, T. W. G. Tisch, and M. H. R. Hutchinson; *Phys. Rev. Lett.* **78**, 3121 (1997).
19. A. McPherson, B. D. Thompson, A. B. Borisov, K. Boyer, and C. K. Rhodes; *Nature* **370**, 631 (1994).
20. S. Dobosz, M. Lezius, M. Schmidt, P. Meynadier, M. Perdrix, D. Normand, J.-P. Rozet, and D. Vernhet; *Phys. Rev. A* **56**, R2526 (1997).
21. T. Ditmire, T. Donnelly, R. W. Falcone, and M. D. Peny; *Phys. Rev. Lett.* **75**, 3122 (1995).
22. Y. L. Shao, T. Ditmire, J. W. G. Tisch, E. Springate, J. P. Marangos, and M. H. R. Hutchinson; *Phys. Rev. Lett.* **77**, 3343 (1996).
23. T. Ditmire, E. Springate, J. W. G. Tisch, Y. L. Shao, M. B. Mason, N. Hay, J. P. Marangos, and M. H. R. Hutchinson; *Phys. Rev. A* **57**, 369 (1998).
24. C. Deiss, N. Rohringer, and J. Burgdörfer; *Phys. Rev. Lett.* **96**, 013203 (2006).
25. S. Micheau, C. Bonte, F. Dorchies, C. Fourment, M. Harmand, H. Jouin, O. Peyrusse, B. Pons, and J. J. Santos; *HEDP* **3**, 191 (2007).
26. T. Fennel, T. Döppner, J. Passig, C. Schaal, J. Tiggesbäumker, and K.-H. Meiwes-Broer; *Phys. Rev. Lett.* **98**, 143401 (2007).
27. F. Calvayrac, P.-G. Reinhard, E. Suraud, and C. A. Ullrich; *Phys. Rep.* **337**, 493 (2000).
28. J. Köhn, R. Redmer, K. - H. Meiwes-Broer, and T. Fennel; *Phys. Rev. A* **77**, 033202 (2008).
29. U. Saalmann, I. Georgescu, and J. M. Rost; *New J. Phys.* **10**, 25014 (2008).
30. P. Hilse, M. Schlanges, T. Bornath, and D. Kremp; *Phys. Rev. E* **71**, 56408 (2005).
31. H. Reinholz, T. Raitza, and G. Röpke; *Int. J. Mod. Phys. B* **21**, 2460 (2007).
32. H. Reinholz, T. Raitza, G. Röpke, and I. Morozov; *Int. J. Mod. Phys. B* **22**, 4627 (2008).
33. T. Raitza, H. Reinholz, G. Röpke, and I. Morozov; *J. Phys. A* **42**, 214048 (2009).
34. T. Raitza, H. Reinholz, G. Röpke, I. Morozov, and E. Suraud; *Contrib. Plasma Phys.* **49**, 498 (2009).
35. M. Belkacem, F. Megi, P.-G. Reinhard, E. Suraud, and G. Zwicknagel; *Eur. Phys. J. D* **40**, 247 (2006).
36. L. Verlet; *Phys. Rev.* **159**, 98 (1967).
37. L. Arndt; *PhD Thesis*, Köln (2006).
38. L. Ramunno, C. Jungreuthmayer, H. Reinholz, and T. Brabec; *J. Phys. B* **39**, 4923 (2006).

RENORMALIZED BOSONS AND FERMIONS

K. A. GERNOTH

Institut für Theoretische Physik, Johannes-Kepler-Universität Linz
Altenbergerstr. 69, A–4040 Linz, Austria
and
The School of Physics and Astronomy, The University of Manchester
Manchester M13 9PL, United Kingdom

M. L. RISTIG

Institut für Theoretische Physik, Universität zu Köln
D–50937 Köln, Germany

Received 27 October 2009

Correlated Density Matrix (CDM) theory permits formal analyses of microscopic properties of strongly correlated quantum fluids and liquids at nonzero temperatures. Equilibrium properties, thermodynamic potentials, correlation and structure functions can be studied formally as well as numerically within the CDM algorithm. Here we provide the essential building blocks for studying the radial distribution function and the single-particle momentum distribution of the ingredients of the quantum systems. We focus on the statistical properties of correlated fluids and introduce the concept of renormalized bosons and fermions. These entities carry the main statistical features of the correlated systems such as liquid ^{4}He through their specific dependence on temperature, particle number density, and wavenumber encapsulated in their effective masses. The formalism is developed for systems of bosons and of fermions. Numerical calculations for fluid ^{4}He in the normal phase demonstrate the power of the renormalization concept. The formalism is further extended to analyze the Bose-Einstein condensed phases and gives a microscopic understanding of Tisza's two-fluid model for the normal and superfluid density components.

Keywords: Strongly correlated quantum fluids; correlated density matrix theory; liquid ^{4}He.

1. Introduction

The present paper reports on recent progress in the physical understanding of thermodynamic equilibrium properties and microscopic structures of strongly correlated quantum fluids and liquids at nonzero temperatures. The *ab initio* theory applied is a natural extension of Correlated Basis Functions (CBF) theory of correlated ground and excited states. Instead of dealing with specific wave functions and analyzing their specific properties it becomes advantageous to concentrate at finite temperatures on a study of the underlying density matrix of the N-body quantum system in question. This correlated density matrix (CDM) theory is now sufficiently well developed to achieve important and interesting new insights as well as to permit reliable quantitative numerical results that can be compared with experimental data and results of stochastic calculations.

The formal development of CDM has heavily drawn on ideas and techniques from the very successful CBF approach, such as generalized Ursell-Mayer cluster expansions, hypernetted-chain and Fermi hypernetted-chain techniques, and Euler-Lagrange optimization procedures. These tools have been introduced or reviewed in a number of early papers and books.[1-7]

The formal development of CDM theory began in the 1980's with the inclusion of direct-direct effects generated by the appearance of collective excitations in Bose liquids such as ^{4}He at nonzero temperatures.[8,9] These and later studies[10] are, however, restricted to Bose-Einstein condensed phases, since the phonons and rotons do not trigger the transition to the normal boson phase. This phase change is generated by "statistical" exchange effects and therefore appears already in free Bose gases. Ref. 9 gives a brief discussion of these shortcomings and indicates the necessary remedy to deal successfully with exchange effects and normal phases.

A systematic development of CDM theory that incorporates the main features of quantum-mechanical exchange of bosons and normal fermions and provides an adequate algorithm has been reported already.[11-13] Also numerical results on liquid ^{4}He (on an exploratory level) are published.[14,15] However, these preliminary applications of the formalism have their shortcomings. They stem from two sources: (a) the employment of the so-called Jackson-Feenberg identity[2] for the evaluation of the internal energy functional at nonzero temperatures. This leads, unfortunately, to a rather complicated energy expression; (b) the corresponding Euler-Lagrange equation that determines the optimal statistical factor has an extremely awkward form and needs an elaborate numerical technique to find numerical solutions.

The present work avoids the Jackson-Feenberg formulation and introduces the concept of renormalized bosons and fermions. As a consequence the internal energy functional has a much simpler form, the optimization of the statistical input can be formulated straightforwardly, and the underlying physics becomes transparent. A number of recent publications[16-21] paved the way for the present achievements.

Section 2 presents the analytic results on the CDM building blocks to study the structure of the one- and two-body reduced density matrix elements for correlated bosons and (spinless) fermions. The renormalization concept is made explicit in Section 3. The renormalized entitities may be interpreted as (stable) constituents of an optimal free background gas of bosons or fermions. The mass of a single renormalized quantity depends in a specific form on temperature, density, and wavenumber. These effective bosons encapsulate the main statistical properties of the correlated many-body system.

The next section reports some of our numerical results on normal fluid ^{4}He in the temperature range $2.17\,\mathrm{K} \leq T \leq 20\,\mathrm{K}$. The renormalization concept may be generalized to deal with anomalous phases of bosons and fermions. Section 5 demonstrates this possibility for the Bose-Einstein condensed phases. The last chapter provides a microscopic interpretation of Tisza's two-fluid model by analyzing a sum rule derived within the CDM algorithm (Sec. 6).

2. CDM Building Blocks

CDM theory begins with an appropriate ansatz for the correlated N-body density matrix that contains the relevant information on the equilibrium properties of quantum matter at nonzero temperatures. References 11 and 13 describe these trial ansätze for normal Bose and Fermi fluids and liquids. They also provide the algorithm that enables us to analyze the structure of the one-body and two-body reduced density matrix elements. They are needed to determine the static equilibrium properties of strongly correlated quantum many-body systems such as fluid ^{4}He and liquid *para*-hydrogen[16,22] and other quantum liquids.

CDM theory then provides the structural results for normal Bose and (spinless) Fermi fluids and liquids. The one-body reduced density matrix elements possess the general structure[11,13]

$$n(r) = n_c N_0(r) \exp\left[-Q(r)\right] .\tag{1}$$

This function leads us directly via its Fourier transform, $n(k)$, to the theoretical single-particle momentum distribution that can be compared with experimental data available from inelastic scattering experiments.[16,23]

Similarly, we may derive within CDM theory the structure of the radial distribution function $g(r)$ with its Fourier transform, $S(k)$, i.e., the static structure function,

$$g(r) = [1 + G_{dd}(r)] F(r)\tag{2}$$

or

$$g(r) = 1 + G_{dd}(r) + 2G_{de}(r) + G_{ee}(r) .\tag{3}$$

The components (dd,de,ee) comprise the diagonal elements of the two-body reduced density matrix elements of Bose and Fermi fluids. In addition the CDM algorithm yields also physically plausible approximations for the entropy expressions of correlated bosons and fermions. These expressions are of the analytic shape[11]

$$S_b(T) = \sum_k \left\{ [1 + n_{qp}(k)] \ln\left[1 + n_{qp}(k)\right] - n_{qp}(k) \ln\left[n_{qp}(k)\right] \right\}\tag{4}$$

for a system of identical bosons and

$$S_f(T) = -\sum_k \left\{ [1 - n_{qp}(k)] \ln\left[1 - n_{qp}(k)\right] + n_{qp}(k) \ln\left[n_{qp}(k)\right] \right\}\tag{5}$$

for correlated fermion systems.

CDM theory developed the analytic and numerical tools to calculate the functions appearing in Eqs. (1)-(5) for given input functions. These input functions[11] are the corresponding pseudo-potential $u(r)$ and the statistical factor $\Gamma_{cc}(r)$. They may be represented by physically motivated ansätze and can be evaluated by solving associated Euler-Lagrange equations which yield the optimal functions[2,18] $u(r)$ and optimal solutions $\Gamma_{cc}(r)$.

As already pointed out, the ingredients of Eqs. (1)-(5) can be evaluated within CDM theory employing hypernetted-chain (HNC) techniques.[4] For this purpose one introduces nodal, non-nodal, and elementary components[11] called $N(r)$, $X(r)$, and $E(r)$ portions, distinguishable further by additional indices to characterize cyclic statistical (cc) effects, direct-direct (dd), direct-exchange (de), exchange-exchange (ee), and phase (Q) or phase-phase (QQ) correlations.[11,13]

To be more explicit, we list here the full expressions for the functions $N_0(k)$, $Q(k)$, $F(r)$, $n_{qp}(k)$, and the exchange structure function $S_{cc}(k)$ in terms of the corresponding HNC components for correlated Bose and (spinless) Fermi systems.[11] The statistical factor in Eq. (1) is given by (upper signs for bosons, lower signs for spinless fermions)

$$N_0(k) = \Gamma_{cc}(k) \pm [X_{Qcc}(k) \pm \Gamma_{cc}(k)]^2 [1 + S_{cc}(k)] \pm E_{QQcc}(k) . \tag{6}$$

The phase-phase correlation function in **k**-space reads explicitly

$$\begin{aligned} -Q(k) = & \; [X_{Qdd}(k) + X_{Qde}(k)] [X_{Qdd}(k) + N_{Qdd}(k)] \\ & + X_{Qdd}(k) [X_{Qde}(k) + N_{Qde}(k)] + E_{QQdd}(k) . \end{aligned} \tag{7}$$

The statistical factor $F(r)$ appearing in Eq. (2) has the form

$$F(r) = [1 + N_{de}(r) + E_{de}(r)]^2 + [N_{ee}(r) + E_{ee}(r)] \pm F_{cc}^2(r) , \tag{8}$$

where $+$ and $-$ stand for bosons and fermions, respectively. The cyclic exchange function $F_{cc}(r)$ reads

$$F_{cc}(r) = \Gamma_{cc}(r) \pm N_{cc}(r) \pm E_{cc}(r) \tag{9}$$

($+$ for bosons, $-$ for fermions).

The entropy expressions (4) and (5) have the familiar forms known for free Bose and Fermi gases. However, the momentum distributions $n_{qp}(k)$ in Eqs. (4) and (5) incorporate effects of the particle-particle interactions via the exchange structure function $S_{cc}(k)$,

$$n_{qp}(k) = \Gamma_{cc}(k) [1 + S_{cc}(k)] . \tag{10}$$

The structure function $S_{cc}(k)$ deviates therefore from the familiar Bose or Fermi distribution function of a free quantum gas by the direct cyclic exchange generator $X_{cc}(k)$,

$$S_{cc}(k) = [X_{cc}(k) \pm \Gamma_{cc}(k)] \{1 - [X_{cc}(k) \pm \Gamma_{cc}(k)]\}^{-1} \tag{11}$$

derived in Ref. 11 for bosons $(+)$ and spinless fermions $(-)$. Available are also explicit expressions for spin-1/2 fermions to deal, for example, with normal liquid ^{3}He.

3. Renormalization of Normal Bosons and Fermions

We employ the CDM building blocks of the preceding section for an interesting reformulation of the physics of strongly correlated quantum liquids at nonzero temperatures. Here we concentrate on the normal phase of strongly interacting identical bosons and fermions.

The CDM/HNC analysis permits to cast the momentum distribution $n_{qp}(k)$ appearing in the entropy expressions (4) and (5) in the explicit analytic form[11] (upper sign for bosons, lower signs for spinless fermions)

$$n_{qp}(k) = \Gamma_{cc}(k) \{1 - [X_{cc}(k) \pm \Gamma_{cc}(k)]\}^{-1} . \tag{12}$$

Next, we define a *renormalized* statistical factor $\Gamma_{qp}(k)$ that replaces the input factor $\Gamma_{cc}(k)$,

$$\Gamma_{qp}(k) = \Gamma_{cc}(k) [1 - X_{cc}(k)]^{-1} . \tag{13}$$

We insert the definition (13) in the formal expression (12) and obtain

$$n_{qp}(k) = \Gamma_{qp}(k) \{1 \mp \Gamma_{qp}(k)\}^{-1} . \tag{14}$$

Result (14) may be complemented by the relation

$$\Gamma_{qp}(k) = n_{qp}(k) [1 \pm n_{qp}(k)]^{-1} . \tag{15}$$

The upper signs in Eqs. (14) and (15) refer to bosons, the lower signs apply to spinless fermions. The entropy expressions (4) and (5) require the inequalities $n_{qp}(k) \geq 0$ for identical bosons and $0 \leq n_{qp}(k) \leq 1$ for fermions. We may therefore conclude from Eq. (15) that $0 \leq \Gamma_{qp}(k) \leq 1$ holds for bosons and $\Gamma_{qp}(k) \geq 0$ for fermions. Consequently, we may read Eq. (14) as a Bose (Fermi) function for the momentum distribution of non-interacting renormalized bosons (fermions). This momentum distribution characterizes a free Bose (Fermi) gas via the associated renormalized statistical factor (13). The corresponding entropies of these entities are given by Eqs. (4) and (5), respectively.

The aforementioned inequalities evidently permit that the renormalized factor $\Gamma_{qp}(k)$ may be cast in exponential form,

$$\Gamma_{qp}(k) = \Gamma_{qp}(0) \exp\left[-\beta\varepsilon_{qp}(k)\right], \qquad (16)$$

familiar from the standard textbook formulation of the average thermal momentum distribution of free quantum gases.

The renormalized bosons or fermions have an effective single-particle kinetic energy

$$\varepsilon_{qp}(k) = \frac{\hbar^2 k^2}{2m_{qp}} \qquad (17)$$

and move in an effective chemical potential $\mu_{qp}(T)$ given by

$$\mu_{qp}(T) = T \ln\left[\Gamma_{qp}(k=0)\right]. \qquad (18)$$

Note that the latter potential differs in its general form from the ordinary chemical potential defined by the derivative $\mu = -\partial J/\partial N$, where J is the grand canonical thermodynamic potential. The effective mass m_{qp} that characterizes the kinetic energy (17) depends, in general, on temperature T, particle number density ϱ, and wavenumber k, $m_{qp} = m_{qp}(T, \varrho, k)$. This mass encapsulates the essential statistical properties of the strongly correlated quantum system composed of the interacting particles, which have bare mass m.

We emphasize once more the interesting features of the renormalized representation: The renormalized entities are (i) free particles and (ii) the thermodynamic phase boundaries are determined by the particular properties of the effetive mass $m_{qp}(T, \varrho, k)$. For example, for a system of interacting bosons the phase transition to the Bose-Einstein condensed states is signalled by the condition $\Gamma_{qp}(k=0) = 1$ or, equivalently, by $\mu_{qp}(T = T_{BE}) = 0$. Moreover, the type and strength of the singularities in function $n_{qp}(k)$ along the phase boundaries are determined by the specific dependence of the renormalized mass m_{qp} on variables T, ϱ, and k. This is due to the particle sum rule

$$N = \sum_{k} n_{qp}(k) = \sum_{k} \Gamma_{qp}(k) \left[1 - \Gamma_{qp}(k)\right]^{-1} \qquad (19)$$

that holds in CDM theory.[11,15] However, we note that CDM theory at the present stage of development is strictly based on the assumption of homogeneity in the three spatial dimensions. This is not supported by experimental results in the very close vicinity of the transition region in bulk boson matter. These singular properties of the transition region are more appropriately addressed by the renormalization group approach.[24] It would be interesting to explore the theoretical connection between the latter theory of critical exponents with the singularities or divergencies appearing in the results derived in the CDM formalism when approaching the phase boundaries.

4. Numerical Results for Liquid ^{4}He

Numerical calculations of the equilibrium properties and correlation functions of normal liquid ^{4}He within CDM theory are based on the interatomic HFDB2 Aziz-potential[25] $v(r)$ and on the input functions, i.e., on the pseudopotential $u(r)$ and the statistical factor $\Gamma_{qp}(r)$ of the renormalized helium constituents. These functions may be optimally determined by employing a minimum principle for the associated grand canonical thermodynamic potential[26] $J = F - \mu N = E - TS - \mu N$ with the Helmholtz free energy F, the internal energy E, the entropy S, and the chemical potential μ. Variation of the potential J with respect to the pseudopotential $u(r)$ yields an Euler-Lagrange equation for the optimal radial distribution function $g(r)$ conventionally known as the effective Schrödinger equation for function $\sqrt{g(r)}$. This equation reads

$$\left\{ -\frac{\hbar^2}{m}\nabla^2 + v(r) + w(r) + v_{\text{coll}}(r) + v_{qp}(r) \right\} \sqrt{g(r)} = 0 \qquad (20)$$

(cf. Eq. (5) of Ref. 18) with the so-called induced potential $w(r)$ and the potential contribution $v_{\text{coll}}(r)$ from the collective excitations (phonons and rotons). The term $v_{qp}(r)$ describes possible coupling effects of the particle constituents with the collective excitations. However, we ignore such effects in the case of normal fluid ^{4}He.

The optimized kinetic energy $\varepsilon_{qp}(k)$ of the renormalized ^{4}He-bosons follows from a second Euler-Lagrange condition that can be derived by minimizing an appropriate grand canonical potential at fixed temperature. The result can be cast in the form

$$\varepsilon_{qp}(k,T) = \varepsilon_0(k) + \dot{D}(k,T,\mu). \qquad (21)$$

The function $\dot{D}$ may be evaluated by solving a set of so-called dot equations generated from the HNC equations for the radial distribution function $g(r)$ and the one-body density matrix $n(r)$. For a system of non-interacting bosons $\dot{D}(k,T,\mu)$ vanishes identically and we recover, of course, the equality $\varepsilon_{qp}(k,T) = \varepsilon_0(k)$ and $\mu_{qp}(T) = \mu(T)$. However, at present, we assume that the effective mass m_{qp} depends only on temperature T, ignoring any dependence on wavenumber k.

In this case we have to perform only a parameter optimization and minimize the thermodynamic grand canonical potential $J = E - TS - \mu N$ with respect to the mass parameter $m_{qp}(T)$,

$$\frac{\partial J}{\partial m_{qp}} = 0. \qquad (22)$$

The components of potential J are the total internal energy $E = E_{\text{kin}} + E_{\text{pot}}$, the entropy (4), the expression (19) for the total number N of particles, and the external chemical potential μ determined by experiment.

The derivatives $\partial N / \partial m_{\text{qp}}$ and $\partial(TS)/\partial m_{\text{qp}}$ may be easily calculated and written in the analytic forms

$$m_{\text{qp}} \frac{\partial N}{\partial m_{\text{qp}}} = \beta \sum_{\mathbf{k}} n_{\text{qp}}(k) \left[1 + n_{\text{qp}}(k)\right] \left[m_{\text{qp}} \frac{\partial \mu_{\text{qp}}}{\partial m_{\text{qp}}} + \varepsilon_{\text{qp}}(k)\right] \tag{23}$$

and

$$m_{\text{qp}} \frac{\partial(TS)}{\partial m_{\text{qp}}} = -\beta \sum_{\mathbf{k}} n_{\text{qp}}(k) \left[1 + n_{\text{qp}}(k)\right] \left[m_{\text{qp}} \frac{\partial \mu_{\text{qp}}}{\partial m_{\text{qp}}} + \varepsilon_{\text{qp}}(k)\right] \left[\mu_{\text{qp}} - \varepsilon_{\text{qp}}(k)\right] . \tag{24}$$

The effective chemical potential μ_{qp} may be expressed also in terms of the momentum distribution n_{qp} via

$$\mu_{\text{qp}} = T \ln \left[\frac{n_{\text{qp}}(0)}{1 + n_{\text{qp}}(0)}\right] . \tag{25}$$

Equation (22) may be solved numerically to construct the optimal mass $m_{\text{qp}}(T)$. To carry out an efficient iteration procedure one needs an appropriate input that permits sufficiently fast convergence. For this purpose we adopt in this work a plausible expression for the renormalized mass of the shape

$$m_{\text{qp}}(T) = m \left\{1 + \sigma_{\text{qp}} \exp \left(1 - \frac{T}{T_{\text{BE}}}\right)\right\} \tag{26}$$

with a parameter σ_{qp} that is constant. Future calculations should be based on a systematic iteration process to solve condition (22) with ansatz (26) as input.

For liquid ^{4}He the numerical calculations are based on the HFDB2 Aziz potential,[25] the optimal solution of the Euler-Lagrange equation (20), and ansatz (26) with $T_{\text{BE}} = 2.17\,\text{K}$ and $\sigma_{\text{qp}} = 1.89$ at particle number density $\varrho = 0.02185\,\text{Å}^{-3}$. We study the properties of the system in the temperature range $2.17\,\text{K} \leq T \leq 20\,\text{K}$. The dependence of the mass ratio $m_{\text{qp}}(T)/m$ is displayed in Figure 1. At $T \geq 12\,\text{K}$ we have $m_{\text{qp}} \simeq m$ and the effective mass at $T = T_{\text{BE}}$ is about twice the bare mass m.

The associated effective chemical potential $\mu_{\text{qp}}(T)$ of the renormalized ^{4}He bosons is calculated via the sum rule (19). The numerical results are shown in Figure 2. The potential vanishes at the transition temperature T_{BE}, indicating the formation of the superfluid phase of ^{4}He where the bosons condense.

The macroscopic occupation of the zero-momentum state at $T \leq T_{\text{BE}}$ of the renormalized bosons and therewith the nonzero condensate fraction of the ^{4}He atoms[13] is signalled by the divergence of the distribution (14). Figure 3 shows the temperature dependence of $n_{\text{qp}}(T)$ at $k = 0$. It diverges at the transition temperature T_{BE}.

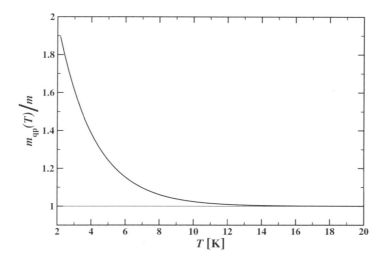

Fig. 1. The renormalized mass ratio $m_{\mathrm{qp}}(T)/m$ adopted in this work.

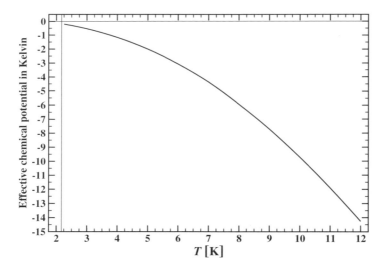

Fig. 2. The effective chemical potential $\mu_{\mathrm{qp}}(T)$ of the renormalized ^{4}He bosons. At $T = T_{\mathrm{BE}} = 2.17\,\mathrm{K}$ the potential vanishes, indicating that the renormalized bosons condense in the Bose-Einstein phase.

Numerical results on the specific heat $c_{\mathrm{v}}(T)$ at constant density corresponding to the entropy (4) are displayed in Figure 4. They are compared to the available data from experimental measurements.[27] At temperatures $T \geq 7\,\mathrm{K}$ theoretical and experimental data agree fairly well but at lower temperatures significant differences appear. They may be caused by the neglect of contributions of collective excitations in the present theoretical study. Improvements may be also expected, once the optimization of the effective mass $m_{\mathrm{qp}}(T, \varrho, k)$ is strictly carried out.

5. Renormalization of Boson Excitations in Condensed Bose-Einstein Phases

A suitable generalization of the ansatz for the N-body correlated density matrix permits a microscopic formulation of CDM theory for the Bose-Einstein condensed phase and subsequent application of the renormalization concept. Refs. 12 and 13 provide the CDM building blocks for this more general case. They report on the analytic results for the radial distribution function $g(r)$, the one-body reduced density matrix $n(r)$, and the entropy $S(T)$. In contrast to the case of bosons in the normal phase the exchange statistical generator is now of long range, i.e. the input factor $\Gamma_{\mathrm{cc}}(r)$ is replaced by $\Gamma_{\mathrm{cc}}(r) + B_{\mathrm{cc}}$ with $0 \leq B_{\mathrm{cc}} \leq 1$ as the relevant order parameter (the condensation strength). Due to this property the correlation functions $g(r)$ and $n(r)$ consist of a normal component and an anomalous portion.[12,13] In contrast, the corresponding entropy of the Bose-Einstein phase has only a normal contribution, since the anomalous superfluid or condensed part does not carry entropy. CDM theory yields the entropy expression that may be split into two normal components[14]

$$S(T) = S_1(T) + S_2(T).   \tag{27}$$

The terms $S_1(T)$ and $S_2(T)$ are given by the explicit forms, respectively,

$$S_1(T) = \frac{1}{2} \sum_{\mathbf{k}} \left\{ \left[1 + n_{\mathrm{qp}}^{(1)}(k) \right] \ln \left[1 + n_{\mathrm{qp}}^{(1)}(k) \right] - n_{\mathrm{qp}}^{(1)}(k) \ln \left[n_{\mathrm{qp}}^{(1)}(k) \right] \right\}   \tag{28}$$

and

$$S_2(T) = \frac{1}{2} {\sum_{\mathbf{k}}}' \left\{ \left[1 + n_{\mathrm{qp}}^{(2)}(k) \right] \ln \left[1 + n_{\mathrm{qp}}^{(2)}(k) \right] - n_{\mathrm{qp}}^{(2)}(k) \ln \left[n_{\mathrm{qp}}^{(2)}(k) \right] \right\}.   \tag{29}$$

The prime in sum (29) indicates that the $\mathbf{k}$-summation extends only over values with $n_{\mathrm{qp}}^{(2)}(k) > 0$. At the phase transition temperature $T = T_{\mathrm{BE}}$ the momentum distributions become identical, $n_{\mathrm{qp}}^{(1)}(k) = n_{\mathrm{qp}}^{(2)}(k)$, and the sum (27) coincides with the result (4) of a normal boson system. We note that our discussion of the entropy does not include the entropy generated by the collective excitations (phonons and rotons). Here we can do without them, since they do not trigger the transition from the normal to the condensed phase in which we are presently interested. The collective contribution to the entropy can be dealt with separately.

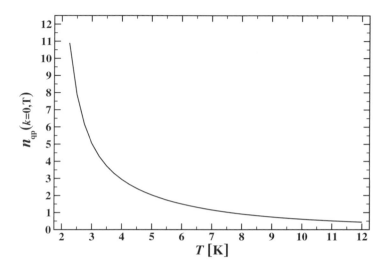

Fig. 3. Displayed are the results on the momentum distribution $n_{qp}(T)$ at zero wavenumber $(k = 0)$ as function of temperature T. It diverges at the transition point.

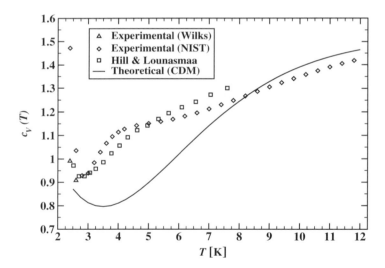

Fig. 4. The result for the specific heat $c_V(T)$ at constant density (volume) compared with experimental results.

The renormalization of the momentum distributions $n_{qp}^{(1)}(k)$ and $n_{qp}^{(2)}(k)$ can be performed in analogy to the concept described in Sec. 3 by starting from the explicit CDM results

$$n_{qp}^{(1)}(k) = \Gamma_{cc}(k) \left[1 - X_{cc}(k) - \Gamma_{cc}(k)\right]^{-1} \qquad (30)$$

and

$$n_{\mathrm{qp}}^{(2)}(k) = n_{\mathrm{qp}}^{(1)}(k) + 2B_{\mathrm{cc}}\Gamma_{\mathrm{cc}}(k)S^{(0)}(k) \,. \tag{31}$$

Expression (31) is limited to **k**-values for which $n_{\mathrm{qp}}^{(2)}(k)$ is positive. This range is determined by the value of the condensation strength $B_{\mathrm{cc}}(T)$ and the behavior of function

$$S^{(0)}(k) = S_{\mathrm{dd}}(k) + 2S_{\mathrm{de}}(k) + S_{\mathrm{cc}}^{(1)}(k) \,, \tag{32}$$

which can become negative for relatively small values of wavenumber k. As in the case of the normal Bose phase all functions appearing in Eqs. (30)–(32) can be evaluated by the CDM/HNC algorithm[12,13] for the condensed phase.

The renormalization concept is now easily carried out. Define

$$\Gamma_{\mathrm{qp}}^{(i)}(k) = n_{\mathrm{qp}}^{(i)}(k)\left[1 + n_{\mathrm{qp}}^{(i)}(k)\right]^{-1} \,, \tag{33}$$

but note that the generator $\Gamma_{\mathrm{qp}}^{(1)}(k)$ is, equivalently, also given by

$$\Gamma_{\mathrm{qp}}^{(1)}(k) = \Gamma_{\mathrm{cc}}(k)\left[1 - X_{\mathrm{cc}}(k)\right]^{-1} \tag{34}$$

already encountered in Eq. (13) of Sec. 3 for the normal phase. Observe further that branch 2 is a function of $\Gamma_{\mathrm{qp}}^{(1)}(k)$. As expected, the renormalized generators $\Gamma_{\mathrm{qp}}^{(i)}(k)$ have the property $0 \le \Gamma_{\mathrm{qp}}^{(i)}(k) \le 1$, provided $n_{\mathrm{qp}}^{(i)}(k) > 0$.

Thus the distribution $n_{\mathrm{qp}}^{(i)}(k)$ may be cast in the familiar Bose form

$$n_{\mathrm{qp}}^{(i)}(k) = \Gamma_{\mathrm{qp}}^{(i)}(k)\left[1 - \Gamma_{\mathrm{qp}}^{(i)}(k)\right]^{-1} \tag{35}$$

with

$$\Gamma_{\mathrm{qp}}^{(i)}(k) = \exp\left\{-\beta\left[\varepsilon_{\mathrm{qp}}^{(i)}(k) - \mu_{\mathrm{qp}}\right]\right\} \tag{36}$$

for non-interacting bosons with effective potential $\mu_{\mathrm{qp}}(T)$ and effective kinetic energy of renormalized entities with effective mass $m_{\mathrm{qp}}^{(i)}(T, \varrho, k)$. The masses $m_{\mathrm{qp}}^{(i)}$ ($i = 1, 2$) are in general different. Only for non-interacting free Bose gases the spectra of the two excitation branches are equal in the whole temperature range $0 \le T \le T_{\mathrm{BE}}$, since function $S^{(0)}(k)$ vanishes identically (Eq. (31)).

6. Tisza's Two-Fluid Model

We conclude with a brief discussion of the CDM result for the particle sum rule of Bose-Einstein condensed phases that generalizes the normal sum rule (19). This particle number conservation law reads[12,13]

$$\sum_{\mathbf{k}} n_{\mathrm{qp}}^{(1)}(k) + B_{\mathrm{cc}} \sum_{\mathbf{k}} n_{\mathrm{qp}}^{(1)}(k) N_{\mathrm{a}}(k) + B_{\mathrm{cc}} N = N \,. \tag{37}$$

The function $N_{\mathrm{a}}(k)$ is defined within CDM theory by

$$N_{\mathrm{a}}(k) = \left[1 + n_{\mathrm{qp}}^{(1)}(k)\right]^{-1} \left[1 - X_{\mathrm{cc}}(k)\right] \left[S_{\mathrm{de}}(k) + S^{(0)}(k)\right] \tag{38}$$

and can be calculated by solving corresponding HNC equations that involve the components appearing on the right-hand side of Eq. (38).

The two series in rule (37) are generated by the momentum distribution of free renormalized bosons of type $i = 1$ excited to the energy level $\varepsilon_{\mathrm{qp}}^{(i)}(k)$ for $i = 1$ and, indirectly, through Eq. (38) to level $\varepsilon_{\mathrm{qp}}^{(i)}(k)$ for $i = 2$. These excitations generate the total entropy (27). Consequently we may identify the two series in Eq. (37) as the total number of renormalized bosons that carry nonzero entropy. This total number, N_{n}, reads explicitly

$$N_{\mathrm{n}} = \sum_{\mathbf{k}} n_{\mathrm{qp}}^{(1)}(k) \left[1 + B_{\mathrm{cc}} N_{\mathrm{a}}(k)\right] \,. \tag{39}$$

The term $B_{\mathrm{cc}} N$ in the sum rule (37) therefore represents the number of renormalized bosons in the Bose condensate with zero entropy. In terms of "normal density" $\varrho_{\mathrm{n}} = N_{\mathrm{n}}/N$ and "superfluid density" $\varrho_{\mathrm{s}} = N_{\mathrm{s}}/N$ we may write sum rule (37) in the shape

$$\varrho_{\mathrm{n}} + \varrho_{\mathrm{s}} = \varrho \,. \tag{40}$$

This is the familiar two-fluid equation postulated phenomenologically by Tisza.[28]

For the free Bose gases we have the identity $n_{\mathrm{qp}}^{(1)}(k) = n_{\mathrm{qp}}^{(2)}(k)$ and $N_{\mathrm{a}}(k) = 0$, i.e., the renormalized bosons of types 1 and 2 are indistinguishable, superfluidity therefore cannot be stable and the density component ϱ_{s} can be interpreted only as the Bose-Einstein condensate fraction of the renormalized bosons with mass $m_{\mathrm{qp}}(T, \varrho, k)$. If the bosonic ingredients of the quantum fluid interact, the excitations of types 1 and 2 can be distinguished and some kind of coupling between them might produce the superfluid behavior of the condensate.

References

1. J. W. Clark and E. Feenberg, *Phys. Rev.* **113**, 388 (1959).
2. E. Feenberg, *Theory of Quantum Fluids* (Academic Press, N.Y., 1969).
3. C. E. Campbell, in "Progress in Liquid Physics," edited by C. A. Croxton (Wiley, N.Y., 1978).
4. J. W. Clark, in "Progress in Particle and Nuclear Physics," edited by D. H. Wilkinson, Vol. 2 (Pergamon, Oxford, 1979).
5. M. L. Ristig, in "From Nuclei to Particles, Proc. Int. School of Physics *Enrico Fermi*," Course LXXIV, edited by A. Molinari (North Holland, Amsterdam, 1981).

6. S. Fantoni and A. Fabrocini, in *Lecture Notes in Physics*, Vol. 510 (Springer, 1998).
7. E. Krotscheck, in *Lecture Notes in Physics*, Vol. 510 (Springer, 1998).
8. C. E. Campbell, K. E. Kürten, M. L. Ristig, and G. Senger, *Phys. Rev. B* **30**, 3728 (1984).
9. G. Senger, M. L. Ristig, K. E. Kürten, and C. E. Campbell, *Phys. Rev. B* **33**, 7562 (1986).
10. B. E. Clements, E. Krotscheck, J. A. Smith, and C. E. Campbell, *Phys. Rev. B* **47**, 5239 (1993).
11. G. Senger, M. L. Ristig, C. E. Campbell, and J. W. Clark, *Ann. Phys. (N.Y.)* **218**, 160 (1992).
12. M. L. Ristig, G. Senger, M. Serhan, and J. W. Clark, *Ann. Phys. (N.Y.)* **243**, 247 (1995).
13. R. Pantfoerder, T. Lindenau, and M. L. Ristig, *J. Low Temp. Phys.* **108**, 245 (1997).
14. M. L. Ristig, T. Lindenau, M. Serhan, and J. W. Clark, *J. Low Temp. Phys.* **114**, 317 (1999).
15. T. Lindenau, M. L. Ristig, J. W. Clark, and K. A. Gernoth, *J. Low Temp. Phys.* **129**, 143 (2002).
16. T. Lindenau, M. L. Ristig, K. A. Gernoth, J. Dawidowski, and F. J. Bermejo, *Int. J. Mod. Phys. B* **20**, 5035 (2006).
17. K. A. Gernoth, M. L. Ristig, and T. Lindenau, *Int. J. Mod. Phys. B* **21**, 2157 (2007).
18. K. A. Gernoth, T. Lindenau, and M. L. Ristig, *Phys. Rev. B* **75**, 174204 (2007).
19. K. A. Gernoth, M. Serhan, and M. L. Ristig, *Phys. Rev. B* **78**, 054513 (2008).
20. K. A. Gernoth, M. Serhan, and M. L. Ristig, *Int. J. Mod. Phys. B* **22**, 4315 (2008).
21. K. A. Gernoth and M. L. Ristig, *Int. J. Mod. Phys. B* **23**, 4096 (2009).
22. J. Dawidowski, F. J. Bermejo, M. L. Ristig, B. Fak, C. Cabrillo, R. Fernandez-Perea, K. Kinugawa, and J. Campo, *Phys. Rev. B* **69**, 014207 (2004).
23. J. Dawidowski, F. J. Bermejo, M. L. Ristig, C. Cabrillo, and S. M. Bennington, *Phys. Rev. B* **73**, 144203 (2006).
24. F. M. Gasparini, M. O. Kimball, K. P. Mooney, and M. Diaz-Avila, *Rev. Mod. Phys.* **80**, 1009 (2008).
25. R. A. Aziz, M. J. Slaman, A. Koide, A. R. Allnat, and J. W. Meath, *Mol. Phys.* **77**, 321 (1992).
26. T. Fließbach, *Statistische Physik* (Akademischer Verlag, Heidelberg, 1995).
27. E. W. Lemmon, M. O. McLinden, and D. G. Friend, in *NIST Chemistry WebBook, NIST Standard Reference Database Number 69*, edited by P. J. Linstrom and W. G. Mallard (2005); http://www.webbook.nist.gov.
28. L. Tisza, *Phys. Rev.* **72**, 838 (1947).

LIGHT CLUSTERS IN NUCLEAR MATTER

GERD RÖPKE

Institut für Physik, Rostock University, Universitätspl. 1
Rostock, 18051, Germany
gerd.roepke@uni-rostock.de

Received 4 December 2009

Nuclei in dense matter are influenced by the medium. In the cluster mean-field approximation, an effective Schrödinger equation for the A-nucleon cluster is obtained accounting for the effects of the surrounding medium, such as self-energy and Pauli blocking. Similar to the single-baryon states (free protons and neutrons), the light elements ($A \leq 4$) are treated as quasiparticles. Fit formulae are given for the quasiparticle shifts as function of temperature, density, asymmetry, and momentum. Composition, quantum condensates and thermodynamic functions are considered. The relevance for nuclear structure, heavy ion collisions and supernova astrophysics is shown.

Keywords: Nuclear matter; equation of state; quantum condensates.

1. Introduction

The equation of state (EoS), the composition and possible occurrence of phase transitions in nuclear matter are widely discussed topics not only in nuclear theory[1], but are also of great interest in astrophysics and cosmology. Experiments on heavy ion collisions, performed over the last decades, gave new insight into the behavior of nuclear systems in a broad range of densities and temperatures. The observed cluster abundances, their spectral distribution and correlations in momentum space can deliver information about the state of dense, highly excited matter. In a recent review[2] constraints to the EoS have been investigated. Different versions have been considered in the high-density region, and comparison with neutron star properties may discriminate between more or less reasonable equations of state.

We restrict ourselves to matter in equilibrium at temperatures $T \leq 30$ MeV and baryon number densities $n_B \leq 0.2$ fm^{-3}, where the quark substructure and the excitation of internal degrees of freedom of the nucleons (protons p and neutrons n) are not of relevance and the nucleon-nucleon interaction can be represented by an effective interaction potential. In this region of the temperature-density plane, we investigate how the quasi-particle picture is improved if few-body correlations are taken into account. The influence of cluster formation on the EoS is calculated for different situations, and the occurrence of phase instabilities is investigated. Derived from the full spectral function, the concept of composition will be introduced as an approximation to describe correlations in dense systems. Another interesting issue is the formation of quantum condensates.

A quantum statistical approach to the thermodynamic properties of nuclear matter can be given using the method of thermodynamic Green functions.[3] In general, within the grand canonical ensemble, the EoS $n_\tau(T, \mu_{\tau'})$ relates the particle number densities n_τ to temperature (T) and the chemical potentials μ_τ of protons (p) or neutrons (n), where the internal quantum number τ can be introduced to describe besides isospin also spin and further quantum numbers. This EoS is obtained from the single-particle spectral function, which can be expressed in terms of the self-energy. Then, thermodynamic potentials such as the pressure $p(T, \mu_\tau)$ or the density of free energy $f(T, n_\tau)$ are obtained by integrations. From these thermodynamic potentials, all other equilibrium thermodynamic properties can be derived. In particular, the stability of the homogeneous system against phase separation has to be considered.

The main quantity to be evaluated is the self-energy. Different approximations can be obtained by partial summations within a diagram representation. The formation of bound states is taken into account considering ladder approximations,[4] leading in the low-density limit to the solution of the Schrödinger equation. The effects of the medium can be included in a self-consistent way within the cluster-mean field approximation (see Ref. 5 for references), where the influence of the correlated medium on the single particle states as well as on the clusters is considered in first order with respect to the interaction. As a point of significance, the quasiparticle energies of the constituents, *i.e.* the bound state energies, are modified besides the single-nucleon self-energy shifts also by the Pauli blocking due to the correlated medium. An extended discussion of the two-particle problem can be found in Ref. 6. Within a generalized Beth-Uhlenbeck formula approach, not only the two-particle properties such as deuteron formation and scattering phase shifts has been used to construct a nuclear matter EoS, but also the influence of the medium has been taken into account which leads to the suppression of correlations at high densities. Similarly, also three and four-particle bound states can be included,[4] and also the medium dependent shift of the cluster binding energies was investigated, see Ref. 7.

The EoS can be applied to different situations. In astrophysics, the relativistic EoS of nuclear matter for supernova explosions was investigated recently.[8] If nuclei are considered as inhomogeneous nuclear matter, within a local density approximation the EoS can serve for comparison to estimate the role of correlations. In nuclear reactions a nonequilibrium theory is needed, but within simple approaches such as the freeze out concept or the coalescence model, the results from the EoS may be used to describe heavy ion reactions. The EoS including the contribution of light clusters has been evaluated recently,[9] and the inclusion of heavier clusters is under consideration.[10] As example, the symmetry energy at subsaturation densities is sensitive to the formation of clusters.[11]

The inclusion of light cluster formation describing nuclear matter has some yet unsolved implications to be discussed here. The account of correlations means besides the contribution of bound states that can be treated in quasiparticle approximation, also the contribution of scattering states. The EoS contains contributions from the scattering phase shifts as seen, for example, from the Beth-Uhlenbeck formula.[6,12] In contrast, cluster yields from HIC do not contain the scattering contributions that quickly decay during the nonequilibrium expansion process. Only the bound states remain because ternary particles are needed for energy and momentum conservation.

There is no unique definition of a bound state and composition in a dense system. From the spectral function in the corresponding A-particle channel, a peak structure can be used to define the quasiparticles, but due to the interactions in dense systems, sharp peaks in the spectral function are broadened and can become resonances in the continuum. We introduce the spectral function in the following section. The quasiparticle approach is introduced to reproduce significant contributions of the clusters to the total nucleon density. The relation to the cluster yields in HIC is not trivial and needs special considerations accounting for non-equilibrium states. Further items to be discussed are the occurrence of quantum condensates, phase instabilities, the role of higher clusters, and the treatment of alpha matter at low temperatures and densities.

2. Quantum Statistical Approach to the Equation of State

2.1. *Single-particle spectral function and quasiparticles*

Using the finite-temperature Green function formalism, a non-relativistic quantum statistical approach can be given to describe the equation of state of nuclear matter including the formation of bound states.[4,6] It is most convenient to start with the nucleon number densities $n_\tau(T, \mu_p, \mu_n)$ as functions of temperature $T = 1/(k_B\beta)$ and chemical potentials μ_τ for protons $(\tau = p)$ and neutrons $(\tau = n)$, respectively,

$$n_\tau(T, \mu_p, \mu_n) = \frac{1}{\Omega} \sum_1 \langle a_1^\dagger a_1 \rangle \delta_{\tau,\tau_1} = 2 \int \frac{d^3 p_1}{(2\pi)^3} \int_{-\infty}^{\infty} \frac{d\omega}{2\pi} f_{1,Z}(\omega) S_1(1, \omega) , \quad (1)$$

where Ω is the system volume, $\{1\} = \{p_1, \sigma_1, \tau_1\}$ denotes the single-nucleon quantum numbers momentum, spin, and isospin. Summation over spin yields the factor 2 and

$$f_{A,Z}(\omega) = \left(\exp\left\{ \beta \left[\omega - Z\mu_p - (A - Z)\mu_n \right] \right\} - (-1)^A \right)^{-1} \quad (2)$$

is the Fermi or Bose distribution function. Instead of the isospin quantum number τ_1 we occasionally use the mass number A and the charge number Z. Both the distribution function and the spectral function $S_1(1, \omega)$ depend on the temperature and the chemical potentials μ_p, μ_n not given explicitly. We work with a grand canonical ensemble and have to use Eq. (1) to replace the chemical potentials by the densities n_p, n_n. For this EoS, expressions such as the Beth-Uhlenbeck formula and its generalizations have been derived. [4,6,12]

We consider both the total number densities of protons and neutrons, n_p^{tot} and n_n^{tot}, and the temperature T as given parameters. Alternatively, the total baryon density $n = n_n^{\text{tot}} + n_p^{\text{tot}}$ and the asymmetry of nuclear matter $\delta = (n_n^{\text{tot}} - n_p^{\text{tot}})/n = 1 - 2Y_p$ are used. Y_p denotes the total proton fraction. Besides the frozen equilibrium where n_p^{tot} and n_n^{tot} are given, we assume homogeneity in space. Thermodynamical stability is considered afterwards. In a further evaluation, allowing for weak interactions, β-equilibrium may be achieved, which is of interest for astrophysical applications. In that case the asymmetry δ is uniquely determined for given n and T.

The spectral function $S_1(1, \omega)$ is related to the self-energy $\Sigma(1, z)$ according to

$$S_1(1, \omega) = \frac{2\text{Im}\,\Sigma(1, \omega - i0)}{[\omega - E(1) - \text{Re}\,\Sigma(1, \omega)]^2 + [\text{Im}\,\Sigma(1, \omega - i0)]^2} \, , \qquad (3)$$

where the imaginary part has to be taken for a small negative imaginary part in the frequency ω. $E(1) = p_1^2/(2m_1)$ is the kinetic energy of the free nucleon. The solution of the relation

$$E_1^{\text{qu}}(1) = E(1) + \text{Re}\,\Sigma[1, E_1^{\text{qu}}(1)] \qquad (4)$$

defines the single-nucleon quasiparticle energies $E_1^{\text{qu}}(1) = E(1) + \Delta E^{\text{SE}}(1)$. Expanding for small $\text{Im}\,\Sigma(1, z)$, the spectral function yields a δ-like contribution. The densities are calculated from Fermi distributions with the quasiparticle energies so that

$$n_\tau^{\text{qu}}(T, \mu_p, \mu_n) = \frac{2}{\Omega} \sum_{k_1} f_{1,z}[E_1^{\text{qu}}(1)] \qquad (5)$$

follows for the EoS in mean field approximation. This result does not contain the contribution of bound states and therefore fails to be correct in the low-temperature, low-density limit where the NSE describes the nuclear matter EoS.

As shown in Refs. 4, 6, the bound state contributions are obtained from the poles of $\text{Im}\,\Sigma(1, z)$ which cannot be neglected in expanding the spectral function with respect to $\text{Im}\,\Sigma(1, z)$. A cluster decomposition of the self-energy has been proposed, see Ref. 5. The self-energy is expressed in terms of the A-particle Green functions which read in bilinear expansion

$$G_A(1...A, 1'...A', z_A) = \sum_{\nu P} \psi_{A\nu P}(1...A) \frac{1}{z_A - E_{A,\nu}^{\text{qu}}(P)} \psi_{A\nu P}^*(1'...A') \, . \qquad (6)$$

The A-particle wave function $\psi_{A\nu P}(1...A)$ and the corresponding eigenvalues $E_{A,\nu}^{\text{qu}}(P)$ result from solving the in-medium Schrödinger equation, see the following subsections. P denotes the center of mass momentum of the A-nucleon system. Besides the bound states, the summation over the internal quantum states ν includes also the scattering states.

The evaluation of the equation of state in the low-density limit is straightforward. Considering only the bound-state contributions, we obtain the result

$$n_p^{\text{tot}}(T, \mu_p, \mu_n) = \frac{1}{\Omega} \sum_{A,\nu,P} Z f_{A,Z}[E_{A,\nu}^{\text{qu}}(P; T, \mu_p, \mu_n)],$$

$$n_n^{\text{tot}}(T, \mu_p, \mu_n) = \frac{1}{\Omega} \sum_{A,\nu,P} (A - Z) f_{A,Z}[E_{A,\nu}^{\text{qu}}(P; T, \mu_p, \mu_n)] \tag{7}$$

for the EoS describing a mixture of components (cluster quasiparticles) obeying Fermi or Bose statistics. The total baryon density results as $n(T, \mu_p, \mu_n) = n_n^{\text{tot}}(T, \mu_p, \mu_n) + n_p^{\text{tot}}(T, \mu_p, \mu_n)$. To derive the extended Beth-Uhlenbeck formula,[4,6] we restrict the summation to $A \leq 2$, but extend the summation over the internal quantum numbers ν not only to excited bound states, but also the scattering states. Note that at low temperatures Bose-Einstein condensation may occur.

The NSE is obtained in the low-density limit if the in-medium energies $E_{A,\nu}^{\text{qu}}(P; T, \mu_p, \mu_n)$ can be replaced by the binding energies of the isolated nuclei $E_{A,\nu}^{(0)}(P) = E_{A,\nu}^{(0)} + P^2/(2Am)$, with $m = 939$ MeV the average nucleon mass. For the cluster contributions, *i.e.* $A > 1$, the summation over the internal quantum numbers is again restricted to the bound states only. We have

$$n_p^{\text{NSE}}(T, \mu_p, \mu_n) = \frac{1}{\Omega} \sum_{A,\nu,P}^{\text{bound}} Z f_{A,Z}[E_{A,\nu}^{(0)}(P)],$$

$$n_n^{\text{NSE}}(T, \mu_p, \mu_n) = \frac{1}{\Omega} \sum_{A,\nu,P}^{\text{bound}} (A - Z) f_{A,Z}[E_{A,\nu}^{(0)}(P)]. \tag{8}$$

The summation over A includes also the contribution of free nucleons, $A = 1$.

In the nondegenerate and non-relativistic case, assuming a Maxwell-Boltzmann distribution, the summation over the momenta P can be performed analytically and the thermal wavelength $\lambda = \sqrt{2\pi/(mT)}$ of the nucleon occurs. As shown below, the medium effects in nuclear matter are negligible below 10^{-4} times the saturation density $n_{\text{sat}} = 0.15$ fm^{-3} for the temperatures considered here.

Interesting quantities are the mass fractions

$$X_{A,Z} = \frac{A}{\Omega n} \sum_{\nu,P} f_{A,Z}[E_{A,\nu}^{\text{qu}}(P; T, \mu_p, \mu_n)] \tag{9}$$

of the different clusters. From the EoS considered here, thermodynamical potentials can be obtained by integration, in particular the free energy per volume F/Ω. In the special case of symmetric nuclear matter, $Y_p^{\text{s}} = 0.5$, the free energy per volume is obtained from the averaged chemical potential $\mu = (\mu_p + \mu_n)/2$ as

$$F(T, n, Y_p^{\text{s}})/\Omega = \int_0^n dn'\, \mu(T, n', Y_p^{\text{s}}). \tag{10}$$

In the quantum statistical approach described above, we relate the EoS to properties of the correlation functions, in particular to the peaks occurring in the A-nucleon spectral function describing the single-nucleon quasiparticle ($A = 1$) as well as the nuclear quasiparticles $(A \geq 2)$. The microscopic approach to these quasiparticle energies can be given calculating the self-energy. Different approaches can be designed which reproduce properties of the nucleonic system known from limiting cases or from empirical data.

2.2. *Medium modification of single nucleon properties*

The single-particle spectral function contains the single-nucleon quasiparticle contribution, $E_1^{\mathrm{qu}}(1) = E_\tau^{\mathrm{qu}}(p)$, given in Eq. (4), where τ denotes isospin of particle 1 and p is the momentum. In the effective mass approximation, the single-nucleon quasiparticle dispersion relation reads

$$E_\tau^{\mathrm{qu}}(p) = \Delta E_\tau^{\mathrm{SE}}(0) + \frac{p^2}{2m_\tau^*} + \mathcal{O}(p^4), \tag{11}$$

where the quasiparticle energies are shifted at zero momentum p by $\Delta E_\tau^{\mathrm{SE}}(0)$, and m_τ^* denotes the effective mass of neutrons $(\tau = n)$ or protons $(\tau = p)$. Both quantities, $\Delta E_\tau^{\mathrm{SE}}(0)$ and m_τ^*, are functions of T, n_p and n_n, characterizing the surrounding matter.

Expressions for the single-nucleon quasiparticle energy $E_\tau^{\mathrm{qu}}(p)$ can be given by the Skyrme parametrization or by more sophisticated approaches such as relativistic mean-field approaches and relativistic Dirac-Brueckner Hartree-Fock calculations. Since there is no fundamental expression for the nucleon-nucleon interaction, a phenomenological form is assumed.

We will use the density-dependent relativistic mean field approach[13] that is designed not only to reproduce known properties of nuclei, but also fits with microscopic calculations in the low density region. It is expected that this approach gives at present an optimal expression to the quasiparticle energies and is applicable in a large interval of densities and temperatures.

The single-nucleon quasiparticle energies (including the rest mass) result as

$$E_n^{\mathrm{qu}}(p) = \sqrt{[m^2 - S(T,n,\delta)]^2 + p^2} + V(T,n,\delta) \tag{12}$$

for neutrons and $E_p^{\mathrm{qu}}(p) = \sqrt{[m^2 - S(T,n,-\delta)]^2 + p^2} + V(T,n,-\delta)$ for protons. In the non-relativistic limit, the shifts of the quasiparticle energies are

$$\Delta E_{n,p}^{\mathrm{SE}}(0) = V(T,n,\pm\delta) - S(T,n,\pm\delta). \tag{13}$$

The effective masses for neutrons and protons are given by

$$m_{n,p}^* = m - S(T,n,\pm\delta). \tag{14}$$

Approximations for the functions $V(T,n,\delta)$ and $S(T,n,\delta)$ are given by Pade expressions

$$S(T,n,\delta) = \frac{s_1 n + s_2 n^2 + s_3 n^3}{1 + s_4 n + s_5 n^2} \tag{15}$$

$$V(T, n, \delta) = \frac{v_1 n + v_2 n^2 + v_3 n^3}{1 + v_4 n + v_5 n^2} \tag{16}$$

with

$$s_i(T, \delta) = \sum_{j=0}^{2} \sum_{k=0,2,4} s_{ijk} \delta^k T^j \qquad v_i(T, \delta) = \sum_{j=0}^{2} \sum_{k=0}^{4} v_{ijk} \delta^k T^j. \tag{17}$$

The parameter values for s_{ijk}, v_{ijk} are given in Tabs. 1, 2.

Table 1. Parameters of the scalar field $S(T, n, \delta)$, Eq (15), in MeV.

k	0	2	4
s_{10k}	4462.35	1.63811	0.293287
s_{11k}	-7.22458	0.92618	-0.679133
s_{12k}	0.00975576	-0.0355021	0.026292
s_{20k}	204334	-11043.9	-46439.7
s_{21k}	7293.23	-49220.9	35263.0
s_{22k}	-209.452	2114.07	-1507.55
s_{30k}	125513.	-64680.5	-4940.76
s_{31k}	1055.3	-19422.6	15842.8
s_{32k}	132.502	572.292	-555.762
s_{40k}	49.0026	-1.76282	-10.6072
s_{41k}	1.70156	-11.1142	7.92604
s_{42k}	-0.0456724	0.473553	-0.337016
s_{50k}	241.935	-19.8568	-48.3232
s_{51k}	6.6665	-52.6306	38.1023
s_{52k}	-0.112997	2.15092	-1.57597

Table 2. Parameters of the vector field $V(T, n, \delta)$, Eq (16), in MeV.

k	0	1	2	3	4
v_{10k}	3403.94	-490.15	-0.0213143	0.00760759	0.0265109
v_{11k}	-0.000978098	-0.000142646	0.00176929	0.00043752	-0.00321724
v_{12k}	0.0000651609	0.0000098168	-0.0000394036	0.0000381407	0.000110931
v_{20k}	-345.863	1521.62	-2658.72	-408.013	-132.384
v_{21k}	29.309	-8.80748	-236.029	13.7447	111.538
v_{22k}	3.63322	0.0163495	6.88256	-0.369704	-3.28749
v_{30k}	33553.8	4298.76	3692.23	-1083.14	-728.086
v_{31k}	-192.395	-52.0101	-141.702	-57.9237	-11.4749
v_{32k}	15.2158	3.86652	-0.785201	1.59625	2.0419
v_{40k}	2.7078	-0.162553	-0.308454	-0.174442	-0.0581052
v_{41k}	0.0161456	-0.00145171	-0.0689643	-0.0000398794	0.0317996
v_{42k}	0.00105179	0.000192765	0.00203728	0.00000561467	-0.000932046
v_{50k}	18.7473	4.0948364	-0.0308012	-0.751981	-0.585746
v_{51k}	-0.102959	-0.044524	-0.308021	-0.0190921	0.0869529
v_{52k}	0.0118049	0.0021141	0.0070548	0.000565564	-0.00182714

These functions reproduce the empirical values for the saturation density $n_{\text{sat}} \approx$ 0.15 fm^{-3} and the binding energy per nucleon $B/A \approx -16$ MeV. The effective mass is somewhat smaller than the empirical value $m^* \approx m(1 - 0.17\, n/n_{\text{sat}})$ for $n < 0.2$ fm^{-3}.

2.3. *Medium modification of cluster properties*

Recent progress of the description of clusters in low density nuclear matter[14,8,15] enables us to evaluate the properties of deuterons, tritons, helions and helium nuclei in a non-relativistic microscopic approach, taking the influence of the medium into account.

In addition to the δ-like nucleon quasiparticle contribution, also the contribution of the bound and scattering states can be included in the single-nucleon spectral function by analyzing the imaginary part of $\Sigma(1, z)$. Within a cluster decomposition, A-nucleon T matrices appear in a many-particle approach. These T matrices describe the propagation of the A-nucleon cluster in nuclear matter. In this way, bound states contribute to $n_\tau = n_\tau(T, \mu_n, \mu_p)$, see Refs. 4, 6. Restricting the cluster decomposition of the nucleon self-energy only to the contribution of two-particle correlations, we obtain the so-called $T_2 G$ approximation. In this approximation, the Beth-Uhlenbeck formula is obtained for the EoS. In the low-density limit, the propagation of the A-nucleon cluster is determined by the energy eigenvalues of the corresponding nucleus, and the simple EoS (7) results describing the nuclear statistical equilibrium (NSE).

For nuclei embedded in nuclear matter, an effective wave equation can be derived.[4,15] The A-particle wave function $\psi_{A\nu P}(1 \ldots A)$ and the corresponding eigenvalues $E_{A,\nu}^{\text{qu}}(P)$ follow from solving the in-medium Schrödinger equation

$$[E^{\text{qu}}(1) + \cdots + E^{\text{qu}}(A) - E_{A,\nu}^{\text{qu}}(P)]\psi_{A\nu P}(1 \ldots A)$$

$$+ \sum_{1' \ldots A'} \sum_{i<j}[1 - \tilde{f}(i) - \tilde{f}(j)]V(ij, i'j') \prod_{k \neq i,j} \delta_{kk'}\psi_{A\nu P}(1' \ldots A') = 0 . \quad (18)$$

This equation contains the effects of the medium in the single-nucleon quasiparticle shifts as well as in the Pauli blocking terms. The A-particle wave function and energy depend on the total momentum P relative to the medium.

The effective Fermi distribution function $\tilde{f}(1) = (\exp\{\beta\, [E^{\text{qu}}(1) - \tilde{\mu}_1]\} + 1)^{-1}$ contains the non-relativistic effective chemical potential $\tilde{\mu}_1$. It is determined by the normalization condition that the total proton or neutron density is reproduced in quasiparticle approximation, $n_\tau^{\text{tot}} = \Omega^{-1} \sum_1 \tilde{f}(1)\delta_{\tau_1, \tau}$ for the particles inside the volume Ω. It describes the occupation of the phase space neglecting any correlations in the medium. This means that the nucleons bound in clusters, e.g. deuterons, also will occupy phase space. The total amount of occupied phase space is correctly given, but the form factor will deviate from the Fermi distribution as given by the bound state wave function.

The solution of the in-medium Schrödinger equation (18) can be obtained in the low-density region by perturbation theory. At higher densities, a variational approach can be used. In particular, the quasiparticle energy of the A-nucleon cluster with Z protons in the ground state follows as

$$E_{A,\nu}^{\text{qu}}(P) = E_{A,Z}^{\text{qu}}(P) = E_{A,Z}^{(0)} + \frac{P^2}{2Am} + \Delta E_{A,Z}^{\text{SE}}(P) + \Delta E_{A,Z}^{\text{Pauli}}(P) + \Delta E_{A,Z}^{\text{Coul}}(P) + \cdots$$
(19)

with various contributions. Besides the cluster binding energy in the vacuum $E_{A,Z}^{(0)}$ and the kinetic term, the self-energy shift $\Delta E_{A,Z}^{\text{SE}}(P)$, the Pauli shift $\Delta E_{A,Z}^{\text{Pauli}}(P)$ and the the Coulomb shift $\Delta E_{A,Z}^{\text{Coul}}(P)$ enter. The latter can be evaluated for dense matter in the Wigner-Seitz approximation,

$$\Delta E_{A,Z}^{\text{Coul}}(P) = \frac{Z^2}{A^{1/3}} \frac{3}{5} \frac{e^2}{r_0} \left[\frac{3}{2} \left(\frac{2n_p}{n_{\text{sat}}} \right)^{\frac{1}{3}} - \frac{n_p}{n_{\text{sat}}} \right] \text{MeV}$$
(20)

with $r_0 = 1.2$ fm. Since the values of Z are small, this contribution is small as well.

The self-energy contribution to the quasiparticle shift is determined by the contribution of the single-nucleon shift

$$\Delta E_{A,Z}^{\text{SE}}(0) = (A - Z)\Delta E_n^{\text{SE}}(0) + Z\Delta E_p^{\text{SE}}(0) + \Delta E_{A,Z}^{\text{SE,eff.mass}} .$$
(21)

The contribution to the self-energy shift due to the change of the effective nucleon mass can be calculated from perturbation theory using the unperturbed wave function of the clusters[8] so that

$$\Delta E_{A,Z}^{\text{SE,eff.mass}} = \left(\frac{m}{m^*} - 1 \right) s_{A,Z} .$$
(22)

Values of $s_{A,Z}$ for $\{A, Z\} = \{i\} = \{d, t, h, \alpha\}$ are given in Tab. 3. Inserting the medium-dependent quasiparticle energies in the distribution functions (2) this contribution to the quasiparticle shift can be included renormalizing the chemical potentials μ_n and μ_p.

The most important effect in the calculation of the abundances of light elements comes from the Pauli blocking terms in Eq. (18) in connection with the interaction potential. This contribution is restricted only to the bound states so that it may lead to the dissolution of the nuclei if the density of nuclear matter increases. The corresponding shift $\Delta E_{A,Z}^{\text{Pauli}}(P)$ can be evaluated in perturbation theory provided the interaction potential and the ground state wave function are known. After angular averaging where in the Fermi functions the mixed scalar product $\vec{p} \cdot \vec{P}$ between the total momentum $\vec{P}$ and the remaining Jacobian coordinates $\vec{p}$ is neglected, the Pauli blocking shift can be approximated as

$$\Delta E_{A,Z}^{\text{Pauli}}(P) \approx \Delta E_{A,Z}^{\text{Pauli}}(0) \exp\left(-\frac{P^2}{2A^2mT} \right) .$$
(23)

Avoiding angular averaging, the full solution gives the result up to the order P^2

$$\Delta E_{A,Z}^{\text{Pauli}}(P) \approx \Delta E_{A,Z}^{\text{Pauli}}(0) \exp\left(-\frac{P^2}{g_{A,Z}} \right)$$
(24)

with the dispersion that is determined by

$$g_i(T, n, Y_p) = \frac{g_{i,1} + g_{i,2}T + h_{i,1}n}{1 + h_{i,2}n} .$$
(25)

The values for $g_{i,1}$ and $g_{i,2}$ can be calculated from perturbation theory using the unperturbed cluster wave functions; the density corrections $h_{i,1}$ and $h_{i,2}$ are fitted to variational solutions of the in-medium wave equation Eq. (18) for given T, n_p, n_n and P. Numerical values of the parameters in symmetric nuclear matter ($Y_p = 0.5$) are given in Tab 3.

The shift of the binding energy of light clusters at zero total momentum which is of first order in density[14] has been calculated recently.[15] The light clusters deuteron ($d = {}^2\mathrm{H}$), triton ($t = {}^3\mathrm{H}$), helion ($h = {}^3\mathrm{He}$) and the α particle (${}^4\mathrm{He}$) have been considered. The interaction potential and the nucleonic wave function of the few-nucleon system have been fitted to the binding energies and the rms radii of the corresponding nuclei.

With the neutron number $N_i = A_i - Z_i$, it can be written as

$$\Delta E_{A_i,Z_i}^{\mathrm{Pauli}}(0; n_p, n_n, T) = -\frac{2}{A_i} \left[Z_i n_p + N_i n_n \right] \delta E_i^{\mathrm{Pauli}}(T, n) , \qquad (26)$$

where the temperature dependence and higher density corrections are contained in the functions $\delta E_i^{\mathrm{Pauli}}(T, n)$. These functions have been obtained with different approximations for the wave function. In case of the deuteron, the Jastrow approach leads to a functional form

$$\delta E_i^{\mathrm{Pauli}}(T, n) = \frac{a_{i,1}}{T^{3/2}} \left[\frac{1}{\sqrt{y_i}} - \sqrt{\pi} a_{i,3} \exp\left(a_{i,3}^2 y_i\right) \mathrm{erfc}\left(a_{i,3}\sqrt{y_i}\right) \right] \frac{1}{1 + [b_{i,1} + b_{i,2}/T]n} \tag{27}$$

with $y_i = 1 + a_{i,2}/T$. For the other clusters $i = t, h, \alpha$, the Gaussian approach is used which gives the simple form

$$\delta E_i^{\mathrm{Pauli}}(T, n) = \frac{a_{i,1}}{T^{3/2}} \frac{1}{y_i^{3/2}} \frac{1}{1 + [b_{i,1} + b_{i,2}/T]n} . \tag{28}$$

The parameters $a_{i,1}$, $a_{i,2}$ and $a_{i,3}$ are determined by low-density perturbation theory from the unperturbed cluster wave functions. The parameters $b_{i,1}$ and $b_{i,2}$ are density corrections and are fitted to the numerical solution of the in-medium wave equation Eq. (18) for given T, n_p, n_n, and $P = 0$. Values are given in Tab 3.

Table 3.   Parameters for the cluster binding energy shifts. For deuterons d, further parameters are $a_{d,3} = 0.2223$, $h_{d,2} = 17.5$ fm^3.

cluster i	s_i [MeV]	$a_{i,1}$ [MeV$^{5/2}$fm^3]	$a_{i,2}$ [MeV]	$b_{i,1}$ [fm^3]	$b_{i,2}$ [MeV fm^3]	$g_{i,1}$ [fm^{-2}]	$g_{i,2}$ [MeV^{-1}fm^{-2}]	$h_{i,1}$ [fm]
d	11.147	38386.4	22.5204	1.048	285.7	0.85	0.223	132
t	24.575	69516.2	7.49232	4.414	43.90	3.20	0.450	37
h	20.075	58442.5	6.07718	4.414	43.90	2.638	0.434	43
α	49.868	164371	10.6701	-	-	8.236	0.772	50

3. Contribution of the Continuum Correlations

Now, the nucleon number densities (7) can be evaluated as in the non-interacting case, with the only difference that the number densities of the particles are calculated with the quasiparticle energies. In the light-cluster quasiparticle approximation, the total densities of neutrons

$$n_n^{\text{tot}} = n_n + \sum_{i=d,t,h,\alpha} N_i n_i \tag{29}$$

and of protons

$$n_p^{\text{tot}} = n_p + \sum_{i=d,t,h,\alpha} Z_i n_i \tag{30}$$

contain the densities of the free neutrons and protons n_n and n_p, respectively, and the contributions due to the correlations in the corresponding few-nucleon channels. The summation over the internal quantum number ν, Eq. (7), covers the bound state part (ground state and possibly excited states) as well as the contribution of the continuum (for example resonances). Thus, the contribution n_d of the deuteron channel is given by the Beth-Uhlenbeck formula that contains besides the bound state contribution also the contribution of the scattering states.[4,6,12]

If we neglect the contribution of the continuum and consider only the formation of bound states, we calculate the densities n_i according to Eq. (7) performing the summation only over the bound state part. The state of the system in chemical equilibrium is completely determined by specifying the corresponding total nucleon density $n^{\text{bound}} = n_n^{\text{tot, bound}} + n_p^{\text{tot, bound}}$, the asymmetry $\delta = (n_n^{\text{tot, bound}} - n_p^{\text{tot, bound}})/n^{\text{bound}}$ and the temperature T as long as no β-equilibrium is considered.

This result is an improvement of the NSE and allows for the smooth transition from the low-density limit up to the region of saturation density. The bound state contributions to the EoS are fading with increasing density because they move as resonances into the continuum of scattering states. Due to the continuously changing Mott momentum, the densities are smooth functions of density at fixed temperature. This improved NSE, however, does not contain the contribution of scattering states explicitly.

We suppose that the densities n_i^{bound} give the cluster yields within a freeze-out approach to heavy ion collisions. In expanding nuclear matter, continuum correlations such as $n-n$ or $p-p$ will decay. The formation or decay of a bound state, e. g. in the deuteron channel, demands a further nucleon as spectator to obey energy and momentum conservation. At decreasing density, such processes become suppressed. The continuum correlations can be included into the free nucleon part n_n, n_p of the total densities $n_n^{\text{tot, bound}}$, $n_p^{\text{tot, bound}}$. A more systematic approach to heavy ion collisions can be given by transport codes based on coupled kinetic equations for the different constituents of nuclear matter (n, p, d, t, h, α in the case considered here).

For the evaluation of the equation of state, the account of scattering states needs further consideration. Investigations on the two-particle level have been performed and extensively discussed.[4,6,12] We use the Levinson theorem to take the contribution of scattering states into account in the lowest-order approximation. Each bound state contribution to the density has to accompanied with a continuum contribution that partly compensates the strength of the bound state correlations. As a consequence, the total proton and neutron densities contain besides the contribution of free nucleons, $A = 1$, considered as quasiparticles with the energy dispersion given by the RMF approach, also the contribution of the different clusters

$$n_i = \frac{1}{\Omega} \sum_{P}^{\text{bound}} \left[f_{A_i,Z_i}[E_i^{\text{qu}}(P; T, \mu_p, \mu_n)] - f_{A_i,Z_i}[E_i^{\text{cont}}(P; T, \mu_p, \mu_n)] \right] + n_i^{\text{scatt}} \quad (31)$$

where E_i^{cont} denotes the edge of the continuum states that is also determined by the single-nucleon self-energy shifts. These expressions guarantee a smooth behavior if bound states merge with the continuum of scattering states.

The summation over P and the subtraction of the continuum contribution is extended only over that region of momentum space where bound states exist. The disappearance of the bound states is caused by the Pauli blocking term; the self-energy contributions to the quasiparticle shifts act on bound as well as on scattering states. Above the so-called Mott density, where the bound states at $P = 0$ disappear, the momentum summation has to be extended only over that region $P > P_{A,\nu}^{\text{Mott}}(T, n, \delta)$ where the bound state energy is lower than the continuum of scattering states.

The contribution of scattering states n_i^{scatt} is necessary to obtain the second virial coefficient according to the Beth-Uhlenbeck equation, see Refs. 6, 12. This leads also to corrections in comparison with the NSE that accounts only for the bound state contributions, neglecting all effects of scattering states. These corrections become important at increasing temperatures for weakly bound clusters. Thus, the corrections which lead to the correct second virial coefficient are of importance for the deuteron system, when the temperature is comparable or large compared with the binding energy per nucleon.

To take the contributions of these continuum correlations into account, we propose the following approximation

$$n_i^{\text{scatt}} = \frac{1}{\Omega} \sum_{P}^{\text{bound}} b_i \frac{E_i^{\text{qu}}(0)}{E_i^{(0)}} \, e^{-P^2/(2A_i mT)} e^{[(A_i-Z_i)(\mu_n - \Delta E_n^{\text{SE}}(0)) + Z_i(\mu_p - \Delta E_p^{\text{SE}}(0))]/T}$$

$$(32)$$

where $b_i(T)$ is the second virial coefficient after subtraction of $e^{-E_i^{(0)}/T} - 1$. We have used the values for the deuteron as given in Ref. 12. Of course, the inclusion of the scattering states can be improved, e.g. by comparing with the Beth-Uhlenbeck formula. To approach the low-density limit of the EoS correctly, one has to reproduce the virial coefficients of the cluster-virial expansion.

Solving Eqs. (29)-(31) for given T, n_p^{tot} and n_n^{tot} we find the chemical potentials μ_p and μ_n. After integration, see Eq. (10), the free energy is obtained, and all

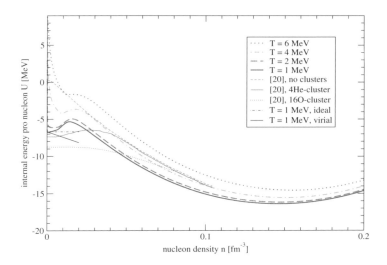

Fig. 1. Internal energy of symmetric matter as function of density for different temperatures. Thin lines: Results of a fully antisymmetrized calculation at $T = 0$ MeV, considering different clusters [20]. For comparison, values for the ideal gas of α particles as well as the virial expansion [12] are also shown.

the other thermodynamic functions are derived from this quantity without any contradictions. Results are shown elsewhere.[9]

4. α Matter

In the low-temperature region, below the Mott density the α particles yield the dominating contribution to the composition of symmetric matter, if we restrict us to clusters with $A \leq 4$. Below densities of the order 10^{-4} fm^{-3}, nuclear matter can be considered as ideal mixture of free nucleons and clusters. The interaction between the constituents can be neglected. The mass action law gives increasing yield of α particles at decreasing temperature for fixed density, but decreasing bound state concentration for decreasing density at fixed temperature (entropy dissociation). The low-density limit at fixed temperature is the ideal classical gas. Corrections are given by the virial expansion.

In particular, within a cluster-virial expansion the empirical scattering phase shifts can be used to evaluate density corrections for the ideal mixture of the different components. Thus, the virial coefficient for $\alpha - \alpha$ interaction is obtained from the corresponding scattering phase shifts.[12] Alternatively, one can use the phase shifts to introduce an effective interaction. The corresponding equation of state are reconsidered recently[21], also taking into account the formation of a quantum condensate. However, such effective interactions become questionable if the density is increasing so that the wave functions of the clusters overlap. Then, Pauli blocking leads to the dissolution of the clusters.

In our approach, these medium effects are included. Results for the internal energy per nucleon are shown in Fig. 1. At very low densities, the internal energy pro nucleon takes the value $U = \frac{3}{2}T$ for the classical ideal gas of free nucleons. As soon as cluster are formed, what happens for low temperatures already at very low densities, the binding energy of the nucleons in clusters determines the internal energy. In the case of α particles this contribution to the internal energy amounts -7.08 MeV. It determines the internal energy at zero temperature in the low-density limit.

With increasing density, the Pauli blocking leads to a reduction of the binding energy of the α particles and its dissolution. This is the reason for the increase of internal energy until the Mott density (≈ 0.006 fm^{-3}) is reached. Above the Mott density, after the bound states are dissolved, the free nucleon RMF approach gives the behavior of the internal energy. A more detailed investigation of the internal energy should also include the virial coefficient for $\alpha - \alpha$ interaction[12,21] as well as the formation of a condensate.[20]

5. Quantum Condensates

If bosonic clusters (A even) are formed, Bose–Einstein condensation may occur at low temperatures. In the low-density limit where the interaction between the clusters can be neglected, the well-known relations for the ideal Bose gas can be applied. With increasing density, the medium effects have to be considered. We consider here symmetric matter.

The critical temperature $T_c^{(\alpha)}$ for the formation of the quantum condensate in the α-particle channel (quartetting) in the low-density limit is given by the relation $n_\alpha \Lambda_\alpha^3 = 2.61$. Assuming that at low temperatures nearly all nucleons are bound in α clusters, $n_\alpha \approx n/4$, we find for the critical nucleon density $n_c^{(\alpha),\text{ideal}}$ for the formation of the quantum condensate in the α-particle channel the relation

$$n_c^{(\alpha),\text{ideal}}(T) = 4 \times 2.61 \left(\frac{4mT}{2\pi} \right)^3 . \tag{33}$$

The corresponding expressions can also be derived for the deuteron channel (spin-triplet pairing).

For the ideal gas of α-particles, at $T = 1$ MeV the value $n_c^{(\alpha),\text{ideal}} = 0.02$ fm^{-3} results, at $T = 10.5$ MeV the value $n_c^{(\alpha),\text{ideal}} = 0.007$ fm^{-3}. However, at that densities we have also to account medium modifications, and the α particles are already dissolved due to Pauli blocking. Within our quasiparticle approach, results for the chemical potential $(\mu_n + \mu_p)/2$ in symmetric matter are shown for low temperatures as function of the nucleon density in Fig. 2. No quantum condensates are obtained above $T = 0.5$ MeV.

The formation of quantum condensates will give further contributions to the EoS. In the region considered here the formation of quantum condensates does not appear. This is in contrast to a recent work employing a quasiparticle gas model[17]

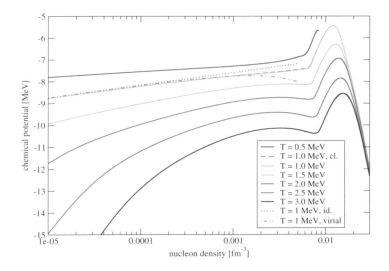

Fig. 2. Averaged chemical potential for different temperatures as function of the nucleon density, symmetric matter. Quantum degeneration is considered. For $T = 1$ MeV, also a classical calculation is shown. The $T = 0.5$ MeV isotherm is only shown in that region where α particles dominate.

where Bose-Einstein condensation of deuterons is observed because the Pauli shift of the deuteron binding energy at high densities is not considered.

Bose-Einstein condensation of α particles will occur below $T = 0.1$ MeV where $n_c^{(\alpha),\text{ideal}} = 0.0006$ fm^{-3}. The question whether also at temperatures around 1 MeV quantum condensates can occur, cannot be answered within the present approach. One has to improve the treatment of the Pauli blocking that is overestimated in our approach. The phase space occupation by strongly bound states such as α particles is smaller than that of free nucleons because the wave functions of the bound states are more extended in momentum representation. Furthermore, a quantum condensate can also be formed from continuum states similar to Cooper pairing in the two-nucleon case.

A more precise treatment of the interaction between α particles needs the full anti-symmetrization of the nucleon wave function.[20] The results for the energy per nucleon at zero temperature is also shown in Fig. 1 and correspond to the calculation for α matter. The interaction potential is taken slightly differently as leading to a different mean-field result.[20] Starting from ^{16}O clusters, the results for the EoS are modified. An interesting feature is the reduction of the quantum condensate fraction due to the interaction between the α particles.[21,22]

6. Conclusion and Outlook

We presented a quantum statistical approach to evaluate the EoS of warm nuclear matter at subsaturation densities. Investigating the self-energy within a cluster decomposition, a quasiparticle approach is given for the free nucleons as well as for the nuclei. The account of medium effects allows to obtain a general approach that combines different well-known cases in the low-density region and near saturation density. Parameters are given. The role of continuum correlations is discussed, that contribute to the cluster yields in heavy ion collisions and to the EoS in different ways.

The approximation of the uncorrelated medium can be improved considering the cluster mean-field approximation.[4,5,15] This would also improve the correct inclusion of α matter as discussed here. The formation of quantum condensates (quartetting) and its disappearance with increasing density demands further investigations.

We do not consider the formation of heavy clusters here. This limits the parameter range n_n^{tot}, n_p^{tot}, T in the phase diagram to that area where the abundances of heavier clusters are small. For a more general approach to the EoS which takes also the contribution of heavier cluster into account, see Refs. 10, 16. Future work will include the contribution of the heavier clusters.

Phase separation occurs when thermodynamic stability $\partial \mu_b / \partial n_b \geq 0$ is violated. For given T, Ω and particle numbers $N_\tau = n_\tau \Omega$, the minimum of the free energy F has to be found. This thermodynamic potential follows from integration, e.g., $F(T, n_p, n_n) = \int_0^{n_n} \mu_n(T, 0, n_n') dn_n' + \int_0^{n_p} \mu_p(T, n_p', n_n) dn_p'$. For stability against phase separation, the curvature matrix $\mathcal{F}_{\tau,\tau'} = \partial F / \partial n_\tau \partial n_{\tau'}|_T$ has to be positive, i.e. Tr $[\mathcal{F}_{\tau,\tau'}] \geq 0$, Det $[\mathcal{F}_{\tau,\tau'}] \geq 0$.

If phase instability occurs, droplet formation has to be considered. Due to the Coulomb interaction, a neutralizing charged background has to be taken into account to avoid divergences in the EoS. Demanding local neutrality, the Wigner-Seitz approximation (20) can be used for the Coulomb energy. The determination of the optimal droplet can be performed in the Thomas-Fermi approximation or the local density approach, but will not detailed here. It is expected that the formation of droplets corresponds to the account of higher clusters. The critical point for the phase transition from the nuclear quantum liquid to nuclear gas is influenced by the formation of clusters.[9] Due to the Coulomb interaction, the phase transition becomes smooth in the thermodynamic limit.

Phase transitions are of high importance for the evolution the early universe[23,24], because they can produce inhomogeneities. To calculate the primordial distribution of elements, medium effects have to be included if densities near the saturation density occur. This would allows for the primordial production of heavy elements.

References

1. M. Baldo, ed., *Nuclear Methods and the Nuclear Equation of State* (World Scientific, Singapore, 1999).
2. T. Klähn *et al.*, *Phys. Rev.* **C 74**, 035802 (2006).
3. A. L. Fetter, J. D. Walecka, *Quantum Theory of Many-Particle systems* (McGraw-Hill, New York, 1971); A. A. Abrikosov, L. P. Gorkov, I. E. Dzyaloshinski, *Methods of Quantum field Theory in Statistical Mechanics* (Dover, New York, 1975).
4. G. Röpke, L. Münchow, and H. Schulz, Nucl. Phys. **A 379**, 536 (1982); G. Röpke,

M. Schmidt, L. Münchow, and H. Schulz, *Phys. Lett.* **B 110**, 21 (1982); G. Röpke, M. Schmidt, L. Münchow, and H. Schulz, *Nucl. Phys.* **A 399**, 587 (1983).

5. G. Röpke, T. Seifert, H. Stolz, and R. Zimmermann, *Phys. Stat. Sol. (b)* **100**, 215 (1980); G. Röpke, M. Schmidt, L. Münchow, and H. Schulz, *Nucl. Phys.* **A 399**, 587 (1983); G. Röpke, in: *Aggregation Phenomena in Complex Systems*, eds. J. Schmelzer et al. (Wiley-VCH, Weinheim, New York, 1999), Chaps. 4, 12; Dukelsky, G. Röpke, and P. Schuck, *Nucl. Phys.* **A 628**, 17 (1998).

6. M. Schmidt, G. Röpke, and H. Schulz, *Ann. Phys. (NY)* **202**, 57 (1990).

7. M. Beyer, W. Schadow, C. Kuhrts, and G. Röpke, *Phys. Rev.* **C 60**, 034004 (1999); M. Beyer, S. A. Sofianos, C. Kuhrts, G. Röpke, and P. Schuck, *Phys. Letters* **B 488**, 247 (2000).

8. K. Sumiyoshi and G. Röpke, *Phys. Rev.* **C 77**, 055804 (2008), arXiv:0801.0110 [astro-ph].

9. S. Typel, G. Röpke, T. Klähn, D. Blaschke, and H. H. Wolter, arXiv:0908.2344

10. M. Hempel and J. Schaffner-Bielich, arXiv:0911.4073

11. S. Kowalski *et al.*, *Phys. Rev.* **C 75**, 014601 (2007), arXiv:nucl-ex/0602023.

12. C. J. Horowitz and A. Schwenk, *Nucl. Phys.* **A 776**, 55 (2006).

13. S. Typel, *Phys. Rev.* **C 71**, 064301 (2005), arXiv:nucl-th/0501056.

14. G. Röpke, A. Grigo, K. Sumiyoshi and Hong Shen, Phys. Part. Nucl. Lett. **2**, 275 (2005).

15. G. Röpke, Phys. Rev. C **79**, 014002 (2009), arXiv:0810.4645 [nucl-th].

16. G. Röpke, M. Schmidt and H. Schulz, Nucl. Phys. A **424**, 594 (1984).

17. S. Heckel, P. P. Schneider and A. Sedrakian, arXiv:0902.3539 [astro-ph.SR].

18. H. Stein, A. Schnell, T. Alm, and G. Röpke, *Z. Phys.* **A 351**, 295 (1995).

19. G. Röpke, A. Schnell, P. Schuck, and P. Nozières, *Phys. Rev. Lett.* **80**, 3177 (1998).

20. H. Takemoto, M. Fukushima, S. Chiba, H. Horiuchi, Y. Akaishi, and A. Tohsaki, *Phys. Rev.* **C 69**, 035802 (2004).

21. F. Carstoiu, S. Misicu, arXiv:0810.4645 [nucl-th].

22. Y. Funaki, H. Horiuchi, G. Ropke, P. Schuck, A. Tohsaki, and T. Yamada, *Phys. Rev.* **C 677**, 064312 (2008).

23. G. Röpke, *Clusters in Nuclear Matter*, in: *Condensed Matter Theories*, vol. 16, ed. S. Hernandez and J. W. Clark (Nova Sciences Publ., New York, 2001).

24. G. Röpke, *Phys. Lett.* **B 185**, 281 (1987).

Part B
Quantum Magnets, Quantum Dynamics and Phase Transitions

MAGNETIC ORDERING OF ANTIFERROMAGNETS ON A SPATIALLY ANISOTROPIC TRIANGULAR LATTICE

R. F. BISHOP and P. H. Y. LI

School of Physics and Astronomy, Schuster Building, The University of Manchester,
Manchester, M13 9PL, UK

D. J. J. FARNELL

Health Methodology Research Group, School of Community-Based Medicine,
Jean McFarlane Building, University Place, The University of Manchester, M13 9PL, UK

C. E. CAMPBELL

School of Physics and Astronomy, University of Minnesota, 116 Church Street SE,
Minneapolis, MN 55455, USA

We study the spin-1/2 and spin-1 J_1–J_2' Heisenberg antiferromagnets (HAFs) on an infinite, anisotropic, two-dimensional triangular lattice, using the coupled cluster method. With respect to an underlying square-lattice geometry the model contains antiferromagnetic ($J_1 > 0$) bonds between nearest neighbours and competing ($J_2' > 0$) bonds between next-nearest-neighbours across only one of the diagonals of each square plaquette, the same diagonal in each square. In a topologically equivalent triangular-lattice geometry the model has two sorts of nearest-neighbour bonds, with $J_2' \equiv \kappa J_1$ bonds along parallel chains and with J_1 bonds providing an interchain coupling. The model thus interpolates between an isotropic HAF on the square lattice at one extreme ($\kappa = 0$) and a set of decoupled chains at the other ($\kappa \to \infty$), with the isotropic HAF on the triangular lattice in between at $\kappa = 1$. For the spin-1/2 J_1–J_2' model, we find a weakly first-order (or possibly second-order) quantum phase transition from a Néel-ordered state to a helical state at a first critical point at $\kappa_{c_1} = 0.80 \pm 0.01$, and a second critical point at $\kappa_{c_2} = 1.8 \pm 0.4$ where a first-order transition occurs between the helical state and a collinear stripe-ordered state. For the corresponding spin-1 model we find an analogous transition of the second-order type at $\kappa_{c_1} = 0.62 \pm 0.01$ between states with Néel and helical ordering, but we find no evidence of a further transition in this case to a stripe-ordered phase.

Keywords: Quantum magnet; quantum phase transition; coupled cluster method.

1. Introduction

Arrays of strongly interacting quantum spins arranged on a regular periodic lattice are among the conceptually simplest to describe via a Hamiltonian, but are amongst the most challenging many-body systems in terms of a theoretical derivation of their ground- and excited-state properties from first principles. They have become the subject of intense interest in recent years, especially since many such models give a good description of real magnetic materials that can be studied experimentally.

Particular attention has focussed on two-dimensional (2D), spin-1/2 Heisenberg antiferromagnets (HAFs), Heisenberg antiferromagnets (HAFs)! 2D spin-1/2 for which the effects of quantum fluctuations are usually larger than for their counterparts with higher spatial dimensionality or with a greater value of the spin quantum number, $s > 1/2$. Of special interest has been the interplay between frustration (either geometric or dynamic in origin) and quantum fluctuations in determining the ground-state (gs) phase diagram for such models. While simple models of this kind are well understood in the absense of frustration,[1] their frustrated counterparts present a much greater theoretical challenge. This is particularly so when there are two or more terms in the interaction Hamiltonian that compete with one another to produce different types of ordering. In particular the zero-temperature ($T = 0$) phase transitions between magnetically ordered quasiclassical phases and novel (magnetically disordered) quantum paramagnetic phases[2,3] have attracted huge interest in recent years, both as prototypical examples of quantum phase transitions of various types and because of the possible fabrication of real magnets that exhibit such phases, and which may have novel properties. Quantum phases! paramagnetic A particularly well studied such model is the frustrated spin-1/2 J_1–J_2 model on the square lattice with nearest-neighbour (NN) bonds (J_1) and next-nearest-neighbour (NNN) bonds (J_2), for which it is now well accepted that there exist two phases exhibiting magnetic long-range order (LRO) at small and at large values of $\alpha \equiv J_2/J_1$ respectively, separated by an intermediate quantum paramagnetic phase without magnetic LRO in the parameter regime $\alpha_{c_1} < \alpha < \alpha_{c_2}$, where $\alpha_{c_1} \approx 0.4$ and $\alpha_{c_2} \approx 0.6$. For $\alpha < \alpha_{c_1}$ the gs phase exhibits Néel magnetic LRO, whereas for $\alpha > \alpha_{c_2}$ it exhibits collinear stripe LRO. We have recently studied this 2D spin-1/2 model exhaustively by extending it to include anisotropic interactions in either real (crystal lattice) space[4] or in spin space.[5] We showed in particular how the coupled cluster method (CCM) provided for this highly frustrated model what is perhaps now the most accurate microscopic description. The interested reader is referred to Refs.[4,5] and references cited therein for further details of the model and the method. Quantum phases! magnetic long-range order (LRO) Quantum phases! collinear stripe LRO

2. The Model

Since the CCM has proven so successful in the above applications we now apply it to the seemingly similar 2D J_1–J_2' model that has been studied recently by other means for the spin-1/2 case.[6-11] Its Hamiltonian is written as

$$H = J_1 \sum_{\langle i,j \rangle} \mathbf{s}_i \cdot \mathbf{s}_j + J_2' \sum_{[i,k]} \mathbf{s}_i \cdot \mathbf{s}_k \tag{1}$$

where the operators $\mathbf{s}_i \equiv (s_i^x, s_i^y, s_i^z)$ are the spin operators on lattice site i with $\mathbf{s}_i^2 = s(s+1)$, and where we study both the cases $s = 1/2$ and $s = 1$ here. On the square lattice the sum over $\langle i, j \rangle$ runs over all distinct NN bonds, but the sum

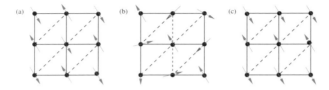

Fig. 1. J_1–J_2' model; — J_1; - - - J_2'; (a) Néel state, (b) spiral state, (c) stripe state.

over $[i, k]$ runs only over one half of the distinct NNN bonds with equivalent bonds chosen in each square plaquette, as shown explicitly in Fig. 1. (By contrast, the J_1–J_2 model discussed above includes *all* of the diagonal NNN bonds.) We shall be interested here only in the case of competing (or frustrating) antiferromagnetic bonds $J_1 > 0$ and $J_2' > 0$, and henceforth for all of the results shown we set $J_1 \equiv 1$. Clearly, the model may be described equivalently as a Heisenberg model on an anisotropic triangular lattice in which each triangular plaquette contains two NN J_1 bonds and one NN J_2' bond. The model thus interpolates continuously between HAFs on a square lattice ($J_2' = 0$) and on a triangular lattice ($J_2' = J_1$). Similarly, when $J_1 = 0$ (or $J_2' \to \infty$ in our normalization with $J_1 \equiv 1$) the model reduces to uncoupled 1D chains (along the chosen diagonals on the square lattice). The case $J_2' \gg 1$ thus corresponds to weakly coupled 1D chains, and hence the model also interpolates between 1D and 2D. Heisenberg model! on anisotropic triangular lattice In the classical case (correponding to the asympototic limit where the spin quantum number $s \to \infty$) the J_1–J_2' model has only two classical gs phases. For $J_2' < \frac{1}{2}J_1$ the gs phase is Néel ordered, as shown in Fig. 1(a), whereas for $J_2' > \frac{1}{2}J_1$ it has spiral order, as shown in Fig. 1(b), wherein the spin direction at lattice site (i, j) points at an angle $\alpha_{ij} = \alpha_0 + (i + j)\alpha_{cl}$, with $\alpha_{cl} = \cos^{-1}(-\frac{J_1}{2J_2'}) \equiv \pi - \phi_{cl}$. The pitch angle $\phi_{cl} = \cos^{-1}(\frac{J_1}{2J_2'})$ thus measures the deviaton from Néel order, and it varies from zero for $2J_2'/J_1 \leq 1$ to $\frac{1}{2}\pi$ as $J_2'/J_1 \to \infty$, as shown in Fig. 3 below. When $J_2' = J_1$ we regain the classical 3-sublattice ordering on the triangular lattice with $\alpha_{cl} = \frac{2}{3}\pi$. The classical phase transition at $J_2' = \frac{1}{2}J_1$ is of continuous (second-order) type, with the gs energy and its derivative both continuous.

In the limit of large J_2'/J_1 the above classical limit represents a set of decoupled 1D HAF chains (along the diagonals of the square lattice) with a relative spin orientation between neighbouring chains that approaches 90°. In fact, of course, there is complete degeneracy at the classical level in this limit between all states for which the relative ordering directions of spins on different HAF chains are arbitrary. Clearly the exact spin-1/2 limit should also be a set of decoupled HAF chains, as given by the exact Bethe ansatz solution.[12] Bethe ansatz solution Quantum fluctuations! order by disorder However, one might expect that this degeneracy could be lifted by quantum fluctuations, via the well-known phenomenon of *order by disorder*.[13] Just such a phase is known to exist in the J_1–J_2 model[4,5] for values of $J_2/J_1 \gtrsim 0.6$, where it is the so-called collinear stripe phase in which, on the square lattice, spins along (say) the rows in Fig. 1 order ferromagnetically while spins along the columns and diagonals order antiferromagnetically, as shown in Fig. 1(c). We note that a recent renormalization-group (RG)[10] analysis of the spin-1/2 J_1–J_2' model considered here has predicted that just such a collinear stripe phase also exists in this case for values of J_2'/J_1 greater than some critical value which was not calculated in that analysis. Our own results are presented below.

3. The Coupled Cluster Method Formalism

The CCM (see, e.g., Refs. [14–16] and references cited therein) that we employ here is one of the most powerful and most versatile modern techniques in quantum many-body Coupled Cluster Method (CCM)! formalism theory. It has been applied very successfully to various quantum magnets (see Refs. [4,5,16–22] and references cited therein), as we now briefly outline below. The starting point for any CCM calculation is to select a normalized model or reference state $|\Phi\rangle$. It is often convenient to take a classical ground state (GS) as a model state for CCM calculations of quantum spin systems. Accordingly our model states here include the Néel and spiral states, although we also employ the nonclassical stripe state. It is very convenient to treat each site on an equal footing, and in order to do so we perform a mathematical rotation of the local axes on each lattice site such that all spins in every reference state align along the negative z-axis. The Schrödinger ground-state ket and bra equations are $H|\Psi\rangle = E|\Psi\rangle$ and $\langle\tilde{\Psi}|H = E\langle\tilde{\Psi}|$ respectively. The CCM parametrizes these exact quantum gs wave functions in the forms $|\Psi\rangle = e^S|\Phi\rangle$ and $\langle\tilde{\Psi}| = \langle\tilde{\Phi}|\tilde{S}e^{-S}$. The correlation operators S and $\tilde{S}$ are expressed as $S = \sum_{I\neq 0} S_I C_I^+$ and $\tilde{S} = 1 + \sum_{I\neq 0} \tilde{S}_I C_I^-$, where $C_0^+ \equiv 1$, the unit operator, and $C_I^+ \equiv (C_I^-)^\dagger$ is one of a complete set of multispin creation operators with respect to the model state (with $C_I^-|\Phi\rangle = 0 = \langle\Phi|C_I^+; \forall I \neq 0$), generically written as $C_I^+ \equiv s_{i_1}^+ s_{i_2}^+ \cdots s_{i_n}^+$, in terms of the spin-raising operators $s_i^+ \equiv s_i^x + s_i^y$ on lattice sites i. We note that $\langle\tilde{\Psi}|\Psi\rangle = \langle\Phi|\Psi\rangle = \langle\Phi|\Phi\rangle \equiv 1$.

The ket- and bra-state correlation coefficients $(S_I, \tilde{S}_I)$ are calculated by requiring the expectation value $\bar{H} \equiv \langle\tilde{\Psi}|H|\Psi\rangle$ to be a minimum with respect to each of them. This immediately yields the coupled set of equations $\langle\Phi|C_I^- e^{-S}He^S|\Phi\rangle = 0$ and $\langle\Phi|\tilde{S}(e^{-S}He^S - E)C_I^+|\Phi\rangle = 0; \forall I \neq 0$, which we solve for the correlation coefficients $(S_I, \tilde{S}_I)$. We may then calculate the gs energy from the relation $E = \langle\Phi|e^{-S}He^S|\Phi\rangle$, and the gs staggered magnetization M from the relation $M \equiv -\frac{1}{N}\langle\tilde{\Psi}|\sum_{i=1}^N s_i^z|\Psi\rangle$ which holds in the local (rotated) spin coordinates on each lattice site i. We note that we work from the outset in the $N \to \infty$ limit, where N is the number of spins on the lattice.

The CCM formalism is exact if a complete set of multispin configurations $\{I\}$ with respect to the model state is included in the calculation. However, it is necessary in practice to use approximation schemes to truncate the expansions in configurations $\{I\}$ of the correlation operators S and $\tilde{S}$. For the case of $s = 1/2$ we employ here, as in our previous work,[4,5,16−22] the localized LSUBm scheme LSUBm!

Table 1.   Numbers of fundamental configurations ($\sharp$ f.c.)
for $s = 1/2$ and $s = 1$ in various CCM approximations.

	$s = 1/2$			$s = 1$	
Scheme	$\sharp$ f.c.		Scheme	$\sharp$ f.c.	
	stripe	spiral		stripe	spiral
LSUB2	2	3	SUB2–2	2	4
LSUB3	4	14	SUB3–3	4	26
LSUB4	27	67	SUB4–4	60	189
LSUB5	95	370	SUB5–5	175	1578
LSUB6	519	2133	SUB6–6	2996	14084
LSUB7	2617	12878	SUB7–7	11778	131473
LSUB8	15337	79408	–	–	–

scheme in which all possible multi-spin-flip correlations over different locales on the lattice defined by m or fewer contiguous lattice sites are retained. The numbers of such fundamental configurations (viz., those that are distinct under the symmetries of the Hamiltonian and of the model state $|\Phi\rangle$) that are retained for the stripe and spiral states of the current model in various LSUBm approximations on the triangular lattice are shown in Table 1. LSUB$m!$ approximations triangular lattice We note next that the number of fundamental LSUBm configurations for $s = 1$ becomes appreciably higher than for $s = 1/2$, since each spin on each site i can now be flipped twice by the spin-raising operator s_i^+. Thus, for the $s = 1$ model it is more practical to use the alternative SUBn–m scheme, where m is the size of the locale on the lattice and n is the maximum number of spin-flips. Hence all correlations involving up to n spin flips spanning a range of no more than m adjacent lattice sites are retained.[16] In our case we set $m = n$, and hence employ the SUBm–m scheme as in our previous work (see, e.g., Refs. [16,22] and references cited therein). More generally, the LSUBm scheme is thus equivalent to the SUBn–m scheme for $n = 2sm$. Hence, LSUB$m \equiv$ SUB$2sm$–m. For s=1/2, LSUB$m \equiv$ SUBm–m; whereas for $s = 1$, LSUB$m \equiv$ SUB$2m$–m. The numbers of fundamental configurations retained at various SUBm–m levels for the $s = 1$ model are also shown in Table 1.

In order to solve the corresponding coupled sets of CCM bra- and ket-state equations we use parallel computing.[23] The highest CCM level that we can reach here, even with massive parallelization and the use of supercomputing resources, is LSUB8 for $s = 1/2$ and SUB7–7 for $s = 1$. For example, to obtain a single data point (i.e., for a given value of J_2', with $J_1 = 1$) for the spiral phase at the LSUB8 level for the $s = 1/2$ case typically required about 0.3 h computing time using 600 processors simultaneously. It similarly required about 0.25 h for the spiral phase at the SUB7–7 level for the $s = 1$ case.

It is important to note that we never need to perform any finite-size scaling, since all CCM approximations are automatically performed from the outset in the infinite-lattice limit, $N \to \infty$, where N is the number of lattice sites. However, we do need as a last step to extrapolate to the $m \to \infty$ limit in the LSUBm or SUBm–m truncation index m. We use here the well-tested[17,18] empirical scaling laws

$$E/N = a_0 + a_1 m^{-2} + a_2 m^{-4} , \tag{2}$$

$$M = b_0 + b_1 m^{-1} + b_2 m^{-2} . \tag{3}$$

4. Results and Discussion

We report here on CCM calculations for the spin-1/2 and spin-1 J_1–J_2' model Hamiltonians of Eq. (1) for given parameters $(J_1 = 1, J_2')$, based respectively on the Néel, Néel states! spiral and stripe spiral and stripe states as CCM model states. We note that, as has been well documented in the past,[24] the LSUBm (and SUBm–m) data for both the gs energy per spin E/N and the on-site magnetization M converge differently for the even-m and the odd-m sequences, similar to what is frequently observed in perturbation theory.[25] Since, as a general rule, it is desirable to have at least $(n + 1)$ data points to fit to any fitting formula that contains n unknown parameters, we prefer to have at least 4 results to fit to Eqs. (2) and (3). Hence, for most of our extrapolated results below we use the even LSUBm sequence with $m = \{2, 4, 6, 8\}$ for the spin-1/2 case. For the spin-1 case we use the SUBm–m sequences with $m = \{2, 4, 6\}$ and $m = \{3, 5, 7\}$.

Firstly, the results obtained using the spiral model state are reported. While classically we have a second-order phase transition from Néel order (for $\kappa < \kappa_{cl}$) to helical Quantum phase transition! from Néel to helical order order (for $\kappa > \kappa_{cl}$), where $\kappa \equiv J_2'/J_1$, at a value $\kappa_{cl} = 0.5$, using the CCM we find strong indications of a shift of this critical point to values $\kappa_{c_1} \approx 0.80$ in the spin-1/2 quantum case and $\kappa_{c_1} \approx 0.62$ in the spin-1 quantum case, as shown in Figs. 2 and 3.

Thus, for example, curves such as those shown in Fig. 2 show that the Néel model state ($\phi = 0$) gives the minimum gs energy for all values of $\kappa < \kappa_{c_1}$, where κ_{c_1} depends on the level of SUBm–m approximation used, as we also observe in Fig. 3. By contrast, for values of $\kappa > \kappa_{c_1}$ the minimum in the energy is found to occur at a value $\phi \neq 0$. If we consider the pitch angle ϕ itself as an order parameter (i.e., $\phi = 0$ for Néel order and $\phi \neq 0$ for spiral order) a typical scenario for a phase transition would be the appearance of a two-minimum structure for the gs energy for values of $\kappa > \kappa_{c_1}$, exactly as observed in Fig. 2 for both the spin-1/2 and spin-1 models in the SUB4–4 approximation. Very similar curves occur for other SUBm–m approximations.

We note from Fig. 2 that for certain values of J_2' (or, equivalently, κ) CCM solutions at a given SUBm–m level of approximation (viz., SUB4–4 in Fig. 2) exist only for certain ranges of spiral angle ϕ. For example, for the pure $s = 1/2$ square-lattice HAF ($\kappa = 0$) the CCM LSUB4 solution based on a spiral model state only exists for $0 \leq \phi \lesssim 0.20\pi$. In this case, where the Néel solution is the stable ground state, if we attempt to move too far away from Néel collinearity the CCM equations

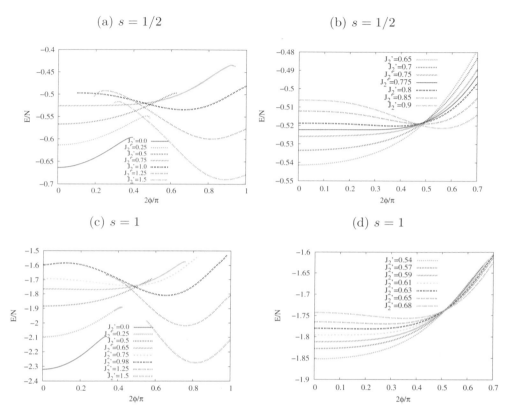

Fig. 2. Ground-state energy per spin of the spin-1/2 and spin-1 J_1–J_2' Hamiltonians of Eq. (1) with $J_1 = 1$, using the SUB4-4 approximation of the CCM with the spiral model state, versus the spiral angle ϕ. For the case of $s = 1/2$, for $J_2' \lesssim 0.774$ the only minimum is at $\phi = 0$ (Néel order), whereas for $J_2' \gtrsim 0.774$ a secondary minimum occurs at $\phi = \phi_{\text{LSUB4}} \neq 0$, which is also a global minimum, thus indicating a phase transition at $J_2' \approx 0.774$ in this approximation. Similarly for the case of $s = 1$, for $J_2' \lesssim 0.610$ the only minimum is at $\phi = 0$ (Néel order), whereas for $J_2' \gtrsim 0.610$ a secondary minimum occurs at $\phi = \phi_{\text{SUB4-4}} \neq 0$, which is also a global minimum.

themselves become "unstable" and simply do not have a real solution. Similarly, we see from Fig. 4 again that for $\kappa = 1.5$; the CCM LSUB4 solution for the $s = 1/2$ case exists only for $0.15\pi \lesssim \phi \leq 0.5\pi$. In this case the stable ground state is a spiral phase, and now if we attempt to move too close to Néel collinearity the real solution terminates.

Such terminations of CCM solutions are very common and are very well documented.[16] In all such cases a termination point always arises due to the solution of the CCM equations becoming complex at this point, beyond which there exist two branches of entirely unphysical complex conjugate solutions.[16] In the region where the solution reflecting the true physical solution is real there actually

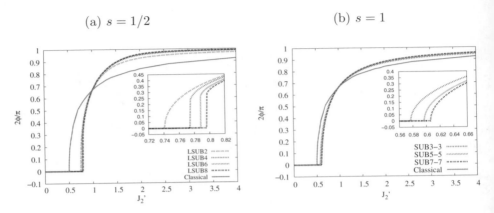

Fig. 3.   The angle $\phi = \phi_{\mathrm{SUB}m-m}$ that minimizes the energy $E_{\mathrm{SUB}m-m}(\phi)$ of the spin-1/2 and spin-1 J_1–J_2' Hamiltonians of Eq. (1) with $J_1 = 1$, in the SUBm–m approximations with $m = \{2, 4, 6, 8\}$ (for $s = 1/2$) and $m = \{2, 4, 6\}$ (for $s = 1$), using the spiral model state, versus J_2'. The corresponding classical result $\phi = \phi_{\mathrm{cl}}$ is shown for comparison. For $s = 1/2$, we find in the LSUBm quantum case with $m > 2$ a seemingly first-order phase transition (e.g., for LSUB8 at $J_2' \approx 0.796$ where ϕ_{LSUB8} jumps abruptly from zero to about 0.14π). By contrast, in the classical case there is a second-order phase transition at $J_2' = 0.5$. For $s = 1$, we find in the SUBm–m quantum case strong evidence of a second-order phase transition (e.g., for SUB6–6 at $J_2' \approx 0.612$), although a very weakly first-order transition cannot be entirely ruled out.

also exists another (unstable) real solution. However, only the (shown) upper branch of these two solutions reflects the true (stable) physical ground state, whereas the lower branch does not. The physical branch is usually easily identified in practice as the one which becomes exact in some known (e.g., perturbative) limit. This physical branch then meets the corresponding unphysical branch at some termination point (with infinite slope in Fig. 2) beyond which no real solutions exist. The LSUBm (or SUBm–m) termination points are themselves also reflections of the quantum phase transitions in the real system, and may be used to estimate the position of the phase boundary,[16] although we do not do so for this first critical point since we have more accurate criteria discussed below. LSUBm! termination points Thus, in Figs. 4 and 5 we show the CCM results for the gs energy and gs on-site magnetization, respectively, where the helical state has been used as the model state and the angle ϕ chosen as described above.

Ground state! magnetic order parameter

Firstly, the gs energy (in Fig. 4) for the $s = 1/2$ case shows signs of a (weak) discontinuity in slope at the critical values κ_{c_1} discussed above. Secondly, however, the gs magnetic order parameter in Fig. 4 for $s = 1/2$ shows much stronger and much clearer evidence of a phase transition at the corresponding LSUBm κ_{c_1} values previously observed in Fig. 4. The extrapolated value of M shows clearly its steep drop towards a value very close to zero at $\kappa_{c_1} = 0.80 \pm 0.01$, which is hence our best estimate of the phase transition point. From the Néel side ($\kappa < \kappa_{c_1}$) the

(a) $s = 1/2$ (b) $s = 1$

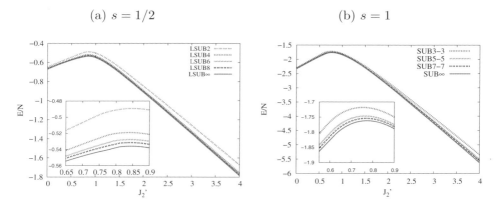

Fig. 4. Ground-state energy per spin versus J_2' for the Néel and spiral phases of the spin-1/2 and spin-1 J_1–J_2' Hamiltonians of Eq. (1) with $J_1 = 1$. The CCM results using the spiral model state are shown for various SUBm–m approximations ($m = \{2,4,6,8\}$ for $s = 1/2$ and $m = \{2,4,6\}$ for $s = 1$) with the spiral angle $\phi = \phi_{\text{SUB}m-m}$ that minimizes $E_{\text{SUB}m-m}(\phi)$.

(a) $s = 1/2$ (b) $s = 1$

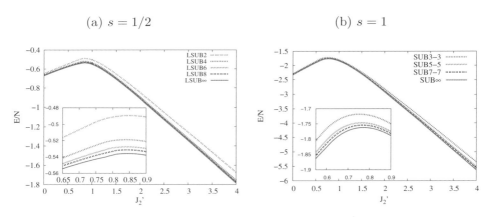

Fig. 5. Ground-state magnetic order parameter (i.e., the on-site magnetization) versus J_2' for the Néel and spiral phases of the spin-1/2 and spin-1 J_1–J_2' Hamiltonians of Eq. (1) with $J_1 = 1$. The CCM results using the spiral model state are shown for various SUBm–m approximations ($m = \{2,4,6,8\}$ for $s = 1/2$ and $m = \{2,4,6\}$ for $s = 1$) with the spiral angle $\phi = \phi_{\text{SUB}m-m}$ that minimizes $E_{\text{SUB}m-m}(\phi)$.

magnetization seems to approach continuously a value $M = 0.025 \pm 0.025$, whereas from the spiral side ($\kappa > \kappa_{c_1}$) there appears to be a discontinuous jump in the magnetization as $\kappa \to \kappa_{c_1}$. The transition at $\kappa = \kappa_{c_1}$ thus appears to be (very) weakly first-order but we cannot exclude it being second-order, since we cannot rule out the possibility of a continuous but very steep drop to zero of the on-site magnetization as $\kappa \to \kappa_{c_1}$ from the spiral side of the transition, for the same reasons

Table 2.   Ground-state energy per spin, E/N, and magnetic order parameter, M (i.e., the on-site magnetization) for the spin-1/2 HAF on the square and triangular lattices. We show CCM results obtained for the J_1–J_2' model with $J_1 > 0$, using the spiral model state in various LSUBm approximations defined on the triangular lattice geometry, for the two cases $\kappa \equiv J_2'/J_1 = 0$ (square lattice HAF, $\phi = 0$) and $\kappa = 1$ (triangular lattice HAF, $\phi = \frac{\pi}{3}$).

Method	E/N	M	E/N	M
	square ($\kappa = 0$)		triangular ($\kappa = 1$)	
LSUB2	-0.64833	0.4207	-0.50290	0.4289
LSUB3	-0.64931	0.4182	-0.51911	0.4023
LSUB4	-0.66356	0.3827	-0.53427	0.3637
LSUB5	-0.66345	0.3827	-0.53869	0.3479
LSUB6	-0.66695	0.3638	-0.54290	0.3280
LSUB7	-0.66696	0.3635	-0.54502	0.3152
LSUB8	-0.66816	0.3524	-0.54679	0.3018
Extrapolations				
LSUB∞ [a]	-0.66978	0.3148	-0.55113	0.2219
LSUB∞ [b]	-0.66974	0.3099	-0.55244	0.1893
LSUB∞ [c]	-0.67045	0.3048	-0.55205	0.2085
QMC [d,e]	-0.669437(5)	0.3070(3)	-0.5458(1)	0.205(10)
SE [f,g]	-0.6693(1)	0.307(1)	-0.5502(4)	0.19(2)

[a] Based on LSUBm results with $m = \{2,4,6,8\}$
[b] Based on LSUBm results with $m = \{4,6,8\}$
[c] Based on LSUBm results with $m = \{3,5,7\}$
[d] QMC (Quantum Monte Carlo) for square lattice (Ref. [26])
[e] QMC for triangular lattice (Ref. [27])
[f] SE (Series Expansion) for square lattice (Ref. [28])
[g] SE for triangular lattice (Ref. [29])

as enunciated above in connection with our discussion of Fig. 2 and 3. These results may be compared with those for the same model of Weihong et al.[9] who used a linked-cluster series expansion technique. They found that while a nonzero value of the Néel staggered magnetization exists for $0 \le \kappa \lesssim 0.7$, the region $0.7 \lesssim \kappa \lesssim 0.9$ has zero on-site magnetization, and for $\kappa \gtrsim 0.9$ they found evidence of spiral order. Nevertheless, their results came with relatively large errors, especially for the spiral phase, and we believe that our own results are probably intrinsically more accurate than theirs.

By contrast, the gs energy shown in Fig. 4 for the $s = 1$ case appears to be quite continuous in slope at the critical values κ_{c_1} for each SUBm–m level of approximation, thereby indicating a second-order transition between the Néel and helical phases. Similarly, the gs magnetic order parameter M shown in Fig. 4 also strongly indicates a second-order transition. The extrapolated value of M provides our best estimate of $\kappa_{c_1} = 0.62 \pm 0.01$ for the $s = 1$ J_1–J_2' model.

As a further indication of the accuracy of our results we show in Tables 2 and 3 data for the two cases of the spin-1/2 and spin-1 HAFs on the square lattice ($\kappa = 0$) and on the triangular lattice ($\kappa = 1$). For the $s = 1/2$ case we present

Table 3. Ground-state energy per spin, E/N, and magnetic order parameter, M (i.e., the on-site magnetization) for the spin-1 HAF on the square and triangular lattices. We show CCM results obtained for the J_1–J_2' model with $J_1 > 0$, using the spiral model state in various SUBm–m approximations defined on the triangular lattice geometry, for the two cases $\kappa \equiv J_2'/J_1 = 0$ (square lattice HAF, $\phi = 0$) and $\kappa = 1$ (triangular lattice HAF, $\phi = \frac{\pi}{3}$).

Method	E/N	M	E/N	M
	square ($\kappa = 0$)		triangular ($\kappa = 1$)	
SUB2–2	-2.29504	0.9100	-1.77400	0.9069
SUB3–3	-2.29763	0.9059	-1.80101	0.8791
SUB4–4	-2.31998	0.8702	-1.82231	0.8405
SUB5–5	-2.32049	0.8682	-1.82623	0.8294
SUB6–6	-2.32507	0.8510	-1.83135	0.8096
SUB7–7	-2.32535	0.8492	-1.83288	0.8006
Extrapolations				
SUB∞ [a]	-2.32924	0.8038	-1.83860	0.7345
SUB∞ [b]	-2.32975	0.7938	-1.83968	0.7086
CCM [c]	-2.3291	0.8067		
SWT–3 [d]	-2.3282	0.8043		
SE [e]	-2.3279(2)	0.8039(4)		

[a] Based on SUBm–m results with $m = \{2, 4, 6\}$
[b] Based on SUBm–m results with $m = \{3, 5, 7\}$
[c] CCM (SUB∞ extrapolation for square lattice based on SUBm–m results with $m = \{2, 4, 6\}$) (Ref. [32])
[d] SWT–3 (Third-order Spin-Wave Theory) for square lattice (Ref. [31])
[e] SE (Series Explansion) for square lattice (Ref. [28])

our CCM results in Table 2 in various LSUBm appoximations (with $2 \leq m \leq 8$) based on the triangular lattice geometry, and using the spiral model state, with $\phi = 0$ for the square lattice and $\phi = \frac{\pi}{3}$ for the triangular lattice. Results are given for the gs energy per spin E/N, and the magnetic order parameter M. We also display our extrapolated ($m \to \infty$) results using the schemes of Eqs. (2) and (3) with the three data sets $m = \{2, 4, 6, 8\}$, $m = \{4, 6, 8\}$ and $m = \{3, 5, 7\}$. The results are seen to be very robust and consistent. For comparison we also show the results obtained for the two lattices using quantum Monte Carlo (QMC) methods[26,27] and linked-cluster series expansions (SE).[28,29] For the square lattice there is no dynamic frustration and the Marshall-Peierls sign rule[30] applies, so that the QMC "minus-sign problem" may be circumvented. In this case the QMC results[26] are extremely accurate, and indeed represent the best available for the spin-1/2 square-lattice HAF. Our own extrapolated results are in complete agreement with these QMC benchmark results, as found previously (see, e.g., Ref. [24] and references cited therein), even though the LSUBm configurations are defined here on the triangular lattice geometry. Thus, we note that whereas the individual LSUBm results for the spin-1/2 square-lattice HAF do not coincide with previous results for

(a) $s = 1/2$ (b) $s = 1$

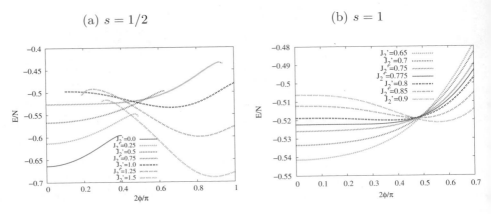

Fig. 6.   Ground-state energy per spin versus J_2' for the stripe-ordered phase of the spin-1/2 and spin-1 J_1–J_2' Hamiltonians of Eq. (1) with $J_1 = 1$. The CCM results using the stripe model state are shown for various SUBm–m approximations ($m = \{2, 4, 6, 8\}$ for $s = 1/2$ and $m = \{2, 4, 6\}$ for $s = 1$).

this model (see, e.g., Ref. [24]) because previous results have been based on defining the fundamental LSUBm configurations on a square-lattice geometry rather than on the triangular-lattice geometry used here, the corresponding LSUB∞ extrapolations in the two geometries are in complete agreement with each other.

For the $s = 1$ case, our CCM results are presented in various SUBm–m approximations (with $2 \leq m \leq 7$) in Table 3. The extrapolated results ($m \rightarrow \infty$) using Eqs. (2) and (3) with $m = \{2, 4, 6\}$ and $m = \{3, 5, 7\}$ are also presented. For comparison we also show the results obtained for the square lattice (i.e., $\kappa = 0$) using third-order spin-wave theory (SWT–3),[31] a linked-cluster series expansion (SE)[28] and previous CCM extrapolated ($m \rightarrow \infty$) results based on SUBm–m calculations on the square lattice.[32]

We turn finally to our CCM results based on the stripe state as CCM gs model state $|\Phi\rangle$. The LSUBm and SUBm–m configurations are again defined with respect to the triangular lattice geometry, exactly as before. Results for the gs energy and magnetic order parameter based on the collinear stripe phase are shown in Figs. 6 and 7.

We see from Fig. 4 that some of the LSUBn solutions based on the stripe state for the $s = 1/2$ case show a clear termination point κ_t of the sort discussed previously, such that for $\kappa < \kappa_t$ no real solution for the stripe phase exists. In particular the LSUB6 and LSUB8 solutions terminate at the values shown in Table 4. As is often the case the LSUB2 solution does not terminate, while the LSUB4 solution shows a marked change in character around the value $\kappa \approx 0.880$ that is not exactly a termination point (but, probably, rather reflects a crossing with another unphysical solution). In any event, the LSUB4 data are not shown below this value in Figs. 4 and 4. Similar termination points are seen for the $s = 1$ case in Figs. 4 and 4.

(a) $s = 1/2$ (b) $s = 1$

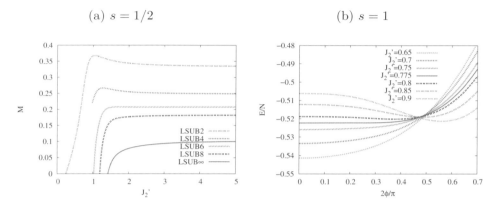

Fig. 7. Ground-state magnetic order parameter (i.e., the on-site magnetization) versus J_2' for the stripe-ordered phase of the spin-1/2 and spin=1 J_1–J_2' Hamiltonians of Eq. (1) with $J_1 = 1$. The CCM results using the stripe model state are shown for various SUBm–m approximations ($m = \{2,4,6,8\}$ for $s = 1/2$ and $m = \{2,4,6\}$ for $s = 1$).

Table 4. The parameters κ_e (the crossing point of the energy curves for the stripe and spiral phases) and κ_t (the termination point of the stripe state solution) in various LSUBm approximations defined on the triangular lattice geometry, for the spin-1/2 J_1–J_2' model, with $\kappa \equiv J_2'/J_1$, $J_1 > 0$. The "LSUB∞" extrapolations are explained in the text.

Method	J_2'	
	κ_e	κ_t
LSUB2	∞	-
LSUB4	4.555	(0.880)
LSUB6	3.593	0.970
LSUB8	3.125	1.150
"LSUB∞"	1.69 ± 0.03	1.69

The large κ limit of the energy per spin results of Fig. 4 again agrees well with the exact $s = 1/2$ 1D chain result of $E/N = -0.4431J_2'$ from the Bethe ansatz solution,[12] just as in Fig. 4 for the spiral phase. However, the most important observation is that for all LSUBm approximations with $m > 2$ the curves for the energy per spin of the $s = 1/2$ stripe phase cross with the corresponding curves (i.e., for the same value of m) for the energy per spin of the $s = 1/2$ spiral phase at a value that we denote as κ_e (as can clearly be seen in Fig. 4, which shows the energy difference between the stripe and spiral states for various LSUBm calculations. The crossing values κ_e are shown in Table 4. Thus, for $\kappa < \kappa_e$ the spiral phase is

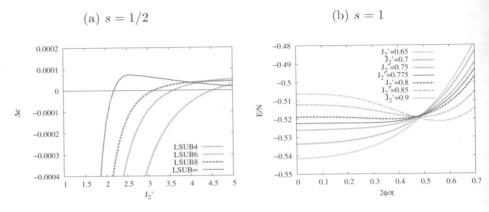

Fig. 8.   Difference between the ground-state energies per spin ($e \equiv E/N$) of the spiral and stripe phases ($\Delta e \equiv e^{\mathrm{spiral}} - e^{\mathrm{stripe}}$) versus J_2' for the spin-1/2 and spin-1 J_1–J_2' Hamiltonians of Eq. (1) with $J_1 = 1$. The CCM results for the energy difference using the stripe and spiral model states for various SUBm–m approximations ($m = \{4, 6, 8\}$ for $s = 1/2$ and $m = \{2, 4, 6\}$ for $s = 1$) are shown. We also show the $m \to \infty$ extrapolated results from using Eq. (2) for the two phases separately.

predicted to be the stable phase (i.e., lies lowest in energy), whereas for $\kappa > \kappa_e$ the stripe phase is predicted to be the stable ground state. We thus have a clear first indication of another (first-order) quantum phase transition in the spin-1/2 J_1–J_2' model at a value $\kappa = \kappa_{c_2}$.

In summary, although it is difficult to put firm error bars on our results for our predicted second critical point, our best current estimate, based on all the above results, is $\kappa_{c_2} = 1.8 \pm 0.4$.

For $s = 1$, the large κ limit of the energy per spin results of Fig. 4 agrees well with the known 1D chain result of $E/N = -1.4015$ obtained from a density-matrix renormalization group analysis,[33] just as in Fig. 4 for the spiral phase. Unlike for their spin-1/2 counterpart, the stripe phase is not a stable state for the spin-1 case because its energy lies higher than that of the spiral state for all values of J_2', as shown in Fig. 4. Hence for the $s = 1$ case, there is only one quantum critical point κ_{c_1} which drives the Néel phase to the helical phase.

In conclusion, we have studied the spin-1/2 and spin-1 J_1–J_2' HAF models. We have observed that as the quantum spin number s is increased, the quantum critical point κ_{c_1} is brought closer to the classical ($s \to \infty$) critical point for the phase transition from the Néel phase to the helical phase. The phase transition from the Néel phase to the helical phase at $\kappa = \kappa_{c_1}$ for the $s = 1/2$ case is observed to be a weakly first-order transition (although a second-order transition cannot be entirey ruled out), whereas the spin-1 case is observed to have a second-order phase transition there. There is only one quantum critical point for the $s = 1$ case at $\kappa_{c_1} = 0.62 \pm 0.01$, in contrast with the $s = 1/2$ case where there is a second quantum critical point for the phase transition from the helical phase to the stripe phase. The two quantum critical point for the spin-$1/2$ case are at $\kappa_{c_1} = 0.80 \pm 0.01$ and $\kappa_{c_2} = 1.8 \pm 0.4$. This latter result provides quantitative verification of a recent qualitative prediction of Starykh and Balents[10] using a renormalization group analysis of the J_1–J_2' model that did not, however, evaluate the actual critical point.

Acknowledgment

We thank the University of Minnesota Supercomputing Institute for Digital Simulation and Advanced Computation for the grant of supercomputing facilities, on which we relied heavily for the numerical calculations reported here.

References

1. S. Sachdev, in *Low Dimensional Quantum Field Theories for Condensed Matter Physicists*, ed. Y. Lu, S. Lundqvist, and G. Morandi (World Scientific, Singapore 1995).
2. J. Richter, J. Schulenburg, and A. Honecker, in *Quantum Magnetism*, Lecture Notes in Physics Vol. 645, ed. U. Schollwöck, J. Richter, D.J.J. Farnell, and R.F. Bishop, (Springer-Verlag, Berlin, 2004), p. 85.
3. G. Misguich and C. Lhuillier, in *Frustrated Spin Systems*, ed. H.T. Diep (World Scientific, Singapore, 2005), p. 229.
4. R. F. Bishop, P. H. Y. Li, R. Darradi, and J. Richter, *J. Phys.: Condens. Matter* **20**, 255251 (2008).
5. R. F. Bishop, P. H. Y. Li, R. Darradi, J. Schulenburg, and J. Richter, *Phys. Rev. B* **78**, 054412 (2008).
6. C. J. Gazza and H. A. Ceccatto, *J. Phys.: Condens. Matter* **5**, L135 (1993).
7. A. E. Trumper, *Phys. Rev. B* **60**, 2987 (1999).
8. J. Merino, R. H. McKenzie, J. B. Marston, and C. H. Chung, *J. Phys.: Condens. Matter* **11**, 2965 (1999).
9. Zheng Weihong, R. H. McKenzie, and R. R. P. Singh, *Phys. Rev. B* **59**, 14367 (1999).
10. O. A. Starykh and L. Balents, *Phys. Rev. Lett.* **98**, 077205 (2007).
11. T. Pardini and R. R. P. Singh, *Phys. Rev. B* **77**, 214433 (2008).
12. H. A. Bethe, *Z. Phys.* **71**, 205 (1931).
13. J. Villain, *J. Phys. (Paris)* **38**, 385 (1977); J. Villain, R. Bidaux, J. P. Carton, and R. Conte, *ibid.* **41**, 1263 (1980); E. Shender, *Sov. Phys. JETP* **56**, 178 (1982).
14. R. F. Bishop, *Theor. Chim. Acta* **80**, 95 (1991).
15. R. F. Bishop, in *Microscopic Quantum Many-Body Theories and Their Applications*, edited by J. Navarro and A. Polls, *Lecture Notes in Physics* **510** (Springer-Verlag, Berlin, 1998), p.1.
16. D. J. J. Farnell and R. F. Bishop, in *Quantum Magnetism*, edited by U. Schollwöck, J. Richter, D. J. J. Farnell, and R. F. Bishop, *Lecture Notes in Physics* **645** (Springer-Verlag, Berlin, 2004), p.307.
17. D. J. J. Farnell, R. F. Bishop, and K. A. Gernoth, *Phys. Rev. B* **63**, 220402(R) (2001).
18. S. E. Krüger, J. Richter, J. Schulenburg, D. J. J. Farnell, and R. F. Bishop, Phys. Rev. B **61**, 14607 (2000).
19. D. Schmalfuß, R. Darradi, J. Richter, J. Schulenburg, and D. Ihle, Phys. Rev. Lett. **97**, 157201 (2006).
20. C. Zeng, D. J. J. Farnell, and R. F. Bishop, *J. Stat. Phys.* **90**, 327 (1998).
21. R. Darradi, J. Richter, and D. J. J. Farnell, *Phys. Rev. B* **72**, 104425 (2005).

22. D. J. J. Farnell, R. F. Bishop, and K. A. Gernoth, *J. Stat. Phys.* **108**, 401 (2002).
23. We use the program package "Crystallographic Coupled Cluster Method" (CCCM) of D. J. J. Farnell and J. Schulenburg, see http://www-e.uni-magdeburg.de/jschulen/ccm/index.html.
24. D. J. J. Farnell and R. F. Bishop, *Int. J. Mod. Phys. B* **22**, 3369 (2008).
25. P. M. Morse and H. Feshbach, *Methods of Theoretical Physics*, Part II (McGraw-Hill, New York, 1953).
26. A. W. Sandvik, *Phys. Rev. B* **56**, 11678 (1997).
27. L. Capriotti, A. E. Trumper, and S. Sorella, *Phys. Rev. Lett.* **82**, 3899 (1999).
28. Zheng Weihong, J. Oitmaa, and C. J. Hamer, *Phy. Rev. B* **43**, 8321 (1991).
29. W. Zheng, J. O. Fjaerestad, R. R. P. Singh, R. H. McKenzie, and R. Coldea, *Phys. Rev. B* **74**, 224420 (2006).
30. W. Marshall, *Proc. R. Soc. London, Ser. A* **232**, 48 (1955).
31. C. J. Hamer, W. H. Zheng, and P. Arndt, *Phys. Rev. B* **46**, 6276 (1992).
32. D. J. J. Farnell, K. A. Gernoth, and R. F. Bishop, *Phys. Rev. B* **64**, 172409 (2001).
33. S. R. White and D. A. Huse, *Phys. Rev. B* **48**, 3844 (1993).

THERMODYNAMIC DETECTION OF QUANTUM PHASE TRANSITIONS

M. K. G. KRUSE* and H. G. MILLER[†]

Department of Physics, University of Pretoria - 0002 Pretoria, South Africa
kruse@up.ac.za
[†]*hmiller@maple.up.ac.za*

A. PLASTINO

National University La Plata, UNLP-IFLP-CCT-Conicet, C.C. 727, 1900 La Plata, Argentina
plastino@fisica.unlp.edu.ar

A. R. PLASTINO

Instituto Carlos I de Física Teórica y Computacional, Universidad de Granada, Granada, Spain
National University La Plata, UNLP-CREG, C.C. 727, 1900 La Plata, Argentina
arplastino@maple.up.ac.za

Received 21 October 2009

We show that, by recourse to thermodynamics's third law, quantum phase transitions in a system of fermions can be detected, in the sense that thereby one automatically finds the values of the external parameter (here coupling strengths of the pertinent Hamiltonian) at which these transitions occur. We illustrate our considerations with reference to an exactly solvable model of Plastino and Moszkowski [Il Nuovo Cimento **47**, 470 (1978)].

Keywords: Quantum phase transitions; critical coupling; Third Law of Thermodynamics.

1. Introduction

The frontier between different phases of matter at zero temperature is called a quantum phase transition (QPT) is[1]. QPTs, opposite to classical phase transitions, can only take place by varying a physical parameter, such as a magnetic field or pressure, at absolute zero temperature. An abrupt modification in the ground state of a many-body system ensues from the changeover. We have first-order phase transitions or continuous ones. Instead, classical phase transitions (CPT) (also called thermal phase transitions), describe a cusp in the thermodynamic properties of a system, signaling a reorganization of the constituent particles. A classical system does not have entropy at zero temperature and therefore no phase transition driven in such a fashion can occur. In finite systems a QPT can occur as well, but strictly speaking, classical phase transitions can not, since at finite temperatures the partition function and all related quantities are analytic and at best only the remnant of a classical phase transition may exist[2]. Thermal fluctuations about equilibrium values are large[3], particularly in the region where this remnant occurs. For example,

143

studies of their effect on an order parameter have concluded that, in atomic nuclei, the super-conducting to normal phase transition is washed out[4,5]. However, in spite of these problems, a phase diagram has been constructed from the remnants in an exactly solvable model[2], by studying the behavior of the specific heat C.

In the case of a system described by a Hamiltonian, $H(\lambda) = H_0 + \lambda H_1$, that changes as a function of the coupling constant, λ, the presence of a QPT can be explained in terms of level crossings so that the ground state (gs) energy is no longer analytic nor monotonic. There are other valid reasons that lead to the loss of analyticity, but the above explanation suffices for our present purposes.

Here we investigate the critical behavior of the gs energy, $\mathcal{E}(\beta, \lambda)$, with focus in the limit $T \to 0$, where β is the inverse temperature. Notice that

$$\lim_{T \to 0} (\frac{\partial \mathcal{E}}{\partial \beta})_\lambda = 0, \tag{1}$$

since, according to the third law, the specific heat must vanish in this limit for any value of λ. Out goal is to show that the $\lim_{T \to 0}(\frac{\partial \mathcal{E}}{\partial \lambda})_\beta = 0$ could be employed to ascertain the critical coupling constant λ_c for which a QPT occurs place. which would imply that enforcing fulfillment of thermodynamics' third law (TTL), for instance using the canonical ensemble formalism, would enable us to automatically determine the values of the coupling constants at which QPTs take place.

2. Formalism

We envisage a system described by the Hamiltonian

$$\hat{H} = \hat{H}_0 + \lambda \hat{H}_1 \tag{2}$$

where $[\hat{H}_0, \hat{H}_1] = 0$. At finite temperatures the Maximum Entropy Principle of Jaynes[6,7] determines the statistical operator, $\hat{\rho}$ by maximizing the entropy subject to appropriate constraints. Thus,

$$S(\hat{\rho}) = Tr[\hat{\rho} \log \hat{\rho}] \quad \text{together with} \quad \delta_\rho S(\hat{\rho}) = 0 \tag{3}$$

subject to the constraints

$$< \hat{H} >= Tr[\hat{\rho}\hat{H}] = \mathcal{E} \quad and \quad Tr[\hat{\rho}] = 1 \tag{4}$$

yield the canonical ensemble operator

$$\hat{\rho} = \frac{\exp^{-\beta \hat{H}}}{\mathcal{Z}} \tag{5}$$

where the partition function reads

$$\mathcal{Z} = Tr[e^{-\beta \hat{H}}]. \tag{6}$$

Usually, in statistical mechanics, the coupling constant λ is regarded as a constant and equation (3) is thus used to determine the Lagrange multiplier β. However, in the case of a QPT, λ is no longer a constant. It stands to reason that a functional relation between β and λ should be obtained by judiciously employing some of the equations of the formalism. Our main point here is that we can choose to that effect the expression for the specific heat in the $T \to 0$ limit, which yields a relation of the type $\lambda = f(\beta)$. Of course, in a real system the "true" coupling constant λ does not depend on the temperature. However, in practice, you may not know the λ-value but can detect phase transitions and can always measure T.

Now, since the specific heat (SH) C is

$$C = -\beta^2 (\frac{\partial <\hat{H}>}{\partial \beta})_\lambda, \tag{7}$$

and, using the function f above,

$$C = -\beta^2 \frac{\partial \lambda}{\partial \beta} (\frac{\partial <\hat{H}>}{\partial \lambda})_\beta, \tag{8}$$

a necessary and sufficient condition for it to vanish at $T = 0$ is

$$(\frac{\partial}{\partial \beta} < \hat{H} >)_\lambda = 0, \tag{9}$$

or equivalently

$$\frac{\partial \lambda}{\partial \beta} (\frac{\partial}{\partial \lambda} < \hat{H} >)_\beta = 0. \tag{10}$$

We will show that λ_c, the critical value of the coupling constant for which, according to the third law, the SH vanishes at T=0, can be determined from equation (9). Of course, C will vanish if

$$\frac{\partial \lambda}{\partial \beta} = 0 \tag{11}$$

(see equation (10)) . We suggest (and will actually show below, for an exactly solvable model of nuclear physics), that information about a QPT is present both in the specific heat and in the factor $(\frac{\partial <\hat{H}>}{\partial \lambda})_{T=0}$ as well. Note, however, that

$$(\frac{\partial < \hat{H} >}{\partial \lambda})_{T=0} = \frac{\partial E_{gs}}{\partial \lambda}. \tag{12}$$

since only the ground state is populated at $T = 0$.

Our principal contribution here will be to demonstrate that

$$\lambda_c^{3rd.\ Law} = \lambda_{critical}^{QPT}, \tag{13}$$

i.e., the critical coupling constant for which the specific heat vanishes at $T = 0$ equals the critical-QPT value of the coupling constant. Only familiar statistical mechanics' tools will be employed. Additionally, it will not be necessary to begin at finite temperatures to ascertain where (in "λ-space") a QPT occurs. For finite systems at finite temperatures ($T \neq 0$), C is analytic and structures in $\frac{\partial <E>}{\partial \beta}$ should be indicative of the remnant of a phase transition. Eq. (9) will allow one to correctly determine the λ-position of the QPT. Alternatively $\frac{\partial E_{gs}}{\partial \lambda}$ can also be used in the manner outlined above to effect such a prediction (see illustrative graphs in the examples discussed below). Thus, the two procedures that we wish to validate here should be equivalent. At this point we conjecture that, if a QPT occurs at a level crossing, then two possibilities are open, namely, 1) a discontinuous derivative

$$G(\lambda) = \left(\frac{\partial E_{gs}(\lambda)}{\partial \lambda}\right)_{\beta=\infty}, \tag{14}$$

if $\frac{\partial E_{gs}}{\partial \lambda}$ does not change sign when passing through λ_c, or, instead, 2) a null derivative, if $\frac{\partial E_{gs}}{\partial \lambda}$ does change sign when passing through λ_c. The conjecture will be investigated below.

3. The Plastino-Moszkowski Model

The PM model is an exactly solvable N-body, SU(2) two-level one.[8] Each level can accommodate N particles, i.e., is $N-$fold degenerate. There are two levels separated by an energetic gap $\mathbb{E}$ occupied by N particles. In the model the angular momentum-like operators J^2, J_x, J_y, J_z, with $J(J+1) = N(N+2)/4$ are used. The Hamiltonian is

$$\hat{H} = \mathbb{E}\hat{J}_z - \lambda[\hat{J}^2 - \hat{J}_z^2 - N/2], \tag{15}$$

its eigenstates being usually referred to as Dicke-states.[9] For convenience we set $\mathbb{E} = 1$, while

$$\hat{J}_z = (1/2)\sum_{i=1}^{N}\sum_{\sigma=1}^{2} a_{i,\sigma}^+ a_{i,\sigma}, \tag{16}$$

with corresponding expressions for $\hat{J}_x$, $\hat{J}_y$. This is a simple yet nontrivial case of the Lipkin model.[10] For now, we will only discuss the model in the zero-temperature regime. The operators appearing in the model Hamiltonian form a commuting set of observables and are thus simultaneously diagonalizable.

The ground state of the unperturbed system (i.e., when $\lambda = 0$) is $|J, J_z\rangle = \left|\frac{N}{2}, -\frac{N}{2}\right\rangle$ with the eigenenergy $E_0 = -\frac{1}{2}N$. When the interaction is turned on ($\lambda \neq 0$) and gradually becomes stronger, the ground state energy will in general be different from the unperturbed system for some critical value of λ, which we will call λ_c. This sudden change of the ground state energy entails the emergence of a quantum phase transition. We emphasize that, for a given value of N, there could be more than one critical point. The critical values of the nth transition, i.e., λ_c at that point, can be evaluated using equation (17) below, provided that $\lambda_c > 0$ and $\lambda_c \neq \infty$.

$$\lambda_{c,n} = \frac{1}{N - (2n - 1)}. \tag{17}$$

4. The $N = 2$ Problem

This simple instance can be tackled analytically without undue effort. The $J = 1$-multiplet for two fermions is $\{J_z = -1; 0; +1\}$. Labelling with the letter i one has, for the three pertinent J_z- eigenstates, the Hamiltonian matrix elements $\{H_{ii} = -1; -\lambda; +1\}$ and $\{\epsilon_i\} = -1; -\lambda; 1\}$, respectively so that the partition function reads

$$Z = e^{-\beta} + e^{\beta\lambda} + e^{\beta} \tag{18}$$

with $\beta-$derivative

$$\frac{\partial Z}{\partial\beta} = e^{\beta} + \lambda e^{\beta\lambda} - e^{-\beta} \tag{19}$$

Accordingly, the mean energy becomes

$$Tr[\rho H] = < E >= Z^{-1}[-e^{\beta} - \lambda e^{\beta\lambda} + e^{-\beta}]. \tag{20}$$

It is of the essence now to enforce the vanishing of the specific heat. This implies implies dealing with the expression

$$Z\frac{\partial < E >}{\partial\beta} = -[e^{\beta} + \lambda^2 e^{\beta\lambda} + e^{-\beta}] - Z^{-1}[-e^{\beta} - \lambda e^{\beta\lambda} + e^{-\beta}][\frac{\partial Z}{\partial\beta}], \tag{21}$$

and setting $Z\frac{\partial < E >}{\partial\beta} = 0$. As a consequence, one is led to

$$0 = -Z[e^{\beta} + \lambda^2 e^{\beta\lambda} + e^{-\beta}] + [e^{\beta} + \lambda e^{\beta\lambda} - e^{-\beta}][e^{\beta} + \lambda e^{\beta\lambda} - e^{-\beta}], \tag{22}$$

that can be recast as

$$[2\cosh\beta + e^{\beta\lambda}][2\cosh\beta + \lambda^2 e^{\beta\lambda}] = [2\sinh\beta + \lambda e^{\beta\lambda}]^2 \tag{23}$$

which is the desired link between λ and β. We go now to the T $= 0$ limit, in which $\beta \to \infty$, $\cosh\beta \to e^{\beta}$, $\sinh\beta \to e^{\beta}$. In such limit (23) becomes

$$(2e^{\beta} + e^{\beta\lambda})(2e^{\beta} + \lambda^2 e^{\beta\lambda}) = (2e^{\beta} + \lambda e^{\beta\lambda})^2 = 4e^{2\beta} + 4\lambda e^{\beta}e^{\beta\lambda} + \lambda^2 e^{2\beta\lambda}, \tag{24}$$

entailing

$$(\lambda - 1)^2 = 0 \Rightarrow \lambda = 1, \tag{25}$$

a relation that happens to yield the exact λ-value at which the QPT takes place, as shown in Ref. 8. Notice, however, that one could alternatively begin with

$$G(\lambda) = \frac{\partial < E >}{\partial\lambda} = \frac{1}{Z^2}[-(1+\beta)e^{\beta\lambda}Z - (-e^{\beta} - \lambda e^{\beta\lambda} + e^{-\beta})\beta e^{\beta\lambda}]. \tag{26}$$

Demanding fulfillment of

$$G(\lambda) = 0, \tag{27}$$

one obtains in the limit $T \to 0$

$$e^{\beta}(1 - \lambda) = 0 \tag{28}$$

or

$$\lambda = 1 \tag{29}$$

again the exact λ-value at which the QPT takes place. Note that at $\lambda = 1$, $G(\lambda)$ changes sign. In accordance with previous considerations revolving around (14), it is clear that, at $T = 0$, the function G above suffers a brutal discontinuity at $\lambda = \lambda_c = 1$, since it is "infinite" everywhere except there, where it vanishes.

5. Other N−values

Let us now discuss the results that one encounters for larger particle-numbers using the Hamiltonian (15). We consider, as an example, the case of six particles. The Hamiltonian is constructed by employing the standard angular momentum matrices in the appropriate $J = 3$-multiplet and is then diagonalized. The resulting $2J+1 = 7$ eigenenergies are in general a function of the coupling constant λ. This dependence on the coupling constant ultimately allows for level crossings to take place at critical values λ_c.

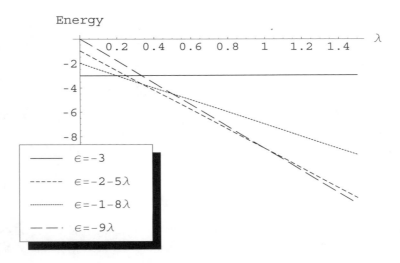

Fig. 1.   The four lowest eigenenergies of the $N = 6$ case have been plotted as a function of the coupling constant λ. There are level crossings at $\lambda_{c,1} = \frac{1}{5}$, $\lambda_{c,2} = 1/3$ and $\lambda_{c,3} = 1$, which is in agreement with equation 17. The eigenenergies of the full system are $\epsilon = \pm 3, \pm 2 - 5\lambda, \pm 1 - 8\lambda, -9\lambda$. The solid, finely-dashed, short-dashed and long-dashed line correspond to the eigenenergies $\epsilon = -3$, $\epsilon = -2 - 5\lambda$, $\epsilon = -1 - 8\lambda$, and $\epsilon = -9\lambda$, respectively.

In figure 1 we have shown the subset of eigenenergies that lead to two level crossings (QPT's) in the $N = 6$−fermions case. One evaluates then the canonical partition function $\mathcal{Z}$ from the full set of eigenvalues and afterwards determine the specific heat as given by the two equations [7-8].

5.1. *The "specific heat"* C_β^*

Once the partition function has been built up from the eigenvalues of the N-particle Hamiltonian, we are able to express the expectation value of the energy as given by the familiar canonical ensemble relation $\mathcal{E} = -\frac{\partial}{\partial\beta}\ln\mathcal{Z}$. The quantity that will be used to map out the phase diagram of the model, which we will call $C_{\beta,\lambda}^*$, is given by the derivative of $\mathcal{E}$, with respect to either β or λ. In this section we will focus our attention on the former case, $C_\beta^* = \frac{\partial}{\partial\beta}\mathcal{E}(\beta,\lambda)$.

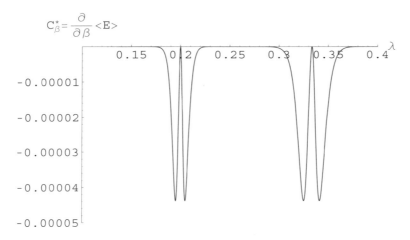

Fig. 2. The quantity C_β^* has been plotted as a function of the coupling constant λ, for the $N = 6$ particle case, with a fixed value of $\beta = 110$ (see text for a discussion on this point). There are two peaks present in the plot, centered around the two critical points λ_c of the system. The peaks coalesce into a single point centered at λ_c as $\beta \to \infty$ (see figure 3). Everywhere else $C_\beta^* = 0$, in agreement with equation 9.

A plot of C_β^* vs. λ is given in figure 2 for a fixed value of $\beta = 110$. The value of β was an arbitrary choice, in order to demonstrate the following point. At finite temperatures, that is when $\beta \neq \infty$, the peaks that are found in figure 2 are a signature of a phase transition taking place. They are smoothed out due to finite temperature and size effects. As the temperature is lowered (β increases), the peaks move together and become smaller in size. This is shown in figure 3. When $\beta \to \infty$, the peaks around each critical point coalesce into a single point, namely λ_c. This is exactly what one would expect at zero temperature; the phase transition takes place where the eigen-energies become degenerate.

5.2. *The "specific heat"* C_λ^*

The above investigation of the quantity C_β^* is one way to characterize the quantum phase transitions. It is also possible to investigate the QPT's from another viewpoint. In this section we will consider the quantity $C_\lambda^* = \frac{\partial}{\partial\lambda}\mathcal{E}(\beta,\lambda)$. From our

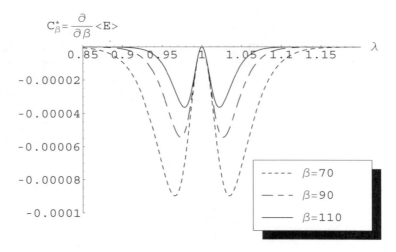

$$C_\beta^* = \frac{\partial}{\partial \beta} <E>$$

Fig. 3.   The temperature dependence of C_β^* in the region of $\lambda_c = 1$ for the $N = 6$ particle case has been plotted. The short-dashed, long-short-dashed and solid peaks correspond to $\beta = 70, 90, 110$ respectively. One can clearly see that as the temperature is lowered ($\beta \to \infty$), that the peaks become smaller in size and narrower in width. In the zero-temperature limit, these peaks would coalesce into a single point situated exactly at the location of the quantum phase transition.

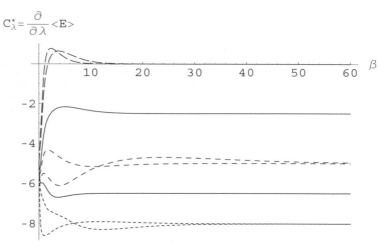

$$C_\lambda^* = \frac{\partial}{\partial \lambda} <E>$$

Fig. 4. C_λ^* as a function of β for various values of λ for the $N = 6$ particle case. The long-dashed curves (top two curves) correspond to $\lambda = 0.05, 0.1$; the medium-dashed curves correspond to $\lambda = 0.25, 0.30$ (in between the two solid curves); the short-dashed curves (lowest two curves) correspond to $\lambda = 0.4, 0.8$. The solid curves correspond to the the critical values of $\lambda_{c,n} = \frac{1}{5}, \frac{1}{3}$. Curves that have same dashing style correspond to the same ground state eigenvalue and in the zero-temperature limit tend to the slope of that corresponding eigenvalue. At the critical points, C_λ^* picks out the average value of the two slopes from the relevant degenerate eigenvalues.

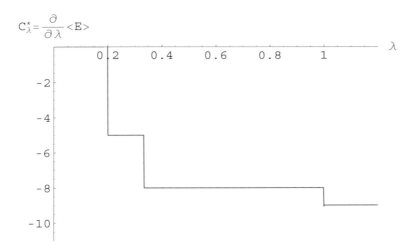

Fig. 5. C_λ^* as a function of λ in the zero-temperature limit for the $N = 6$ particle case. The horizontal segments of the plot correspond to the derivatives (with respect to λ) of the ground state eigenvalue for that particular range of λ. For example, $\frac{1}{5} < \lambda < 1/3$, $C_\lambda^* = -5$, which corresponds to the slope of the ground state eigenvalue $\epsilon = -1-5\lambda$. The discontinuities take place at the critical values of the system, viz $\lambda_{c,n} = \frac{1}{3}, 1$, respectively. At the QPT, the value of C_λ^* is the average value of the two slopes of the relevant degenerate eigenvalues.

previous discussion we would expect that C_λ^* should be related to C_β^* insofar as structure present in the former should manifest itself in the latter. As we shall show this 'duality' is indeed clearly present and substantiates our earlier suggestions (see Sec. 2).

In figure 4 we have plotted the dependence of C_λ^* on β for various values of λ. It can be seen that if the coupling constant is set in a range corresponding to one particular value of the ground state eigenenergy, that at low temperatures C_λ^* tends to the value of the slope of the given eigenenergy. For example, when $0 \leq \lambda < \frac{1}{5}$, $C_\lambda^* \to 0$ as β becomes large. For that range of the coupling constant, the corresponding ground-state eigenvalue is $\epsilon = -3$, which of course has a slope of zero. Similarly for $\frac{1}{5} < \lambda < \frac{1}{3}$, $C_\lambda^* \to -5$, which corresponds to the slope of the ground state eigenvalue $\epsilon = -2 - 5\lambda$. At the critical values λ_c, C_λ^* takes on the average value of the slope of the two degenerate eigenenergies involved. In figure 5, we have plotted the zero temperature limit of C_λ^* as a function of λ. There are three discontinuities in the figure, corresponding to the values of λ_c where the QPT takes place. The horizontal lines in the figure correspond to the slope of the current ground state eigenvalue.

6. Conclusion

A rather unexpected relationship between the third law of thermodynamics (TTL), on the one hand, and quantum phase transitions (QPTs), on the other hand, has been established here, based on the requirement that the specific heat vanish. This demand generates a relationship $\lambda = f(T)$ between a microscopic quantity (the coupling constant) and a macroscopic one (temperature). If one forces obeyance to the 3rd law by setting the expression for the specific heat equal to zero one encounters as a result the critical coupling constants at which the quantum phase transitions occur.

Our treatment relies entirely on the concept of specific heat. It is remarkable that it can be applied for finite systems, for which only the remnant of a classical phase transition exists.

Our considerations were illustrated in an exactly solvable model of Plastino and Moszkowski. We single out the fact that information about quantum phase transitions can be derived from the quantity $\frac{\partial E_{gs}}{\partial \lambda}$, which is tantamount to looking for the zero temperature limit of the specific heat.

References

1. S. Sachdev, Quantum Phase Transitions (Cambridge University Press, Cambridge, 1999).
2. E. D. Davis, H. G. Miller, Phys. Lett. B **196**, 277 (1987).
3. Y. Alhassid, J. Zingman, Phys. Rev C **30**, 684 (1984).
4. J. L. Egido, P. Ring, S. Iwasaki, H. J. Mang, Phys. Lett. B **154**, 1 (1985).
5. A. L. Goodman, Phys. Rev. C **29**, 1887 (1984).
6. E. T. Jaynes, Phys. Rev. **106**, 620 (1957).
7. E. T. Jaynes, Phys. Rev. **108** 171 (1957).
8. A. Plastino, S. A. Moszkowski, Il Nuevo Cimento **47A**, 470 (1978).
9. R. H. Dicke, Phys. Rev. **93**, 99 (1954).
10. H. J. Lipkin, N. Meshkov, A. J. Glick, Nucl. Phys. **62**, 188 (1965).

THE SU(2) SEMI QUANTUM SYSTEMS DYNAMICS AND THERMODYNAMICS

C. M. SARRIS*

Laboratorio de Sistemas Complejos, FIUBA
Facultad de Ingenieria, Universidad de Buenos Aires,
Av. Paseo Colón 850 Buenos Aires, (1063), Argentina
clsasarris@fi.uba.ar

A. N. PROTO

Comisión de Investigaciones Científicas, PBA.
Laboratorio de Sistemas Complejos, FIUBA
Facultad de Ingenieria, Universidad de Buenos Aires,
Av. Paseo Colón 850 Buenos Aires, (1063), Argentina
aproto@fi.uba.ar

Received 30 November 2009

The dynamical description of a semi quantum nonlinear systems whose classical limit is not chaotic is still an open question. These systems are characterized by mixing a classical system with a quantum-mechanical one. As some of them lead to an irregular dynamics, the name "semi quantum chaos" arises. In this contribution we study two different Hamiltonians through the Maximum Entropy Principle Approach (MEP). Taking advantage of the MEP formalism, it can be clearly established that the Hamiltonians belonging to the $SU(2)$ Lie algebra have common properties and a common treatment can be developed for them. These Hamiltonians resemble a quantum spin system coupled to a classical cavity. In the present contribution, we show that all of them share the generalized uncertainty principle as an invariant of the motion and other invariants as well. Two different classical potentials $V(q)$ have been studied. Their specific heat are evaluated in terms of the extensive (mean values) and the intensive (Lagrange multipliers) variables. The main result of the present contribution is to show that the specific heat of these systems can be fixed independently of the temperature by setting only the initial conditions on the extensive or intensive variables, as well as the value of the quantum-classical coupling parameter. It could be possible to infer that this result can be extended to generalized forms for the $V(q)$ classical potential.

Keywords: Semiquantum Systems; Specific Heat; Quantum Dynamics.

1. Introduction

A semiquantum nonlinear time independent Hamiltonian is characterized by the mixing of quantal and classical degrees of freedom as well as by exhibiting a nonlinear term which takes account the interacction between both kind of variables. Such

---

*Ciclo Básico Común, Cátedra de Física, Universidad de Buenos Aires

a Hamiltonian may be written as[1]

$$\hat{H} = \sum_j \sum_{i=1}^{n} a_j(q_i, p_i)\hat{O}_j + \sum_k \sum_{i=1}^{n} f_k(q_i, p_i), \tag{1}$$

where the classical variables (q_i, p_i) are contained in the coefficients $a_j(q_i, p_i)$, and the second term is a purely classical one. Through the maximum entropy principle (MEP) approach we are going to work, through the maximization of the Shannon-Gibbs entropy $S = -Tr(\hat{\rho}\ln\hat{\rho})$, to obtain a general treatment for any non-linear time independent semiquantum system whose quantal degrees of freedom close a partial Lie algebra under commutation with the Hamiltonian (1) through the so-called closure semiquantum condition

$$\left[\hat{H}, \hat{O}_k\right] = i\hbar \sum_{r=0}^{N} \sum_{i=1}^{n} g_{rj}(q_i, p_i)\hat{O}_r \tag{2}$$

with: $k = 1, \ldots, N$, and the $\hat{O}_k$'s are N quantum operators which close a partial Lie algebra under commutation with the Hamiltonian (1)[2]. The $g_{rj}(q_i, p_i)$ are coefficients which may depend upon $2n$ canonically conjugate classical variables (q_i, p_i) and define a $N \times N$ dynamical semiquantuma matrix $G(q_i, p_i)$ which entails the semiquantum dynamics of the mean values of the quantal degrees of freedom defined through (2) as well as the corresponding ones to the Lagrange multipliers associated to them

$$\frac{d\left\langle\hat{O}_k\right\rangle}{dt} = -\sum_{r=1}^{N} \sum_{i=1}^{n} g_{rk}(q_i, p_i)\left\langle\hat{O}_r\right\rangle, \tag{3}$$

$$\frac{d\lambda_k}{dt} = \sum_{r=1}^{N} \sum_{i=1}^{n} g_{kr}(q_i, p_i)\lambda_k(t) \tag{4}$$

To obtain the equations of motion of the classical degrees of freedom, we need the help of the quantum state $\hat{\rho}$: indeed, the energy of the system is taken to coincide with the quantum expectation value of the Hamiltonian (1)

$$\left\langle\hat{H}\right\rangle = Tr\left(\hat{\rho}\hat{H}\right) = \sum_j \sum_{i=1}^{n} a_j(q_i, p_i)\left\langle\hat{O}_j\right\rangle$$

$$+ \sum_k \sum_{i=1}^{n} f_k(q_i, p_i). \tag{5}$$

Equation (5) generates the temporal evolution of the classical variables through

$$\frac{dq_i}{dt} = \left\{q_i, \left\langle\hat{H}\right\rangle\right\}; \quad i = 1, \ldots, n \tag{6}$$

$$\frac{dp_i}{dt} = \left\{ p_i, \left\langle \hat{H} \right\rangle \right\}; \quad i = 1, ..., n \tag{7}$$

where $\{ , \}$ indicates Poisson brackets. Although Eqs. (6) and (7) generates the dynamic evolution of the classical variables, its nature is semiquantum as well because it get involved the mean values of the quantum variables (see Refs. 1 and 3 for more details). The semiquantum dynamics of the system develops in a semiquantum phase space

$$\mathbb{V}_{Sq} = span\{ \left\langle \hat{O}_1 \right\rangle (t), ..., \left\langle \hat{O}_N \right\rangle (t),$$
$$q_1, ..., q_n, \ p_1, ..., p_n \}, \tag{8}$$

whose dimension is $N + 2n$ on account of the N quantum variables are linearly independent and so are the $2n$ classical ones. So, in this semiquantum phase space, the quantum mean values span the quantum manifold of the system $\mathbb{QM} = span\{ \left\langle \hat{O}_1 \right\rangle (t), ..., \left\langle \hat{O}_N \right\rangle (t) \}$ whose dimension is N and, the classical variables span the classical manifold of the system $\mathbb{CM} = span\{ q_1, ..., q_n, p_1, ..., p_n \}$ whose dimension is $2n$, so that, it holds $\mathbb{V}_{Sq} = \mathbb{QM} \oplus \mathbb{CM}$.

1.1. *Metric on the semiquantum phase space and dynamical invariants*

As the classical variables were assumed to have the classical equations of motion (6) and (7), the classical manifold of the system remains as a symplectic one. Meanwhile, on the quantum manifold ($\mathbb{QM}$), it is possible to define a definite positive metric as in the full quantum case[4]: $\bullet : \mathbb{QM} \times \mathbb{QM} \longrightarrow \mathbb{R}$ /

$$\left\langle \hat{O}_i \right\rangle \bullet \left\langle \hat{O}_j \right\rangle = \frac{1}{2} Tr \left(\hat{\rho} \left[\hat{O}_i, \hat{O}_j \right]_+ \right) -$$
$$- Tr \left(\hat{\rho} \hat{O}_i \right) Tr \left(\hat{\rho} \hat{O}_j \right)$$
$$= \frac{1}{2} \left\langle \hat{O}_i \hat{O}_j + \hat{O}_j \hat{O}_i \right\rangle -$$
$$- \left\langle \hat{O}_i \right\rangle \left\langle \hat{O}_j \right\rangle$$
$$= K_{ij}(t) = K_{ji}(t), \tag{9}$$

which, in turn, defines the components, $K_{ij} = K_{ji}$, of the second-rank covariant metric tensor, $K(t)$, in terms of the quantum correlation coefficients between the N quantum degrees of freedom of the system. Because of the positive definiteness requirement of the metric, the Schwarz inequality holds and, as we did in Ref. 4, this inequality enables us to establish the connection between the metric and the uncertainty principle, which is obtained as the summation over the principal minors of order 2 belonging to the covariant metric tensor

$$I^H = \sum_{\substack{j,k=1 \\ j<k}}^{N} \left(\Delta\hat{O}_j\right)^2 \left(\Delta\hat{O}_k\right)^2 -$$

$$- \sum_{\substack{j,k=1 \\ j<k}}^{N} \left[\left\langle \hat{L}_{jk} \right\rangle - \left\langle \hat{O}_j \right\rangle \left\langle \hat{O}_k \right\rangle\right]^2 \geq$$

$$\geq -\frac{1}{4} \sum_{j,k=1}^{N} \left\langle \left[\hat{O}_j, \hat{O}_k\right] \right\rangle^2, \tag{10}$$

$\left(\left\langle \hat{L}_{jk} \right\rangle = \frac{1}{2} \left\langle \hat{O}_j\hat{O}_k + \hat{O}_k\hat{O}_j \right\rangle\right)$. It is possible to demonstrate[1] that an *anti-symmetry of $G(q_i, p_i)$ is a sufficient condition for the generalized uncertainty principle given by Eq. (10) be a dynamical invariant for the semiquantum system (1) as well as the quantity*[1]

$$I_\Lambda = \sum_{r=1}^{N} \left[\lambda_k(t)\right]^2 \tag{11}$$

expressed in term of the N Lagrange multipliers only.

2. Metric and Specific Heat Relationship

To establish this relationship, we follow the prescription given in Ref. 5, it is to say: to include the Hamiltonian of the system into the relevant set of operators, so that, this set is now composed by $RS = \{\hat{O}_0 = \hat{I}, \hat{O}_1 = \hat{H}, \hat{O}_2, \hat{O}_3, ..., \hat{O}_{q+1}\}$. The inclusion of the Hamiltonian, does not modify the dynamics of the system (see Eqs.(2), (3) and (4consequence of this, Eq.(q−dimensional generalized phase space (8), but now, we have well-defined the intensive variable temperature[5]: $\beta = \frac{1}{T} = \left|\frac{\partial S}{\partial \langle \hat{H} \rangle}\right|_{\{\langle \hat{O}_i \rangle\}}$, with: $i = 2, 3, ..., q + 1$. So, we are able to obtain not only the dynamic evolution of the system but the thermodynamical one. The state operator now reads: $\hat{\rho}(\beta, t) = \exp(-\lambda_0 \hat{I} - \beta\hat{H} - \sum_{r=2}^{q} \lambda_r\hat{O}_r)$, and the entropy S reads: $S(\beta) = \lambda_0 + \beta\left\langle \hat{H} \right\rangle + \sum_{r=2}^{q} \lambda_r(t) \left\langle \hat{O}_r \right\rangle(t)$. So, the specific heat results (see ref. citeIJMPB for more details)

$$C = \beta \sum_{j=1}^{q+1} \sum_{r=2}^{q+1} a_r(q_i, p_i) \,\lambda_j(t) K_{jr}(t) \tag{12}$$

in terms of the metric tensor's components $K_{jr}(t)$, the Lagrange multipliers associated to the relevant operators, the $a_r(q_i, p_i)$ coefficients appearing in the Hamiltonian (1) and the temperature β.

3. The SU(2) Case

Let us considerer the following nonlineal semiquantum Hamiltonians , where $1/2$ spin particle interacts with a classical particle[1,7]

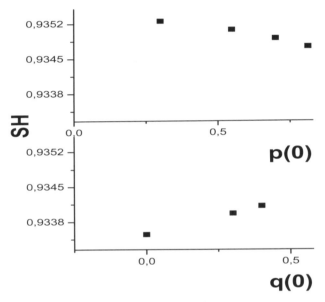

Fig. 1. SH dependence on the initial conditions for $< \hat{H}_1 >$. The values at $t = 0$ are: $< \hat{H}_1 >= 2$, $\langle \hat{\sigma}_x \rangle_{(0)} = 0$, $\langle \hat{\sigma}_y \rangle_{(0)} = 0.5$, $\langle \hat{\sigma}_z \rangle_{(0)} = 0.5$, $\hat{\lambda}_{x(0)} = 0.2$, $\hat{\lambda}_{y(0)} = 0.2$, $\hat{\lambda}_{z(0)} = 0.2$. The parameter values are: $\beta = 1$, $\omega = 1$, $\omega_0 = 3$, $m = 1$, $C = 1$.

$$\hat{H}_1 = -\frac{\omega_0}{2} \, \hat{\sigma}_z + C \, q \, \hat{\sigma}_x + \frac{p^2}{2m} + \frac{m\omega^2 q^2}{2}, \cdot \tag{13}$$

$$\hat{H}_2 = B \, \hat{\sigma}_z + C \, q \, \hat{\sigma}_x + \frac{p^2}{2} + \frac{q^4}{4} - \frac{q^2}{2}, \tag{14}$$

where q and p are canonically conjugate classical variables and $\hat{\sigma}_i$ are $\frac{1}{2}$ spin particles operators. The $-\frac{\omega_0}{2} \, \hat{\sigma}_z$ ($B\hat{\sigma}_z$)term is the spin Hamiltonian, the $\frac{p^2}{2m} + \frac{\eta \, q^\alpha}{\alpha}$ (α an integer, positive) term, according to the $\hat{H}_i$ is the classical particle Hamiltonian and the $C \, q \, \hat{\sigma}_x$ term represents the interaction between them.

By considering the relevant set, $\{O_j\} = \{\hat{\sigma}_x, \hat{\sigma}_y, \hat{\sigma}_z\}$, we see that, through Eq.(2), it define a complete set of non-commuting observables ($CSNCO$) whose

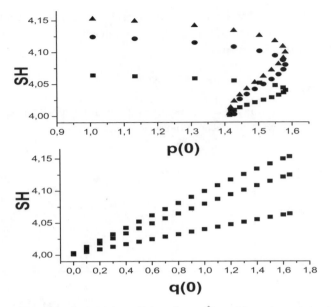

Fig. 2. SH dependence on the initial conditions for $< \hat{H}_2 >$. The values at $t = 0$ are: $< \hat{H}_2 > = 2, \langle \hat{\sigma}_x \rangle_{(0)} = 0.1, \langle \hat{\sigma}_y \rangle_{(0)} = 0.1, \langle \hat{\sigma}_z \rangle_{(0)} = 0.7, \hat{\lambda}_{x(0)} = 0.2, \hat{\lambda}_{y(0)} = 0.2, \hat{\lambda}_{z(0)} = 0.2.$ The parameter values are: $m = 1, B = 2, \beta = 1$. For Square, dot, and triangle points $C = 2, 4, 5$, respectively.

mean values are the quantum degrees of freedom of the system, while q and p are the classical ones. The semiquantum phase space is 5 dimensional and is defined by $\mathbb{V}_{Sq} = span\{\langle \hat{\sigma}_x \rangle, \langle \hat{\sigma}_y \rangle, \langle \hat{\sigma}_z \rangle, q, p\}$. Equation (2) leads to the following commutation relationships: $\left[\hat{H}_1, \hat{\sigma}_x\right] = -i\omega_0 \hat{\sigma}_y$, $\left[\hat{H}_1, \hat{\sigma}_y\right] = iC\omega_0 \hat{\sigma}_x + i2Cq \ \hat{\sigma}_z$, $\left[\hat{H}_1, \hat{\sigma}_z\right] = -i2Cq \ \hat{\sigma}_y$.and to the the 3×3 antysimmetric matrix $G_1(q)$

$$G_1(q) = \begin{pmatrix} 0 & \omega_0 & 0 \\ -\omega_0 & 0 & -2Cq \\ 0 & 2Cq & 0 \end{pmatrix},$$ (15)

From Eqs. (15), (3) and (4) the following system of equations for the $< O_j >$ and the associate Lagrange multipliers are obtained for H_1

$$\frac{d\langle \hat{\sigma}_x \rangle}{dt} = \omega_0 \langle \hat{\sigma}_y \rangle,$$ (16)

$$\frac{d\langle \hat{\sigma}_y \rangle}{dt} = -\omega_0 \langle \hat{\sigma}_x \rangle - 2Cq \langle \hat{\sigma}_z \rangle,$$ (17)

$$\frac{d\langle \hat{\sigma}_z \rangle}{dt} = 2Cq \langle \hat{\sigma}_y \rangle,$$ (18)

$$\frac{d\lambda_x}{dt} = \omega_0 \lambda_y, \tag{19}$$

$$\frac{d\lambda_y}{dt} = -\omega_0 \lambda_x - 2Cq\lambda_z, \tag{20}$$

$$\frac{d\lambda_z}{dt} = 2Cq\lambda_y. \tag{21}$$

Taking

$$\left\langle \hat{H}_1 \right\rangle = -\frac{\omega_0}{2} \left\langle \hat{\sigma}_x \right\rangle + C \, q \, \left\langle \hat{\sigma}_z \right\rangle + \frac{p^2}{2m} + \frac{m\omega^2 q^2}{2}, \tag{22}$$

the equations of motion for p and q are given by Eqs.(6) and (7)

$$\frac{dq}{dt} = \frac{p}{m}, \tag{23}$$

$$\frac{dp}{dt} = -C \left\langle \hat{\sigma}_x \right\rangle - m\omega^2 q. \tag{24}$$

Following similar procedure, for H_2 we can write: $\left[\hat{H}_2, \hat{\sigma}_x\right] = i2B \, \hat{\sigma}_y$, $\left[\hat{H}_2, \hat{\sigma}_y\right] = -i2B \, \hat{\sigma}_x + i2Cq \, \hat{\sigma}_z$, $\left[\hat{H}_2, \hat{\sigma}_z\right] = -i2Cq \, \hat{\sigma}_y$. The antysimmetric matrix $G_2 \, (q)$ is

$$G_2 \, (q) = \begin{pmatrix} 0 & -2B & 0 \\ 2B & 0 & -2qC \\ 0 & 2qC & 0 \end{pmatrix}. \tag{25}$$

From Eqs. (25), (3) and (4) the following system of equations for the $< O_j >$ and the asociate Lagrange multipliers are obtained for H_2

$$\frac{d \left\langle \hat{\sigma}_x \right\rangle}{dt} = -2B \left\langle \hat{\sigma}_y \right\rangle, \tag{26}$$

$$\frac{d \left\langle \hat{\sigma}_y \right\rangle}{dt} = 2B \left\langle \hat{\sigma}_x \right\rangle - 2Cq \left\langle \hat{\sigma}_z \right\rangle, \tag{27}$$

$$\frac{d \left\langle \hat{\sigma}_z \right\rangle}{dt} = 2Cq \left\langle \hat{\sigma}_y \right\rangle, \tag{28}$$

$$\frac{d\lambda_x}{dt} = -2B\lambda_y, \tag{29}$$

$$\frac{d\lambda_y}{dt} = 2B\lambda_x - 2Cq\lambda_z, \tag{30}$$

$$\frac{d\lambda_z}{dt} = 2Cq\lambda_y. \tag{31}$$

Taking

$$\left\langle \hat{H}_2 \right\rangle = B \left\langle \hat{\sigma}_z \right\rangle + C q \left\langle \hat{\sigma}_x \right\rangle + \frac{p^2}{2} + \frac{q^4}{4} - \frac{q^2}{2}, \tag{32}$$

the equations of motion for p and q are given by Eqs.(6) and (7)

$$\frac{dq}{dt} = p, \tag{33}$$

$$\frac{dp}{dt} = -C \left\langle \hat{\sigma}_x \right\rangle + q - q^3. \tag{34}$$

According to Eqs. (9), it is possible to obtain the following invariants for the former Hamiltonian (13) and (14)[4]

$$B = \left\langle \hat{\sigma}_x \right\rangle^2 + \left\langle \hat{\sigma}_y \right\rangle^2 + \left\langle \hat{\sigma}_y \right\rangle^2 = \left\langle \hat{\sigma} \right\rangle^2, \tag{35}$$

$$Tr\left[K\right] = 3 - \left\langle \hat{\sigma} \right\rangle^2, \tag{36}$$

$$I^H = 3 - 2\left\langle \hat{\sigma} \right\rangle^2 \geq \left\langle \hat{\sigma} \right\rangle^2 \tag{37}$$

$$\det\left[K\right] = 1 - \left\langle \hat{\sigma} \right\rangle^2, \tag{38}$$

$$I_\Lambda = |\alpha|^2 = \left(\lambda_x\right)^2 + \left(\lambda_y\right)^2 + \left(\lambda_z\right)^2. \tag{39}$$

because of the positive definiteness requirement of the metric, Eqs. (37) and (38) allow us to express

$$0 < \left\langle \hat{\sigma} \right\rangle^2 = \left\langle \hat{\sigma}_x \right\rangle^2 + \left\langle \hat{\sigma}_y \right\rangle^2 + \left\langle \hat{\sigma}_z \right\rangle^2 < 1. \tag{40}$$

As the metric on the quantum manifold is a positive definite one, the former invariant (10) is positive and it holds that: $I^H = 3 - 2\left\langle \hat{\sigma} \right\rangle^2 \geq \left\langle \hat{\sigma} \right\rangle^2$, with $\left\langle \hat{\sigma} \right\rangle^2 =$

$\langle \hat{\sigma}_x \rangle^2 + \langle \hat{\sigma}_y \rangle^2 + \langle \hat{\sigma}_z \rangle^2$, which implies the well known condition $\langle \hat{\sigma} \rangle^2 < 1$. So, the uncertainty principle (10) can be put into the form

$$0 < \langle \hat{\sigma} \rangle^2 < 1. \tag{41}$$

The metric tensor's components are: $K_{xx} = 1 - \langle \hat{\sigma}_x \rangle^2$, $K_{yy} = 1 - \langle \hat{\sigma}_y \rangle^2$, $K_{zz} = 1 - \langle \hat{\sigma}_z \rangle^2$, $K_{xy} = K_{yx} = - \langle \hat{\sigma}_x \rangle \langle \hat{\sigma}_y \rangle$, $K_{xz} = K_{zx} = - \langle \hat{\sigma}_x \rangle \langle \hat{\sigma}_z \rangle$, $K_{yz} = K_{zy} = - \langle \hat{\sigma}_z \rangle \langle \hat{\sigma}_y \rangle$.

Using Eq.(12) it is possible to obtain the "dynamic" expressions for the specific heat, in terms of the metric tensor's components, which are

$$SH_1 = \beta \left(\lambda_x + \beta Cq \right) \times \left[Cq \left(1 - \langle \hat{\sigma}_x \rangle^2 \right) + \frac{\omega_0}{2} \langle \hat{\sigma}_x \rangle \langle \hat{\sigma}_z \rangle \right] +$$
$$+ \beta \lambda_y \left[-Cq \langle \hat{\sigma}_x \rangle \langle \hat{\sigma}_y \rangle + \frac{\omega_0}{2} \langle \hat{\sigma}_y \rangle \langle \hat{\sigma}_z \rangle \right] +$$
$$+ \beta \left(\lambda_z - \beta \frac{\omega_0}{2} \right) \times \left[-Cq \langle \hat{\sigma}_x \rangle \langle \hat{\sigma}_z \rangle - \frac{\omega_0}{2} \left(1 - \langle \hat{\sigma}_z \rangle^2 \right) \right]. \tag{42}$$

$$SH_2 = \beta \left(\lambda_x + \beta Cq \right) \times \left[Cq \left(1 - \langle \hat{\sigma}_x \rangle^2 \right) - B \langle \hat{\sigma}_x \rangle \langle \hat{\sigma}_z \rangle \right] -$$
$$- \beta \lambda_y \left[Cq \langle \hat{\sigma}_x \rangle \langle \hat{\sigma}_y \rangle + B \langle \hat{\sigma}_y \rangle \langle \hat{\sigma}_z \rangle \right] +$$
$$+ \beta \left(\lambda_z + \beta B \right) \times \left[-Cq \langle \hat{\sigma}_x \rangle \langle \hat{\sigma}_z \rangle + B \left(1 - \langle \hat{\sigma}_z \rangle^2 \right) \right]. \tag{43}$$

Equations (42) and (43) show clearly that the specific heat is a "dynamic" concept because of it depends on the initial conditions imposed on the set $\Lambda(0) = \{ \lambda_{x(0)}, \lambda_{y(0)}, \lambda_{z(0)} \}$ and on the set $\Gamma(0) = \{ \langle \hat{\sigma}_x \rangle_{(0)}, \langle \hat{\sigma}_y \rangle_{(0)}, \langle \hat{\sigma}_z \rangle_{(0)} \}$.

4. Numerical Simulations

We have been done two different kind of numerical simulation in order to show how the SH depends on the initial conditions and in order to see how the SH may be taken as a tool for defining the nature of "semiquantum chaos".

4.1. *The SH depends on the Initial Conditions*

To show how the SH depends on the initial conditions, we have depicted the SH values vs. $p_{(0)}$ and $q_{(0)}$ initial values. In both cases we have always taken the same values at $t = 0$ for the sets $\Gamma(0) = \{ \langle \hat{\sigma}_x \rangle_{(0)}, \langle \hat{\sigma}_y \rangle_{(0)}, \langle \hat{\sigma}_z \rangle_{(0)} \}$ and $\Lambda(0) = \{ \lambda_{x(0)}, \lambda_{y(0)}, \lambda_{z(0)} \}$, and the same values for the parameters $< \hat{H} >$, ω_0, B, C, $m\omega$ in order the $q_{(0)}$ and $p_{(0)}$ values are allowed to vary. Once the $q_{(0)}$ value is fixed, the corresponding $p_{1(0)}$ and $p_{2(0)}$ values are derived from the $< \hat{H}_1 >$, and $< \hat{H}_2 >$ respectively (see Eqs. (22) and (32))

$$p_{1(0)} = \left[2m \left(\left\langle \hat{H}_1 \right\rangle + \frac{\omega_0}{2} \left\langle \hat{\sigma}_z \right\rangle_{(0)} - Cq_{(0)} \left\langle \hat{\sigma}_x \right\rangle_{(0)} - \frac{m\omega^2 q_{(0)}^2}{2} \right) \right]^{1/2} \quad (44)$$

$$p_{2(0)} = \left[2m \left(\left\langle \hat{H}_2 \right\rangle + B \left\langle \hat{\sigma}_z \right\rangle_{(0)} - Cq_{(0)} \left\langle \hat{\sigma}_x \right\rangle_{(0)} + \frac{q_{(0)}^2}{2} - \frac{q_{(0)}^4}{4} \right) \right]^{1/2} \quad (45)$$

In Fig. (1) are shown the *SH* values vs. $p_{(0)}$ and $q_{(0)}$ for $< \hat{H}_1 >$ and, in Fig.(2), are shown the same for $< \hat{H}_2 >$. In both cases we have taken the value $\beta = 1$ ($\beta = 1/KT$). Both figures show how the specific heat depends on the initial conditions.

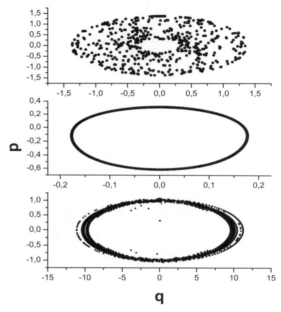

Fig. 3. p versus q Poincare sections for $\left\langle \hat{H}_1 \right\rangle$. The values at $t = 0$ are: $< \hat{H}_1 > = 0.5$, $\langle \hat{\sigma}_x \rangle_{(0)} = 0$, $\langle \hat{\sigma}_y \rangle_{(0)} = 0$, $\langle \hat{\sigma}_z \rangle_{(0)} = 0.9$ (upper), $\langle \hat{\sigma}_z \rangle_{(0)} = 0.1$ (middle), $\langle \hat{\sigma}_z \rangle_{(0)} = 0.5$ (lower). $\lambda_{x(0)} = 0.2$, $\lambda_{y(0)} = 0.2$, $\lambda_{z(0)} = 0.2$, $q_{(0)} = 0$, $p_{(0)} = 0.31623$. The parameter values are: $m = 1$, $C = 1$, $\beta = 1$, $\omega = 1$, $\omega_0 = 2$.

4.2. *The SH as a tool for defining the nature of "semiquantum chaos"*

In order to show how the *SH* could be taken as a tool for defining the nature of "semiquantum chaos" we have depicted the Poincare surfaces p vs. q (always for the cut plane $\langle \hat{\sigma}_x \rangle = 0$) for both examples $\left\langle \hat{H}_1 \right\rangle$ and $\left\langle \hat{H}_2 \right\rangle$. In Fig.(3) we can see it for

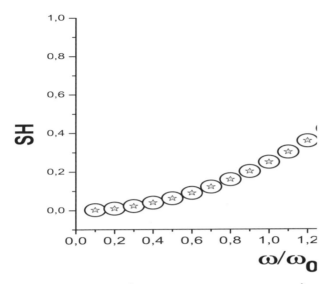

Fig. 4.   *SH* dependence with ω_0 for $< \hat{H}_1 >$. The values at $t = 0$ are: $< \hat{H}_1 > = 0.5$, $\langle \hat{\sigma}_x \rangle_{(0)} = 0$, $\langle \hat{\sigma}_y \rangle_{(0)} = 0$, $\langle \hat{\sigma}_z \rangle_{(0)} = 0.9$ (for $p_{(0)} = 0.31623$) and $\langle \hat{\sigma}_z \rangle_{(0)} = 0.1$(for $p_{(0)} = 0.94869$), $\lambda_{x(0)} = 0.2$, $\lambda_{y(0)} = 0.2$, $\lambda_{z(0)} = 0.2$, $q_{(0)} = 0$. The parameter values are: $m = 1$, $C = 1$, $\beta = 1$, $\omega = 1$, $\omega_0 = 2$.

the $\left\langle \hat{H}_1 \right\rangle$ case, for three different values of $\langle \hat{\sigma}_z \rangle_{(0)}$. These plots suggest that a change in the regime has taken place and, from these three Poincare sections, it could be inferred that the dynamics of the system has changed. However, in Fig.(4), we see that the behavior of the *SH* for different values of the ratio ω/ω_0 (always for the sames values of $\left\langle \hat{H}_1 \right\rangle$, $\langle \hat{\sigma}_x \rangle_{(0)}$, $\langle \hat{\sigma}_y \rangle_{(0)}$, $q_{(0)} = 0$ and $p_{(0)}$) is a very simple one. In Fig.(5) we show the Poincare surfaces p vs. q (always for the cut plane $\langle \hat{\sigma}_x \rangle = 0$) for $\left\langle \hat{H}_2 \right\rangle$ case for three different valuess of $\langle \hat{\sigma}_z \rangle_{(0)}$ and the *SH* dependence vs. $\langle \hat{\sigma}_z \rangle_{(0)}$. Again, the Poincare sections suggest that a change in the system regime take place but the *SH* vs. $\langle \hat{\sigma}_z \rangle_{(0)}$ behavior suggests that no such change has happened on account of this behavior is a very simple one. It is to say that the *SH* behavior does not show any irregularity that could point out to a "chaotic" situation. So, Figs.(4) and (5) suggest us that we are in the presence of a complex but non chaotic dynamics.

5. Conclusion

Hamiltonians which belong to the $SU(2)$ Lie algebra accept a common treatment, share the generalized uncertainty principle as an invariant of the motion and other

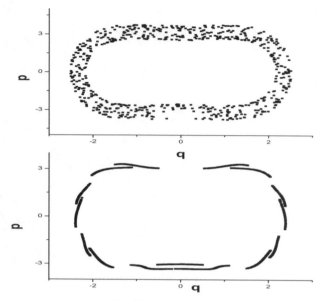

Fig. 5.   Poincare sections p versus q :$\left\langle \hat{H}_2 \right\rangle = 5$, $\langle \hat{\sigma}_x \rangle_{(0)} = \langle \hat{\sigma}_y \rangle_{(0)} = 0.1$, $\langle \hat{\sigma}_z \rangle_{(0)} = 0.9$ (upper), $\langle \hat{\sigma}_z \rangle_{(0)} = 0.5$ (middle), $\langle \hat{\sigma}_z \rangle_{(0)} = 0.1$ (lower),$q_{(0)} = 0$, $p_{(0)} = 2.465$.SH vs. $\langle \hat{\sigma}_z \rangle_{(0)}$, $\lambda_x(0) = 0.2$,$\lambda_y(0) = 0.2$, $\lambda_z(0) = 0.2$, $q_{(0)} = 0$. The parameter values are: $m = 1$, $C = 4$, $\beta = 1$, $B = 2$.

invariants as well. This result can be extended to generalized forms for the $V(q)$ classical potential. Their Specific Heats are evaluated in terms of the extensive (mean values) and the intensive (Lagrange multipliers) variables, and fixed through the IC imposed on the set of nonlinear differential equations. From the numerical results, we propose SH as a new tool to clarify the nature of the dynamics presented by this type of Hamiltonians. As we have shown it is tempting, looking at the Poincare surfaces, to talk about different chaotic-regular dynamics. However, as the SH presents an smooth variation with the IC, it seems highly controversial to do so.

References

1. A. N. Proto and C. M. Sarris, International Journal of Bifurcation and Chaos (to be published)
2. Y. Alhassid, R. D. Levine, J. Chem. Phys. **67** (1977) 4321-4340; Phys. Rev. A **18** (1978) 89-116
3. A. M. Kowalski, A. Plastino, A. N. Proto, Phys. Rev. E, **52** (1995) 165-177
4. C. M. Sarris, A. N. Proto, Phys. A **377** (2007) 33-42
5. J. Aliaga, D. Otero, A. Plastino and A.N. Proto; Phys. Rev. A, 38, 2455 (1988).
6. C. M. Sarris and A. N. Proto, Int. J. of Modern Phys. B, Vol 23, Nos. 20 & 21 (2009) 4170-4185
7. R. Bonci, R. Roncaglia, B. J. West, P. Grigolini, Phys. Rev. A **45** (1992) 8490-8500

Part C
Physics of Nanosystems and Nanotechnology

QUASI-ONE DIMENSIONAL FLUIDS THAT EXHIBIT HIGHER DIMENSIONAL BEHAVIOR

SILVINA M. GATICA*

Department of Physics and Astronomy, Howard University, Washington, DC, 20059, USA
sgatica@howard.edu

M. MERCEDES CALBI

Department of Physics, Southern Illinois University, Carbondale, IL USA

GEORGE STAN

Department of Chemistry, University of Cincinnati, Cincinnati, OH 45221, USA

R. ANDREEA TRASCA

Dieffenbachstr. 58A, 10967 Berlin, Germany

MILTON W. COLE*

Department of Physics, Penn State University, University Park, PA 16802, USA
miltoncole@aol.com

Received 13 October 2009

Fluids confined within narrow channels exhibit a variety of phases and phase transitions associated with their reduced dimensionality. In this review paper, we illustrate the crossover from quasi-one dimensional to higher effective dimensionality behavior of fluids adsorbed within different carbon nanotubes geometries. In the single nanotube geometry, no phase transitions can occur at finite temperature. Instead, we identify a crossover from a quasi-one dimensional to a two dimensional behavior of the adsorbate. In bundles of nanotubes, phase transitions at finite temperature arise from the transverse coupling of interactions between channels.

Keywords: Carbon nanotubes; adsorption; quasi-one dimensional systems.

1. Introduction

One of the most interesting topics within modern condensed matter physics is that of phenomena in reduced dimensionality, resulting from some degree of spatial localization of the particles comprising a system.[1] For example, chemists, materials scientists and physicists have created and explored numerous physical systems in

---

*Corresponding author

which atoms and molecules are confined within quasi-one dimensional (Q1D) environments. The variety of these systems is remarkable, such as the peapod geometry, *i.e.*, a line of buckyballs within a carbon nanotube,[2,3] fluids within artificial materials created by templating[4,5] and Q1D optical lattices created by laser fields.[7] Unfortunately, as far as we know, there exists no comprehensive review of this general problem, although many relevant subfields have been summarized.[8-11] The present paper addresses a small subset of this exciting research field. Specifically, we consider problems involving fluids, both classical and quantum, confined within Q1D channels, the focus of our group's research during the last decade.[12]

Here, the term Q1D refers to a system in which particles move in an external potential field $V(\mathbf{r})$ which is either constant or slowly varying in *one* direction (z), while $V(\mathbf{r})$ is strongly localizing in the two other (transverse) directions. In the case of quantum particles, for which the transverse spectrum is discrete, one expects that the corresponding degrees of freedom are frozen out at low temperature (T); transverse excitation does occur at higher T, as determined by the gaps in the transverse spectrum of states. This plausible expectation is borne out in some cases, but we shall see that there can be dramatic consequences of the transverse degrees of freedom in other cases, even at low T.

One of the many exciting aspects of strictly 1D physics is its susceptibility to weak perturbations. The reason for this behavior arises from the fact that (in all practical situations)[13] no phase transition can exist in a purely 1D system at any finite T, even though the ground state may exhibit symmetry-breaking order. Thus, there do exist transitions at T identically equal to zero. A familiar example is the 1D Ising model, for which the correlation length and susceptibility both diverge as T approaches 0.

Since the purely 1D system has no finite T transition, what brings about more interesting behavior in the Q1D case? As discussed below, the difference can arise from considering a set of parallel 1D systems which are weakly coupled. Alternatively, the behavior can happen because the system is only 1D insofar as the transverse dimensions are finite, unlike the length in the z direction, L, which achieves the thermodynamic limit.....but the transverse dimensions *are* large enough to have an observable effect (e.g. low energy gaps). A third, more surprising, origin of interesting phenomena is when the system is a collection of noninteracting 1D systems, with "quenched" heterogeneity, which are coupled to a particle bath, so they possess a common chemical potential.

The next section discusses the conceptually and computationally simplest case of a Q1D system: a low density, noninteracting gas within a single channel; then, the solution of the one-particle Schrödinger equation determines the physical behavior. Section 3 considers the case of many *interacting* particles within a single channel; such a problem is often used as a model for fluids within regular or irregular porous materials. Section 4 considers the problem of Q1D channels containing fluids that interact with one another as well as with fluids occupying other channels.

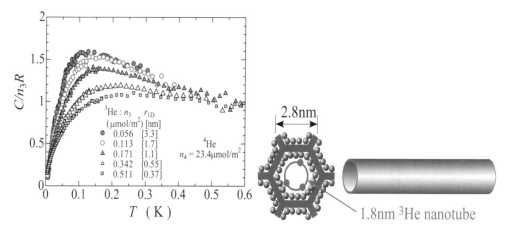

Fig. 1. Left: Experimental heat capacity for ^{3}He at various densities, inside ^{4}He-coated FSM-16, from Ref. 5. Right: schematic depiction of the geometry, from Ref. 6. The tube represents the environment experienced by a ^{3}He atom.

2. Low Density Gas in a Single Q1D Channel

Our first problem is conceptually and calculationally simple: a low density gas confined within a Q1D geometry; nevertheless, it provides interesting, and sometimes surprising, results. One example that has been studied extensively is that of a gas inside single carbon nanotubes, *e.g.* quantum fluids or the "peapod" case of C_{60} molecules.[3] Another is the case of quantum gases inside templated regular pores.[5] A third example is the so-called groove region between two nanotubes, *e.g.* on the outside of a bundle of nanotubes.[14–16]

A *classical* noninteracting gas has a kinetic energy per particle of $(3/2)\,k_BT$ and a mean potential energy $\langle U \rangle$ determined by its interaction with the environment. For the case of a particle localized near the z axis within a channel, $\langle U \rangle = k_BT$, due to two transverse directions of excitation. Hence, the classical specific heat per particle is $[C(T)/N]_{classical} = (5/2)k_B$. For a quantum gas, instead, the transverse degrees of freedom are frozen out at low T, so one expects $[C(T)/N]_{quantum} = (1/2)k_B$. The generalization to D "effective" dimensions yields this expression for the dimensionless specific heat, C^*, of a noninteracting Boltzmann gas:

$$C* \equiv C/(Nk_B) = D/2 \tag{1}$$

Fig. 1 shows experimental results for C^* in the case of ^{3}He inside of FSM-16, a material consisting of straight hexagonal pores of cross-sectional distance of order 2 to 3 nm, pre-coated with a thin film of ^{4}He. At the low densities shown here, the ^{3}He gas can be assumed to be noninteracting, although interaction effects appear at higher density.[17] The behavior observed inside FSM-16 can be understood by

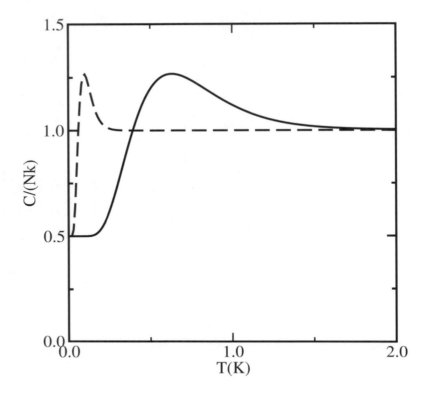

Fig. 2.   Heat capacity for a low density gas of ^{4}He atoms within a single wall carbon nanotube, for cases R=0.8 nm (dashes) and 0.5 nm (full curve). Quantum statistical effects are omitted from the calculation. Adapted from Ref. 18.

analogy to calculations in Fig. 2, for a noninteracting gas of ^{4}He within a carbon nanotube. Note the overall similarity of these two figures.

In Fig. 2,[18] one observes that the low T limit is $C^* = 1/2$, corresponding to a 1D classical gas, as expected. At high T, instead, the limit is $C^* = 1$; this limit is interpreted as that of a 2D gas moving on the inner surface of the nanotube. The bump at intermediate T is a general property found for spectra, such as that of the rigid rotor,[19] for which the inter-level spacing increases with quantum number. In the present case, the relevant spectrum is that arising from the azimuthal kinetic energy,

$$E_\theta = (\hbar\nu)^2/(2m\langle\rho\rangle^2) \equiv \nu^2 k_B \Theta \tag{2}$$

Here $\nu = 0, \pm1, \pm2, \ldots$ is the azimuthal quantum number and ρ is the radial coordinate, while Θ is defined as a temperature characteristic of azimuthal excitation. The peak in the specific heat occurs at a temperature near 3 Θ for both radii considered in Fig. 2, so its position serves as a benchmark from which one can determine the value of $\langle\rho\rangle$. Note that inside a nanotube, the relevant value of $\langle\rho\rangle$ is typically

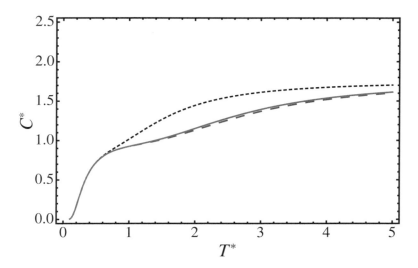

Fig. 3. Reduced specific heat C* for the gases He (dashed-blue), H₂ (solid-red) and Ne (short-dashed-black) between two nearly parallel nanotubes, each with $R = 0.7$ nm, and a divergence half-angle $\gamma = 0.5$ degrees. From Ref. 1.

$R - \sigma$, where σ is the gas-surface hard core interaction length. The key qualitative difference between the behaviors seen in Figs. 1 and 2 is that the experimental data in Fig. 1 plunge to $C^* = 0$ as $T \to 0$. This is an effect of quantum degeneracy, manifested in Nernst's law, as found in explicit calculations which revise Fig. 2 by taking quantum statistics into account.[20]

There have been many theoretical studies of gas adsorption in the presence of nanotubes.[8–12] In most treatments of these systems, one assumes that neighboring tubes are parallel. In that case, there exists a region of space- the so-called "groove"- which is a 1D channel with a strongly attractive potential, created by the adjacent tubes. The adsorbed gas then exhibits Q1D behavior at low N. However, one can also inquire about the case when the tubes are not quite parallel, but instead diverge, leaving a particularly attractive region between them, with a minimum potential energy (V_0) located at equilibrium position r_0. In a forthcoming study,[1] we will report remarkable results for this geometry, as exemplified in Fig. 3. The low T behavior is that of a gas in a 1D harmonic potential, $V \simeq V_0 + k_z(z - z_0)^2/2$, where k_z is the force constant for particle motion parallel to the z axis, midway between and nearly parallel to the tubes' axes. We introduce the characteristic temperature for this motion, $T_z \equiv \hbar(k_z/m)^{1/2}/k_B$ and a reduced temperature $T^* \equiv T/T_z$. For $T^* \ll 1$, the specific heat is of Arrhenius form, as seen in the figure, while for $T^* \sim 1$, $C^* \sim 1$, the specific heat of a 2D gas. That behavior might not have been anticipated, at first glance, because for small divergence half-angle γ, one might have expected 1D behavior, i.e., $C^* = 1/2$. Another surprise is the high T^* limiting behavior, $T^* \to 7/4$. This peculiar result arises because the relevant particles'

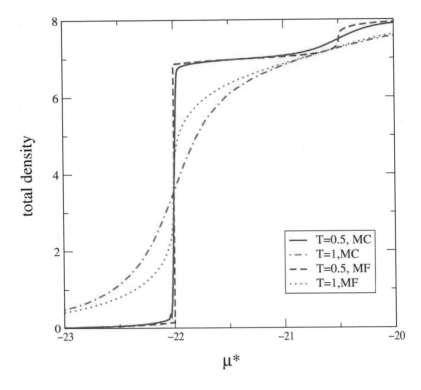

Fig. 4.   Mean field isotherms compared with exact Monte Carlo isotherms as a function of reduced chemical potential and temperature, from Ref. 21.

motions are those in the plane perpendicular to the $x - z$ plane of the nanotubes, for which the potential variation is unusual- proportional to y^4.

3. Many Interacting Particles in a Channel of Finite Width

From the perspective of phase transitions, a channel of finite width is a Q1D system, so that no true thermodynamic singularities can occur. This means that many attempts to explore behavior in porous media with single channel models (such as cylindrical and slit pores) cannot accurately describe phase transition behavior that is seen in genuinely 3D porous media. Nevertheless, these models may provide good semiqualitative predictive power, sufficient for most purposes since the fully 3D geometry is not known. We have explored a variety of such models, with several different goals. These include assessing the accuracy of simplifying models, such as mean-field-theory (MFT) and the use of periodic boundary conditions. Both of these approximations are suspect, at first glance, due to the important role of fluctuations in 1D statistical physics.

Fig. 4 presents adsorption isotherms for uptake in a single cylindrical pore, described by a lattice gas model.[21] In this model, continuous space is discretized into

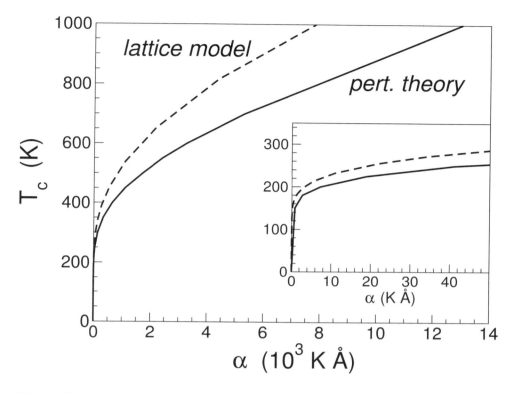

Fig. 5.   Critical temperature T_c as a function of the transverse interaction. Perturbation theory results and predictions from the anisotropic lattice gas model are compared. From Ref. 3.

adsorption sites, which may be either occupied or empty in any microstate of the system. In the specific model used here, the set of sites (within one transverse section) consists of one axial site and seven "cylindrical shell" sites, corresponding to positions near the inner boundary of the nanotube. The ensemble of sites include an infinite sequence of such layers of eight sites. The energy of the system includes interactions between particles occupying these sites, plus interactions between particles and the substrate host.

Fig. 4 compares (numerically) exact results with those obtained from MFT. The results are quite similar, overall, given the highly expanded scale of reduced chemical potential (μ^*). Note that the spurious transition seen in MFT (at reduced temperature $T = 0.5, \mu^* = -22$) is not very different from the nearly discontinuous isotherm seen in the exact results.

4. Real Transitions of Gases within Weakly Coupled Q1D Channels

Consider a geometry consisting of a set of parallel Q1D fluids, as in gases within or between nanotubes comprising a bundle of such tubes. While no transition can occur for an *isolated* Q1D system, once coupling between such systems is present, a finite temperature transition can occur. Fig. 5 presents results for such a geometry.[3] The critical temperature T_c is shown as a function of the transverse coupling α, for the case of buckyballs confined within parallel nanotubes. As indicated, rather similar results were obtained from exact solutions for a lattice-gas model and from a perturbation theory. In the latter case, the unperturbed equation of state was the exact 1D result for a model with Lennard-Jones interactions between the buckyballs and the perturbation was the weak van der Waals interaction between balls in adjacent tubes.

Particularly striking in this figure is the singular behavior for small α, as blown-up in the inset. This behavior is well-known for the lattice-gas, for which the transition temperature satisfies (for weak coupling)[22]

$$k_B T_c = \frac{2J_l}{ln(1/c) - ln[ln(1/c)]} \qquad (3)$$

Here J_l is the longitudinal interaction and $c = J_t/J_l$ is the anisotropy ratio of transverse to longitudinal interaction strengths. The singular behavior reflects the divergent susceptibility of the 1D system at low T; as the correlation length diverges, larger regions of fluid in adjacent channels are coupled, so increases rapidly with increasing J_t.The other notable result in this formula and the figure is that the characteristic energy and T_c scale are given by the *longitudinal* coupling, which is about 500 K for the peapod case, whereas naively one might have expected to be proportional to J_t.

Acknowledgments

We are grateful to DOE and NSF for support of this research and to S. J. Full and N. Wada for help with figures.

References

1. S. J. Full, J. P. McNutt, M. W. Cole, M. Mbaye and S. M. Gatica, Anomalous effective dimensionality of quantum gas adsorption near nanopores, to appear in J. Phys.:Cond. Mat.
2. J. Vavro, M. C. Llaguno, B. C. Satishkumar, D. E. Luzzi, and J. E. Fisher, *Appl. Phys. Lett.* **80**, 1450 (2002); D. E. Luzzi, B. W. Smith, *Carbon* **38**, 1751 (2000); D. J. Hornbaker et al., *Science* **295**, 828 (2002).
3. M. M. Calbi, S. M. Gatica, and M. W. Cole, *Phys. Rev.* **B67**, 205417 (2003).
4. N. Wada and M. W. Cole, *J. Phys. Soc. Japan* **77**, 111012 (2008).
5. J. Taniguchi, et al, *Phys. Rev. Lett.* **94**, 065301 (2005).
6. Y. Matsushita, T. Matsushita, R. Toda, M. Hieda and N. Wada, AIP Conf. Proc. 850, 297 (2006).
7. V.A. Yurovsky, M. Olshanii, and D. S. Weiss, *Adv. Opt At. Mol. Phys.* **55**, 61 (2007).
8. M. M. Calbi, S. M. Gatica, M. J. Bojan, G. Stan and M. W. Cole, *Rev. Mod. Phys.* **73**, 857 (2001).

9. A. C. Dillon and M. J. Heben, *Appl. Phys. A Mater.* **72**, 133 (2001).

10. M. M. Calbi, M. W. Cole, S. M. Gatica, M. J. Bojan and J. K. Johnson, in *Adsorption by Carbons*, ed. E. J. Bottani and Juan M. D. Tascón (Elsevier Science Publishing, 2008), pp. 187-210.

11. J. K. Johnson and M. W. Cole, in *Adsorption by Carbons*, op cit, pp. 369-402.

12. S. M. Gatica, M. M. Calbi, R. D. Diehl and M. W. Cole, *J. Low Temp. Phys.* **152**, 89 (2008).

13. We exclude the case of an infinitely long-range interaction, for which mean-field theory correctly predicts a transition in 1D. See M. Kac, G. E. Uhlenbeck and P. C. Hemmer, *J. Math. Phys.* **4**, 216 (1963).

14. A. D. Migone, in *Adsorption by Carbons*, op cit, Chapter 16.

15. S. M. Gatica, M. J. Bojan, G. Stan, and M. W. Cole, *J. Chem. Phys.* **114**, 3765 (2001).

16. M. M. Calbi, S. M. Gatica, M. J. Bojan and M. W, Cole, *J. Chem. Phys.* **115**, 9975 (2001).

17. R. Toda, M. Hieda, T. Matsushita, N. Wada, J. Taniguchi, H. Ikegami, S. Inagaki and Y. Fukushima, *Phys. Rev. Lett.* **99**, 255301 (2007).

18. G. Stan, and M. W. Cole, *Surf. Sci.* **395**, 280 (1998).

19. R. K. Pathria, *Statistical Mechanics*, second edition (Butterworth-Heinemann, Oxford, 1996), Fig. 6.5.

20. G. Stan, S. M. Gatica, M. Boninsegni, S. Curtarolo, and M. W. Cole, *Am. J. Phys.* **67**, 1170 (1999).

21. R. A. Trasca, M. M. Calbi, M. W.Cole, and J. L. Riccardo, *Phys. Rev.* **E69**, 011605 (2004).

22. M. E. Fisher, *Phys.Rev.* **162**, 480 (1967); T. Graim and D. P. Landau, *Phys. Rev.* **B24**, 5156 (1981).

SPECTRAL PROPERTIES OF MOLECULAR OLIGOMERS. A NON-MARKOVIAN QUANTUM STATE DIFFUSION APPROACH

JAN RODEN

Max-Planck-Institut für Physik Komplexer Systeme,
Nöthnitzer Str. 38, D-01187 Dresden, Germany
roden@mpipks-dresden.mpg.de

WALTER T. STRUNZ

Institut für Theoretische Physik, Technische Universität Dresden,
D-01062 Dresden, Germany
Walter.Strunz@tu-dresden.de

ALEXANDER EISFELD

Max-Planck-Institut fr Physik komplexer Systeme,
Nöthnitzer Str. 38, D-01187 Dresden, Germany
eisfeld@mpipks-dresden.mpg.de

Received 6 December 2009

Absorption spectra of small molecular aggregates (oligomers) are considered. The dipole-dipole interaction between the monomers leads to shifts of the oligomer spectra with respect to the monomer absorption. The line-shapes of monomer as well as oligomer absorption depend strongly on the coupling to vibrational modes. Using a recently developed approach [Roden *et al.*, PRL 103, 058301] we investigate the length dependence of spectra of one-dimensional aggregates for various values of the interaction strength between the monomers. It is demonstrated, that the present approach is well suited to describe the occurrence of the J- and H-bands.

Keywords: J-aggregates; H-band; quantum state diffusion; Frenkel excitons.

1. Introduction

Due to their unique optical and energy-transfer properties molecular aggregates have attracted researchers for decades.[1-3] In recent years there has been a growing interest in one-dimensional molecular aggregates consisting of a small number N (ranging from only two to a few tens) of molecules,[4-10] where finite size effects play an important role. Such systems are e.g. dendrimers,[8] stacks of organic dye molecules[5] or the light harvesting complexes in photosynthesis.[11-13] Usually in these aggregates there is negligible overlap of electronic wavefunctions between neighboring monomers. However, there is often a strong (transition) dipole-dipole interaction, that leads to eigenstates where an electronic excitation is coherently delocalized over many molecules of the aggregate, accompanied by often drastic changes in the aggregate absorption spectrum with respect to the monomer spectrum. Since these changes depend crucially on the conformation of the aggregate, commonly the first step in the investigation of such aggregates is optical spectroscopy. Important properties, like the strength of the dipole-dipole interaction can often already be deduced from these measurements.[4,9,14,15] For small aggregates, often referred to as oligomers, the finite size leads to pronounced effects in their optical properties.

The interpretation of measured absorption spectra is complicated by the fact that the electronic excitation couples strongly to distinct vibrational modes of the molecules and to the environment, leading to a broad asymmetric monomer absorption spectrum that often also shows a vibrational progression.[4,14] Because of this (frequency dependent) interaction with a quasi-continuum of vibrations, the theoretical treatment, even of quite small oligomers (of the order of five to ten monomers), becomes a challenge. Exact diagonalization[5,6,16–18] is restricted to a small (one to five) number of vibrational modes. Furthermore, to compare with experiments, the resulting "stick-spectra" have to be convoluted with some line-shape function to account for the neglected vibrations. For large aggregates the "coherent exciton scattering" (CES) approximation, which works directly with the monomer absorption line-shape as basic ingredient,[19] has been shown to give very good agreement with experiment.[14,20,21] However, one expects problems for small ($N < 10$) oligomers.[6]

To treat oligomers with strong coupling to vibrational modes, recently an efficient approach was put forward[22] which is based on a non-Markovian stochastic Schrödinger equation. In the present paper this approach will be used to investigate the absorption of linear one-dimensional oligomers.

The paper is organized as follows: In section 2 the Hamiltonian of the aggregate is introduced followed by a brief review of our method to calculate the absorption spectrum in section 3. The calculated oligomer spectra are presented and discussed in section 4. We conclude in section 5 with a summary and an outlook.

2. The Model Hamiltonian

In this work we will use a commonly applied model of a molecular aggregate including coupling of electronic excitation to vibrations.[9,23,24] We consider an oligomer consisting of N monomers. The transition energy between the electronic ground and electronically excited state of monomer n is denoted by ε_n. We will investigate absorption from the state where all monomers are in their ground state. A state in which monomer n is excited and all other monomers are in their ground state is denoted by $|\pi_n\rangle$. States in which the whole aggregate has more than one electronic excitation are not considered. We expand the total Hamiltonian

$$H = H_{\text{el}} + H_{\text{int}} + H_{\text{env}} \tag{1}$$

of the interacting monomers and the vibrational environment with respect to the electronic "one-exciton" states $|\pi_n\rangle$. The purely electronic Hamiltonian

$$H_{\text{el}} = \sum_{n,m=1}^{N} \left(\varepsilon_n \delta_{nm} + V_{nm} \right) |\pi_n\rangle\langle\pi_m| \tag{2}$$

contains the interaction V_{nm} which describes exchange of excitation between monomer n and m. The environment of each monomer is taken to be a set of harmonic modes described by the environmental Hamiltonian

$$H_{\text{env}} = \sum_{n=1}^{N} \sum_{j} \hbar\omega_{nj} a_{nj}^{\dagger} a_{nj}. \tag{3}$$

Here a_{nj} denotes the annihilation operator of mode j of monomer n with frequency ω_{nj}. The coupling of a local electronic excitation to its vibrational environment is assumed to be linear and the corresponding interaction Hamiltonian is given by

$$H_{\text{int}} = -\sum_{n=1}^{N} \sum_{j} \kappa_{nj} (a_{nj}^{\dagger} + a_{nj}) |\pi_n\rangle\langle\pi_n|, \tag{4}$$

where the coupling constant κ_{nj} scales the coupling of the excitation on monomer n to the mode j with frequency ω_{nj} of the local vibrational modes. This interaction is conveniently described by the spectral density

$$J_n(\omega) = \sum_{j} |\kappa_{nj}|^2 \delta(\omega - \omega_{nj}) \tag{5}$$

of monomer n. It turns out that the influence of the environment of monomer n is encoded in the bath correlation function,[23] which at temperature T reads

$$\alpha_n(\tau) = \int d\omega \, J_n(\omega) \left(\cos(\omega\tau) \coth\frac{\hbar\omega}{2k_B T} - i\sin(\omega\tau) \right). \tag{6}$$

In this work we restrict the discussion to the case $T = 0$ in which the bath-correlation function $\alpha_n(\tau)$ reduces to

$$\alpha_n(\tau) = \int d\omega \, J_n(\omega) e^{-i\omega\tau} = \sum_{j} |\kappa_{nj}|^2 e^{-i\omega_{nj}\tau}. \tag{7}$$

When the environment has no "memory", i.e. $\alpha_n(\tau) \propto \delta(\tau)$ it is termed Markovian, otherwise it is non-Markovian.

3. Method of Calculation

We use a recently developed method to describe the interaction of the excitonic system with a non-Markovian environment[22] which is based on ideas from a stochastic Schrödinger equation approach to open quantum system dynamics.[25,26] Within this approach it is possible to exactly describe the dynamics governed by a Hamiltonian of the form of Eq. (1) by a stochastic Schrödinger equation in the Hilbert space of the electronic system alone. However, due to the appearance of functional derivatives in this stochastic Schrödinger equation, the general method is of limited practical use. To circumvent this problem in Ref. 22 these functional derivatives were approximated by operators $D^{(m)}(t,s)$ which are independent of the stochastics and can be obtained from a set of coupled differential equations.

Within this approximation the cross-section for absorption of light with frequency Ω at zero-temperature is simply given by

$$\sigma(\Omega) = \frac{4\pi}{\hbar c} \Omega \operatorname{Re} \int_0^\infty dt\, e^{i\Omega t} \langle \psi(0)|\psi(t)\rangle \tag{8}$$

with initial condition

$$|\psi(0)\rangle = \sum_{n=1}^N (\hat{\mathcal{E}} \cdot \vec{\mu}_n)|\,\pi_n\,\rangle, \tag{9}$$

where the orientation of the monomers enters via the transition dipoles $\vec{\mu}_n$ and the polarization of the light $\hat{\mathcal{E}}$. The state $|\psi(t)\rangle$ is obtained from solving a Schrödinger equation in the small Hilbert space of the excitonic system

$$\partial_t|\psi(t)\rangle = -\frac{i}{\hbar}H_{\mathrm{el}}|\psi(t)\rangle + \sum_m |\pi_m\rangle\langle\pi_m|\bar{D}^{(m)}(t)|\psi(t)\rangle \tag{10}$$

with initial condition (9). The operator $\bar{D}^{(m)}(t)$ appearing on the right hand side of equation (10) contains the interaction with the environment upon excitation of monomer m in an approximate way,[22] and is given by

$$\bar{D}^{(m)}(t) = \int_0^t ds\, \alpha_m(t-s)D^{(m)}(t,s), \tag{11}$$

where $\alpha_m(t-s)$ is defined in Eq. (7) and the operator $D^{(m)}(t,s)$ is obtained by solving

$$\partial_t D^{(m)}(t,s) = \left[-\frac{i}{\hbar}H_{\mathrm{el}}, D^{(m)}(t,s)\right] + \sum_l \left[|\pi_l\rangle\langle\pi_l|\bar{D}^{(l)}(t), D^{(m)}(t,s)\right], \tag{12}$$

with initial condition $D^{(m)}(t=s,s) = -|\pi_m\rangle\langle\pi_m|$.

This method allows a numerically fast and simple treatment of an assembly of interacting monomers with a complex environment.

4. Results

Using the method described in the previous section we will now investigate the dependence of oligomer absorption spectra on the number N of monomers and the interaction strength between them. To this end we choose a one dimensional arrangement of the monomers and take the transition energies to be equal, $\varepsilon_n = \varepsilon$. We take only coupling between nearest neighbors into account, which we assume

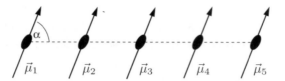

Fig. 1.   Sketch of the geometry considered for the case $N = 5$. The points mark the positions of the monomers and the arrows indicate the direction of the respective transition dipoles.

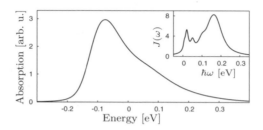

Fig. 2.   Absorption spectrum of a single monomer taken in the calculations. The inset shows the corresponding spectral density (unit 10^{-2} eV/$\hbar$).

to be equal for all neighboring monomers, i.e. $V_{n,n+1} = V$ for all n. For simplicity we also take all transition dipole moments $\vec{\mu}_n$ to be identical and parallel to the polarization of the light. This arrangement is sketched in Fig. 1. Before including the complexity of the vibrations we will briefly discuss the purely electronic case, given by Eq. 2. For the arrangement described above the Hamiltonian Eq. 2 can be diagonalized analytically to obtain the eigenenergies $E_j = \varepsilon + 2V \cos(\pi j/(N + 1))$ where the sign and magnitude of V depend crucially on the angle α between the transition dipoles and the axis of the oligomer. The oscillator strength F_j for absorption from the electronic ground state of the oligomer to an excited state with energy E_j is given by[15,27,28]

$$F_j = \frac{1 - (-1)^j}{N + 1} \operatorname{ctg}^2 \frac{\pi j}{2(N + 1)} \tag{13}$$

where the oscillator strength of a monomer is taken to be unity. From this one sees that the state with $j = 1$, which is located at the edge of the exciton band, carries nearly all the oscillator strength. For large N it carries roughly 81% of the oscillator strength; for $N < 7$ even more than 90%. Depending on the sign of V this state is either shifted to lower energies (for $V < 0$) or to higher energies (for $V > 0$) with respect to the electronic transition energy ε. Note that with increasing number of monomers N this state is shifted further away from the monomer transition energy, approaching for $N \to \infty$ the value $E_1 = 2V$. Note also that the absorption spectra

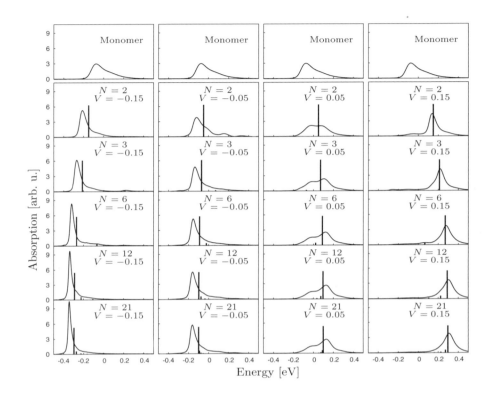

Fig. 3.   Absorption spectra of linear oligomers for different numbers of monomers N and various coupling strength V (in eV). The solid curves are calculated using the method described in section 3 for the spectral density of Fig.2; sticks are obtaind from the purely electronic Hamiltonian.

for V and $-V$ are just mirror images of each other. This will change drastically when one includes coupling to vibrations.

For the vibrational environment we consider a continuum of frequencies so that the spectral density $J_n(\omega)$ becomes a smooth function. Exemplarily we will consider the monomer spectrum shown in Fig. 2. This spectrum is obtained from a spectral density (shown in the inset of Fig. 2) which is composed of a sum of Lorentzians. The zero of energy is located at the transition energy ε. Due to the strong interaction with the vibrations this spectrum is considerably broadened. It also shows a pronounced asymmetry (it is much steeper at the low energy side). This strong asymmetry is mainly due to coupling to high energy vibrations with vibrational energies around $0.18\,\mathrm{eV}$, which is roughly the value of a C=C stretch mode in organic molecules.

We will now take a closer look at the dependence of the oligomer spectra on the number N of monomers for various values of the interaction strength V. Each column of Fig. 3 shows, for a distinct value of the dipole-dipole interaction V, a sequence of absorption spectra with increasing length N of the oligomer from top to bottom. The values of V and N are indicated in the individual plots. On top of each column the monomer spectrum from Fig. 2 is shown again for comparison. In addition to the spectra calculated including all vibrational modes also "stick-spectra" resulting from the purely electronic theory are shown.

The values chosen for the interaction strength V between neighboring monomers are $V = \pm 0.15$ eV (column 1 and 4) and $V = \pm 0.05$ eV (column 2 and 3). These values are of the order of magnitude as what is found in experiment.[4,5]

As explained above, the purely electronic stick spectra for opposite sign (but same magnitude) of V are mirror images with respect to each other. However, as can clearly be seen, this is no longer the case for the spectra including coupling to the vibrations. Although, in accordance with sum rules,[19,24,29] the mean of each absorption spectrum is centered at the mean of the corresponding purely electronic spectrum, the line-shape is completely different for negative and positive interaction V.

We will now discuss the dependence of the absorption spectra on N and V in more detail. For large negative interaction $V = -0.15$ eV (first column) the absorption line-shape considerably narrows upon increasing the number N of monomers of the oligomer. This exemplifies the appearance of the famous so-called J-band,[2] a very narrow red-shifted peak found in the absorption spectra of large dye-aggregates. It shows a strong asymmetry, with a steep slope at the low energy side and a long tail extending to higher energies. Due to this asymmetry of the line-shape the peak maximum appears shifted to lower energies with respect to the purely electronic "stick". For weaker interaction $V = -0.05$ eV (second column) the narrowing can already be seen, but is less pronounced than for $V = -0.15$ eV. Furthermore, when increasing the number of monomers from 3 to 6 there is little change in the absorption spectrum for $V = -0.05$ eV, while for $V = -0.15$ eV there is still a noticeable narrowing. In contrast to the narrowing found for negative values of V, for the same absolute value $|V|$ but positive V, the spectra are much broader and show little resemblance with their negative counterparts. This is the typical case of the so-called H-band, which is blue-shifted with respect to the monomer absorption.

Note that for the present values of V there is only minor change in the absorption spectrum for $N > 6$, which indicates that for larger N finite size effects will no longer be of great importance.

5. Summary and Outlook

In this work we have investigated the properties of one-dimensional linear oligomers using a recently developed approach based on a stochastic Schrödinger equation[22] to handle coupling of the electronic excitation to vibrations. Considering the length dependence of the oligomer spectra as well as the dependence on the dipole-dipole interaction between the monomers, we found that this approach, although approximate, is well suited to describe the basic features like the narrow J-band and the broad H-band. While in this work discussion was restricted to zero temperature, we plan to study the temperature dependence with the present method and compare to experiment.

Acknowledgments

A. E. acknowledges discussions with S. Trugman and thanks Gerd for the great hospitality in Quito.

References

1. J. Franck and E. Teller. *J. Chem. Phys.* **6**, 861–872 (1938).
2. T. Kobayashi, editor. *J-Aggregates.* World Scientific (1996).
3. D. M. Eisele, J. Knoester, S. Kirstein, J. P. Rabe, and D. A. Vanden Bout. *Nature Nanotechnology* **4**, 658–663 (2009).
4. B. Kopainsky, J. K. Hallermeier, and W. Kaiser. *Chem. Phys. Lett.* **83**, 498–502 (1981).
5. J. Seibt, T. Winkler, K. Renziehausen, V. Dehm, F. Würthner, H.-D. Meyer, and V. Engel. *J. Phys. Chem. A* **113**, 13475–13482 (2009).
6. J. Roden, A. Eisfeld, and J. S. Briggs. *Chemical Physics* **352**, 258–266 (2008).
7. M. Wewer and F. Stienkemeier. *Phys. Rev. B* **67**, 125201 (2003).
8. C. Supritz, V. Gounaris, and P. Reineker. *J. Luminescence* **128**, 877 – 880 (2008).
9. P. O. J. Scherer. Molecular Aggregate Spectra. In T. Kobayashi, editor, *J-Aggregates.* World Scientific (1996).
10. J. Guthmuller, F. Zutterman, and B. Champagne. *J. Chem. Theory Comput.* **4**, 2094–2100 (2008).
11. R. van Grondelle and V. I. Novoderezhkin. *Phys. Chem. Chem. Phys.* **8**, 793 – 807 (2006).
12. P. Rebentrost, R. Chakraborty, and A. Aspuru-Guzik. *J. Chem. Phys.* **131**, 184102 (2009).
13. T. Renger, V. May, and O. Khn. *Physics Reports* **343**, 137 – 254 (2001).
14. A. Eisfeld and J. S. Briggs. *Chem. Phys.* **324**, 376–384 (2006).
15. H. Fidder, J. Knoester, and D. A. Wiersma. *J. Chem. Phys.* **95**, 7880–7890 (1991).
16. R. L. Fulton and M. Gouterman. *J. Chem. Phys.* **41**, 2280–2286 (1964).
17. J. Bonča, S. A. Trugman, and I. Batistić. *Phys. Rev. B* **60**, 1633–1642 (1999).
18. F. C. Spano. *J. Chem. Phys.* **116**, 5877–5891 (2002).
19. J. S. Briggs and A. Herzenberg. *J.Phys.B* **3**, 1663–1676 (1970).
20. A. Eisfeld and J. S. Briggs. *Chem. Phys. Lett.* **446**, 354–358 (2007).
21. A. Eisfeld, R. Kniprath, and J. S. Briggs. *J. Chem. Phys.* **126**, 104904 (2007).
22. J. Roden, A. Eisfeld, W. Wolff, and W. T. Strunz. *Phys. Rev. Lett.* **103**, 058301 (2009).
23. V. May and O Khn. *Charge and Energy Transfer Dynamics in Molecular Systems.* WILEY-VCH (2000).
24. H. van Amerongen, L. Valkunas, and R. van Grondelle. *Photosynthetic Excitons.* World Scientific, Singapore (2000).
25. L. Diósi and W. T. Strunz. *Phys. Lett. A* **235**, 569–573 (1997).
26. L. Diósi, N. Gisin, and W. T. Strunz. *Phys. Rev. A* **58**, 1699–1712 (1998).
27. E. W. Knapp. *Chem. Phys.* **85**, 73–82 (1984).
28. V. Malyshev and P. Moreno. *Phys. Rev. B* **51**, 14587–14593 (1995).
29. A. Eisfeld. *Chem. Phys. Lett.* **445**, 321–324 (2007).

QUANTUM PROPERTIES IN TRANSPORT THROUGH NANOSCOPIC RINGS: CHARGE-SPIN SEPARATION AND INTERFERENCE EFFECTS

K. HALLBERG

Instituto Balseiro and Centro Atómico Bariloche, CNEA, CONICET,
8400 Bariloche, Argentina
karen@cab.cnea.gov.ar

JULIAN RINCON

Instituto Balseiro and Centro Atómico Bariloche, CNEA, CONICET,
8400 Bariloche, Argentina

S. RAMASESHA

Solid State and Structural Chemistry Unit, Indian Institute of Science, Bangalore 560 012, India

Received 2 December 2009

Many of the most intriguing quantum effects are observed or could be measured in transport experiments through nanoscopic systems such as quantum dots, wires and rings formed by large molecules or arrays of quantum dots. In particular, the separation of charge and spin degrees of freedom and interference effects have important consequences in the conductivity through these systems.

Charge-spin separation was predicted theoretically in one-dimensional strongly interacting systems (Luttinger liquids) and, although observed indirectly in several materials formed by chains of correlated electrons, it still lacks direct observation. We present results on transport properties through Aharonov-Bohm rings (pierced by a magnetic flux) with one or more channels represented by paradigmatic strongly-correlated models. For a wide range of parameters we observe characteristic dips in the conductance as a function of magnetic flux which are a signature of spin and charge separation.

Interference effects could also be controlled in certain molecules and interesting properties could be observed. We analyze transport properties of conjugated molecules, benzene in particular, and find that the conductance depends on the lead configuration. In molecules with translational symmetry, the conductance can be controlled by breaking or restoring this symmetry, e.g. by the application of a local external potential.

These results open the possibility of observing these peculiar physical properties in anisotropic ladder systems and in real nanoscopic and molecular devices.

Keywords: Charge-spin separation; quantum interference; strong correlations.

1. Introduction

New artificial structures made using nanotechnology allow for the possibility of tailoring systems with novel physical properties. For example, the Kondo effect was achieved in a system consisting of one quantum dot (QD) connected to leads;[1-3]

systems of a few QD's have been proposed theoretically as realizations of the two-channel Kondo model,[4,5] the ionic Hubbard model,[6] and the double exchange mechanism.[7] Also, the correlation-driven metal-insulator transition has been studied in a chain of QD's.[8]

In addition, molecular systems also pose a challenging scenario as seen from various successful attempts.[9–11] The possibility of achieving controlled quantum transport has been studied, for example in electronic conductance through single π-conjugated molecules using theoretical[12–16] and experimental[17–20] techniques.

Several interesting properties arise from these artificial and molecular systems. On one hand, artificial nanorings of nanoladders with strong interacting electrons could be assembled to observe the elusive property of charge-spin separation (SCS), characteristic of strongly correlated low-dimensional systems. Among the theoretical methods for detecting and visualizing SCS, direct calculations of the real-time evolution of electronic wave packets in finite Hubbard rings revealed different velocities in the dispersion of spin and charge densities as an immediate consequence of SCS.[21] Also, Kollath and coworkers[22] have repeated this calculation for larger systems using the Density Matrix Renormalization Group (DMRG) technique[23] and observed distinct features of SCS in a model for one-dimensional cold Fermi gases in a harmonic trap, proposing quantitative estimates for an experimental observation of SCS in an array of atomic wires.

Other calculations were done in Refs. 24 and 25, where the authors analyzed the transmission through infinite Aharonov-Bohm (AB) rings. The motion of the electrons in the ring was described by a Luttinger liquid (LL) propagator with different charge and spin velocities, v_c and v_s, included explicitly. With this assumption the flux-dependence of the transmission has, in addition to the periodicity in multiples of the flux quantum $\Phi_0 = hc/e$, new structures which appear at fractional values of the flux. Numerical calculations of the transmittance through finite AB rings described by the $t - J$ and Hubabrd models show clear dips at the fluxes that correspond to Eq. (4).[26–29] The extension of these results to ladders of two legs was also considered, as a first step to higher dimensions.[30] In essence, these structures arise because transmission requires the separated spin and charge degrees of freedom of an injected electron to recombine at the drain lead after traveling through the ring in the presence of the AB flux. In this work we will review these results for rings and ladder rings.

On the other hand, quantum interference might play a crucial role in transport measurement through molecules and could be used as a handle to control conductance through such systems. In two recent theoretical works,[14,15] the idea of a Quantum Interference Effect Transistor (QuIET) based on single annulene molecules, including benzene, was proposed. The equilibrium conductance at zero bias and gate voltage (V_g) was calculated in both papers for the strong coupling limit using the non-equilibrium Green's function and the Landauer-Büttiker formalism. However, annulenes have a gap at the energy corresponding to $V_g = 0$ due mainly to the strong Coulomb interactions present in the molecule ($N = 4n + 2$

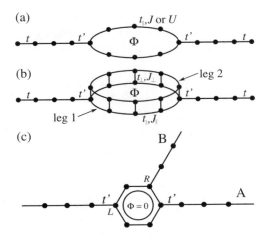

Fig. 1.   Schematic representation of the systems considered for electronic transport. a) An interacting ring modelled by the $t - J$ and Hubbard Hamiltonians; b) a ladder ring represented by the
$t - J$ model and c) conjugated molecules with different number of Carbon atoms and two different
lead configurations.

annulenes like benzene, are closed-shell molecules with no level at zero energy even
in the absence of interactions). For zero gate voltage the conductance is finite, albeit
small, only for strong coupling to the leads. For weak coupling the conductance will
be appreciable only through the main channels of the molecule which are a few eV
away from the Fermi energy. By analyzing the QuIET at the main transmittance
channels the switching effect will be much more pronounced and robust as we will
show below.

In this paper we will show results for the conductance through the three systems
depicted in Fig. 1: a) an interacting ring, b) an interacting ladder ring and c)
conjugated molecules.

2. The Method

Before introducing the models we will describe the method used for the calculation
of the transmittance.

The systems considered are sketched in Fig. 1. They consist of interacting rings
weakly coupled to two non-interacting leads. Our model Hamiltonian reads $H =
H_{\text{leads}} + H_{\text{link}} + H_{\text{ring}}$, where H_{leads} describes free electrons in the left and right
leads,

$$H_{\text{link}} = -t' \sum_{\sigma} (a^{\dagger}_{-1,\sigma} c_{L,\sigma} + a^{\dagger}_{1,\sigma} c_{R,\sigma} + \text{H.c.}) \tag{1}$$

describes the exchange of quasiparticles between the leads ($a_{i,\sigma}$) and particular sites
of the ring, and H_{ring} depends on the modelling of the ring.

Following Ref. 24, the transmission from the left to the right lead can be calculated to second order in t' where the ring is integrated out. The resulting effective Hamiltonian is equivalent to a one-particle model for a non-interacting chain
with two central sites modified by the interacting ring, with effective on-site energy
$\epsilon(\omega) = t'^2 G^R_{L,L}(\omega)$ and effective hopping between them $\tilde{t}(\omega) = t'^2 G^R_{L,R}(\omega)$. $G^R_{i,j}(\omega)$
denotes the Green's function of the isolated ring.

The transmittance and conductance of the system at zero temperature may then
be computed using the effective impurity problem. The transmittance $T(\omega)$ is given
by[24]

$$T(\omega, V_g, \phi) = \frac{4t^2 \sin^2 k \, |\tilde{t}(\omega)|^2}{\left| [\omega - \epsilon(\omega) + t e^{ik}]^2 - |\tilde{t}^2(\omega)| \right|^2}, \tag{2}$$

where $\omega = -2t \cos k$ is the tight-binding dispersion relation for the free electrons in
the leads which are incident upon the impurities. These equations are exact for a
non-interacting system.

From Eq. (2), $T(\omega, V_g, \phi)$ may be calculated from the Green functions of the
isolated ring. We consider holes incident on a ring of L' sites and $N = N_e + 1$
electrons in the ground state, obtaining the Green functions from the ground state
of the ring modelled by Eqs. (3), (5) and (6) below, and calculated using numerical
diagonalization[31] Then we substitute these in Eq. (2). We fix the energy $\omega = 0$
to represent half-filled leads and explore the dependence of the transmittance on
the threading flux, obtained by integration over the excitations in a small energy
window, which accounts for possible voltage fluctuations and temperature effects.[26]

3. Charge-spin Separation

3.1. *Conductance through rings*

In this section we will calculate the transmittance through a ring (Fig. 1a) for which
the Hamiltonian reads:

$$H_{\text{ring}} = -eV_g \sum_{l,\sigma} c^\dagger_{l,\sigma} c_{l,\sigma} - t_\| \sum_{l,\sigma} (c^\dagger_{l,\sigma} c_{l+1,\sigma} e^{-i\phi/L'} + \text{H.c.}) + H_{\text{int}}. \tag{3}$$

where the flux is given in units of the flux quantum $\phi = 2\pi\Phi/\Phi_0$ and the system is
subjected to an applied gate voltage V_g.

For the Hubbard model $H_{\text{int}} = \sum_{l,\sigma\neq\sigma'} U n_{l,\sigma} n_{l,\sigma'}$, with $n_{l,\sigma} = c^\dagger_{l,\sigma} c_{l,\sigma}$. This
model is related to the also well known $t - J$ model for very large interactions by
$J = 4t_\|^2/U$ where a similar behaviour to the one described below is also found.[27]

In the case of infinite on-site repulsion U (or equivalently $J = 0$), the wave function can be factorized into a spin and a charge part.[26,32,33] Therefore, charge-spin
separation becomes evident for finite systems and independent of system size. For
each spin state, the system can be mapped into a spinless model with an effective
flux which depends on the spin. Considering a non-degenerate ground state containing $N = N_e + 1$ particles and analyzing the part of the Green function that

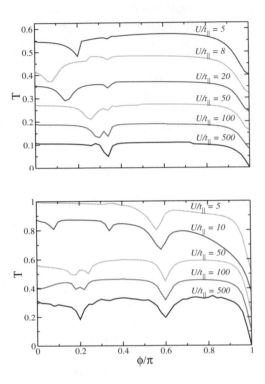

Fig. 2.   Transmittance vs flux for the Hubbard model in the 1D ring for $L = 6$ sites, and a) $N_e + 1 = 4$ and b) $N_e + 1 = 6$ particles in the ground state and several values of U (curves are shifted vertically for better visualization).

enters the transmittance when a particle is destroyed, it is shown that the dips occur when two intermediate states cross at a given flux and interfere destructively. These particular fluxes depend on the spin quantum numbers and are located at[27]

$$\phi_d = \pi(2n + 1)/N_e, \qquad (4)$$

with n integer. If the integration energy window includes these levels, a dip in the conductance arises.

In Fig. 2 we show results for the transmittance using the Hubbard model in the ring. For large interactions, $U \gg t_{\parallel}$, dips are found at the positions given by Eq. (4). However, when U is reduced increasing the mixture between different spin sectors, the spin-charge separation is affected and we observe the appearance of new dips and a shift of some of them. This is because the destructively interfering level crossings which lead to the reduction in the conductance occur at different values of the flux, as explained in Ref. 27.

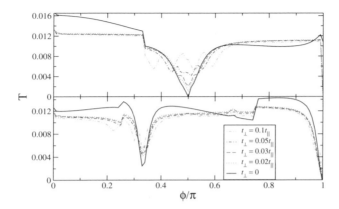

Fig. 3.   Flux-dependent transmittance for the anisotropic $t-J$ model with $J=0$, $t_\parallel = 0.1$, $L=6$ rungs, $t' = 0.05t_\parallel$, several values of $t_\perp$ and $N=6$ (top) and $N=8$ (bottom) electrons in the ground state.

3.2. *Conductance through ladder rings*

In this subsection we analyze the conductance through rings formed by two-leg ladder systems described by the $t-J$ model as a first step towards two dimensions (Fig. 1b). For this case the Hamiltonian of the interacting system reads:

$$H_{\text{ring}} = -eV_g \sum_{i,l,\sigma} c^\dagger_{i_l,\sigma} c_{i_l,\sigma} - t_\parallel \sum_{i,\sigma} (c^\dagger_{i_l,\sigma} c_{i_l+1,\sigma} e^{-i\phi/L'} + \text{H.c.})$$

$$- t_\perp \sum_{i,\sigma} (c^\dagger_{i_1,\sigma} c_{i_2,\sigma} + \text{H.c.}) + H_{\text{int}} \qquad (5)$$

The fermionic operators $c^\dagger_{i_l,\sigma}$ create an electron at site $i = 1, L'$ of leg $l = 1, 2$ with spin σ. The AB ring has L' rungs, is threaded by a flux ϕ ($\phi = 2\pi\Phi/\Phi_0$), and is subjected to an applied gate voltage V_g.

We first show results for weakly coupled chains ($t_\perp \ll t_\parallel$) and $J=0$, where we know that in 1D there is complete SCS.[26,32,33] In Fig. 3 we show the results for several small values of $t_\perp$ and in fact, observe clear dips at certain fractional values of the magnetic flux.

To understand the position of the dips for this case of weakly coupled rings we must consider the following: For the ladder with $t_\perp = 0$ and a total even number of electrons N in the ground state, the lowest-lying state has $N/2$ electrons in each leg. Since we are calculating the transmittance through one leg only, and the intermediate state has one particle less, one expects to see dips at $\phi_d = \pi \frac{2n+1}{N/2-1}$ from the condition for ϕ_d with $N_e = N/2 - 1$. We find this behaviour in Fig. 3: for the top figure there will be $N_e + 1 = 3$ electrons in each leg, leading to a dip at $\phi = \pi/2$ and for the bottom figure there will be 4 electrons in each leg leading to a dip at $\phi = \pi/3$. When $t_\perp \neq 0$, we find that the dips remain and are quite robust,

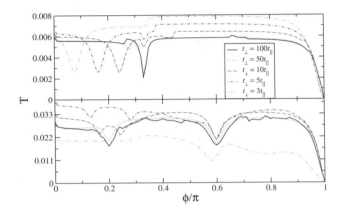

Fig. 4.   Transmittance as a function of flux for $J = 0$. Top: $t' = 0.05t_{\parallel}$ and $N = 4$ electrons. Bottom: $t' = 0.09t_{\parallel}$ and $N = 6$ electrons.

even for values of $t_{\perp}/t_{\parallel}$ as high as 0.1.

For strongly coupled chains, $i.e$ with a large coupling between the legs, $t_{\perp} \gg t_{\parallel}$, the bands corresponding to each leg are very far apart in the non-interacting case and one might expect the reappearance of SCS. This is, in fact, the case as can be seen in Fig. 4, where we plot the transmittance for a ladder with several values of $t_{\perp}$ and fillings.

For this case, the total number of electrons in the lower band corresponds to the total filling N (for a less than half filled band) and the transmittance will involve $N_e = N - 1$ electrons. Hence, the dips will be found at the fluxes $\phi_d = \pi \frac{2n+1}{N-1}$ as seen in Fig. 4 for large values of $t_{\perp}/t_{\parallel}$. For smaller values of $t_{\perp}$, we find a shift in the location of the minima and sometimes a splitting of the dips. We have understood this numerically for a smaller system by realizing that the dips correspond to certain level crossings which occur for particular values of the magnetic flux.[30]

For intermediate values of the perpendicular hopping the dips disappear. It is interesting to see that, for constant filling, we find dips at different values of the flux for weak and strong coupling. Finite J introduces an extra spin shuffling in the system, reducing the depth and shifting the position of the dips.

4. Quantum Interference

We consider here an annular π-conjugated molecules with N sites, weakly connected to non-interacting leads in the A or B configurations (Fig. 1c). Now the interacting Hamiltonian H_{ring} describes the isolated π-conjugated molecule, modelled by the

PPP Hamiltonian,[34] with on-site energy given by a gate voltage V_g:

$$H_{\text{ring}} = -eV_g \sum_{i=1,\sigma}^{N} c_{i\sigma}^{\dagger} c_{i\sigma} - t \sum_{\langle ij \rangle, \sigma} c_{i\sigma}^{\dagger} c_{j\sigma} + \sum_i U_i \left(n_{i\downarrow} - \frac{1}{2} \right) \left(n_{i\uparrow} - \frac{1}{2} \right)$$

$$+ \sum_{i>j} V_{ij} (n_i - 1)(n_j - 1) \tag{6}$$

where the operators $c_{i\sigma}^{\dagger}$ ($c_{i\sigma}$) create (annihilate) an electron of spin σ in the π orbital of the Carbon atom at site i, n are the corresponding number operators and $\langle \cdots \rangle$ stands for bonded pairs of Carbon atoms. The intersite interaction potential V_{ij} is parametrized so as to interpolate between U and e^2/r_{ij} in the limit $r_{ij} \longrightarrow \infty$.[35] In the Ohno interpolation, V_{ij} is given by

$$V_{ij} = U_i \left(1 + 0.6117 \, r_{ij}^2 \right)^{-1/2} \tag{7}$$

where the distance r_{ij} is in Å. The standard Hubbard parameter for sp^2 Carbon is $U_i = 11.26$ eV and hopping parameter t for $r = 1.397$ Å is 2.4 eV,[36] and all energies are in eV.

For the isolated benzene molecule, which has translational symmetry, the allowed total momentum quantum numbers are $k = 2r\pi/(4n+2) = r\pi/(2n+1)$, with r an integer. For a two-terminal set up like the one considered here, wave functions travelling through both branches of the molecule will interfere producing different interference patterns depending on the positions of the leads. The phase difference will be momentum times the difference in the lengths of the two trajectories (in units of the C-C separation): $\Delta\phi = k\Delta x$. For leads in the "para" position, $\Delta x = 0$ and the waves are in phase, interfering constructively (Fig. 1c, A configuration). However, in the "meta" (B configuration) position ($\Delta x = 2$), the interference will depend on the k value of the particular channel. For the highest occupied molecular orbital (HOMO) and lowest unoccupied molecular orbital (LUMO) the phase differences will be $\Delta\phi = 2\pi/3$ and $\Delta\phi = 4\pi/3$ respectively and the interference will reduce the amplitude in these channels.

In Fig. 5 we show results for the conductance through benzene molecules in the "para" and "meta" configurations of the leads. The nearly destructive interference effect for the latter case is clearly visible.

In view of the reduction of the HOMO (or LUMO) peaks for the "meta" configuration, one can ask whether disrupting translational invariance would lead to a larger conductance. This question was first addressed in Ref. 14 by introducing a local energy (Σ) at one site in benzene, the real part of which would produce elastic scattering and its imaginary part, decoherence. These authors focused on the Fermi energy of the leads (set to zero) where the observed effect is small. In 37 we studied the effect of external perturbations on the main transmittance channels such as the HOMO and LUMO and we found a much larger response. These results are shown in the bottom part of Fig. 5, where an additional diagonal energy is added to the site to the right of the B ("meta") position (the effect is not qualitatively dependent

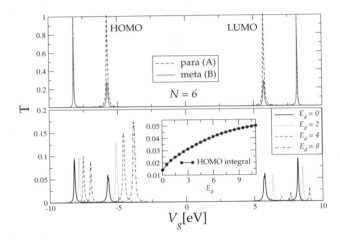

Fig. 5.   Top: Transmittance vs. gate voltage (measured from the Fermi energy) through a benzene molecule for the "para" (A substitution in Fig. 1c) and "meta" (B substitution in Fig. 1c) configurations for $t' = 0.4$ and a finte Lorentzian width $\eta = 0.03$. Bottom: Same for the "meta" configuration in the presence of different on-site potentials which break the translational invariance. Inset: evolution of the weight of the transmission peak through the HOMO as a function of the on-site potential.

on this position). It is clearly seen that, in this case, the small peak corresponding to the HOMO level develops and grows as the local energy is increased, disrupting the translational symmetry responsible for the destructive interference (see inset). We also find that this effect is much more striking for larger annulenes. These results are not affected by molecular vibrations at room temperature since modes that can cause decoherence are excited at temperatures higher than $500K$.[14]

5. Conclusion

We have reviewed results on the existence of dips in the conductance through finite strongly correlated low-dimensional systems which arise as a consequence of nontrivial destructive interference effects at fractional values of the flux quantum Φ_0. This feature is a strong indication of the existence of SCS in these systems. We have presented results for the transmittance through interacting one-dimensional and anisotropic ladder rings. In all cases the dip structure is robust against finite interactions (small J's or large U for the $t - J$ or Hubbard models respectively). However, we find new dips and shifts of their positions with respect to the ideal scenario of complete charge-spin separation. We also find that the dip structure, originally predicted for 1D systems, is still present for two transmission channels modelled by a ladder system in the anisotropic limit. For a wide range of parameters, in particular for weak and strong hoppings across the rungs $t_\perp$, the dips remain, but they disappear for intermediate values of this parameter. These findings open the possibility of measuring this peculiar phenomenon in real nanoscopic systems or artificial structures, such as rings of quantum dots on the sub-μm scale, where the magnetic fields needed for this kind of experiments become accessible.

In addition, we have analyzed the resonant conductance through the HOMO and LUMO channels of benzene in the weak lead-molecule coupling regime. We find a strong dependence on the source-drain configuration and on the molecular geometry due to quantum interference.

Acknowledgments

This investigation was sponsored by PIP 5254 of CONICET and PICT 2006/483 of the ANPCyT. We are partially supported by CONICET.

References

1. D. Goldhaber-Gordon, H. Shtrikman, D. Mahalu, D. Abusch-Magder, U. Meirav, and M.A. Kastner, Nature **391**, 156 (1998).
2. S. M. Cronenwet, T. H. Oosterkamp, and L. P. Kouwenhoven, Science **281**, 540 (1998).
3. W.G. van der Wiel, S. De Franceschi, T. Fujisawa, J.M. Elzerman, S. Tarucha and L.P. Kouwenhoven, Science **289**, 2105 (2000).
4. Y. Oreg and D. Goldhaber-Gordon, Phys. Rev. Lett. **90**, 136602 (2003).
5. R. Žitko and J. Bonča, Phys. Rev. B **74**, 224411 (2006).
6. A. A. Aligia, K. Hallberg, B. Normand, and A. P. Kampf, Phys. Rev. Lett. **93**, 076801 (2004).
7. G. B. Martins, C. A. Bsser, K. A. Al-Hassanieh, A. Moreo, and E. Dagotto, Phys. Rev. Lett. **94**, 026804 (2005).
8. L. P. Kouwenhoven, F. W. J. Hekking, B. J. van Wees, C. J. P. M. Harmans, C. E. Timmering and C. T. Foxon, Phys. Rev. Lett. **65**, 361 (1990).
9. A. Aviram, M. A. Ratner, Chem. Phys. Lett. **29**, 277 (1974).
10. M. A. Reed, M. A. Reed, C. Zhou, C. J. Muller, T. P. Burgin, J. M. Tour, Science **278**, 252 (1997).
11. G. Cuniberti, G. Fagas, K. Richter (eds), Introducing Molecular Electronics, Springer, Berlin **2005**.
12. A. Nitzan, M. A. Ratner, Science **300**, 1384 (2003).
13. N. Tao, Nature Nanotech. **1**, 173-181 (2006).
14. D. Cardamone, C. Stafford, S. Mazumdar, Nano Lett. **6**, 2422 (2006).
15. S-H. Ke, W. Yang, H. Baranger, Nano Lett. **8**, 3257 (2008).
16. S. Yeganeh, M. Ratner, M. Galperin, A. Nitzan, Nano Lett. **9**, 1770 (2009).
17. J. Park, J.Park, A.N.Pasupathy, J.I.Goldsmith, C.Chang,. Y.Yaish, J.R.Petta, M.Rinkoski, J.P.Sethna, H.D.Abruna, P.L.McEuen and D.C.Ralph, Nature **417**, 722-725 (2002).
18. L. Venkataraman, J. E. Klare, C. Nuckolls, M. S. Hybertsen and M. L. Steigerwald, Nature **442**, 904 (2006).
19. A. V. Danilov, S. Kubatkin, S. Kafanov, P. Hedegard, N. Stuhr-Hansen, K. Moth-Poulsen, and T. Bjornholm, Nano Lett. **8**, 1 (2008).
20. T. Dadosh, T. Dadosh, Y. Gordin, R. Krahne, I. Khivrich, D. Mahalu, V. Freydman, J. Sperling, A. Yacoby, I. Bar, Nature **436**, 677 (2005).
21. E. A. Jagla, K. Hallberg, and C. A. Balseiro, Phys. Rev. B **47**, 5849 (1993).
22. C. Kollath, U. Schollwoeck and W. Zwerger, Phys. Rev. Lett. **95**, 176401 (2005)

23. S. White, Phys. Rev. Lett. **69**, 2863 (1992);K. Hallberg, New Trends in Density Matrix Renormalization, Adv. in Phys. **55**, 477 (2006); U. Schollwck, The density-matrix renormalization group, Rev. Mod. Phys. **77**, 259 (2005)
24. E. A. Jagla and C. A. Balseiro, Phys. Rev. Lett. **70**, 639 (1993).
25. S. Friederich and V. Meden, Phys. Rev. B **77**, 195122 (2008).
26. K. Hallberg, A. A. Aligia, A. Kampf and B. Normand, Phys. Rev. Lett. **93**, 067203 (2004).
27. J. Rincón, A. A. Aligia and K. Hallberg, Phys. Rev. B **79**, 035112 (2009).
28. J. Rincón, A. A. Aligia and K. Hallberg, Physica B **404**, 2270 (2009).
29. J. Rincón, K. Hallberg, and A. Aligia, Physica B **404**, 3147 (2009).
30. J. Rincón, K. Hallberg and A. A. Aligia, Phys. Rev. B **78**, 125115 (2008).
31. E. R. Davidson, J. Comput. Phys. **17**, (1975) 87; Comput. Phys. Comm. **53**, 49 (1989).
32. M. Ogata and H. Shiba, Phys. Rev. B **41**, 2326 (1990).
33. W. Caspers and P. Ilske, Physica A **157**, 1033 (1989); A. Schadschneider, Phys. Rev. B **51**, 10386 (1995).
34. R. Pariser, R. Parr, J. Chem. Phys. **21**, 466 (1953); J. A. Pople, Trans. of the Faraday Soc., **49**, 1375 (1953).
35. K. Ohno, Theor. Chim. Acta **2**, 219 (1964).
36. Z. G. Soos, and S. Ramasesha, Phys. Rev. B **29**, 5410 (1984).
37. J. Rincón, K. Hallberg and S. Ramasesha, Phys. Rev. Lett. in print (2009).

COOPERATIVE LOCALIZATION-DELOCALIZATION IN THE HIGH T_c CUPRATES

JULIUS RANNINGER

Institut Néel, CNRS and Université Joseph Fourier, 25 rue des Martyrs, BP 166
38042 Grenoble cedex 9, France
julius.ranninger@grenoble.cnrs.fr

Received 17 December 2009

The intrinsic metastable crystal structure of the cuprates results in local dynamical lattice instabilities, strongly coupled to the density fluctuations of the charge carriers. They acquire in this way simultaneously both, delocalized and localized features. It is responsible for a partial fractioning of the Fermi surface, i.e., the Fermi surface gets hidden in a region around the anti-nodal points, because of the opening of a pseudogap in the normal state, arising from a partial charge localization. The high energy localized single-particle features are a result of a segregation of the homogeneous crystal structure into checker-board local nano-size structures, which breaks the local translational and rotational symmetry. The pairing in such a system is dynamical rather than static, whereby charge carriers get momentarily trapped into pairs in a deformable dynamically fluctuating ligand environment. We conclude that the intrinsically heterogeneous structure of the cuprates must play an important role in this type of superconductivity.

Keywords: Localization-delocalization; Fermi surface fractioning; local symmetry breaking; crystalline metastability.

1. Introduction

The long standing efforts to synthesize superconductors with critical temperatures higher than about 25 K gradually faded away as the decades past. Three quarters of a century after the first discovery of superconductivity by Kamerlingh-Onnes[1] in 1911 (Hg with a T_c 4.25 K), a ceramic material of not particularly good materials qualities surfaced in 1986 and Bednorz and Mueller produced this long searched after goal.[2] The inhomogeneities of these materials, which seem to be intrinsic, are related to local dynamical lattice instabilities, and could be a prime factor to bypass the limitation of T_c[3] in classical low temperature BCS type phonon-mediated superconductors.[4] Local dynamical lattice instabilities are known to trigger diamagnetic fluctuations[5] without leading to any global translational symmetry breaking, which would kill the superconducting state and end up in a charge ordered phase. Maximal values of T_c can be expected for materials, which are synthesized at the highest temperature in the miscibility phase diagram,[6] see Fig. 1. They present thermodynamically stable single-phase solutions at the boundary of insulating and metallic regions, which upon cooling condense into single-phase solid solutions. Their high kinetic stability prevents them from decomposing into the different compositions of the mixture one started with, due to a freezing-in of the high entropy mismatch of thermodynamically stable phases in the synthesization process. Microscopically, their metastability arises from adjacent cation-ligand complexes with incompatible inter-atomic distances.

Following such a strategy in material preparations, $BaBi_xPb_{1-x}O_3$,[7] $Ba_{1-x}K_xBiO_3$ [8,9] and $Pb_{1-x}Tl_xTe$,[10,11] just to mention a few representative examples, were synthesized and exhibited superconducting phases with $T_c = 13, 30, 1.5$ K. These are relatively high values, considering their small carrier density of around 10^{-20}/cm. Their parent compounds are respectively $BaBiO_3$, which is a diamagnetic insulating charge ordered state and PbTe, which is a small gap semiconductor. Their non-metallic phases are turned into superconductors upon partially substituting Bi by Pb, Ba by K and Te by Tl. According to band theory, $BaBiO_3$ should be metallic, composed of exclusively Bi^{4+}. But the high polarizibility of the oxygens[12] makes it synthesize in a mixture of Bi^{5+} and Bi^{3+}, rendering Bi^{4+} an unstable valence state. Such materials are sometimes referred to as "valence skippers". Bi^{5+} occurs in a regular octahedral ligand environment, with a Bi–O distance of 2.12 Å. Bi^{3+}, on the contrary, occurs in a pseudo-octahedral ligand environment, with one of the oxygen ions in the octahedral being displaced to such an extent (with a corresponding Bi–O distance of 2.28 Å) that it effectively becomes O^{2-}, after having transferred an electron to Bi^{4+}.[13] This charge disproportionation results in negative U centers - the Bi^{3+} sites on which electrons pair up. The undoped parent compound stabilizes in a translational symmetry broken state of alternating Bi^{3+} and Bi^{5+} ions. Upon doping (partially replacing Bi by Pb), the system becomes superconducting with locally fluctuating $[Bi^{3+} \Leftrightarrow Bi^{5+}]$, see Fig. 2. The situation is similar in $Pb_{1-x}Tl_xTe$. Its parent compound, PbTe, involves unstable Te^{2+} valence cations, which disproportionate into Te^{1+} and Te^{3+} ions, with the first again acting as negative U centers. Upon doping PbTe $\to Pb_{1-x}Tl_xTe$ it changes into a superconductor, driven by those negative U centers.

2. The Scenario

Cuprates High T_c materials bare some similarity to such socalled valence skipping compounds. Upon doping, their anti-ferromagnetic half-filled band Mott-insulating parent compound, composed of Cu^{II} - O - Cu^{II} bonds in the CuO_2 planes, becomes dynamically locally unstable and leads to dynamically fluctuating Cu^{II} - O - $Cu^{II} \Leftrightarrow Cu^{III}$ - O - Cu^{III} bonds.[14,15] Different from the socalled valence skippers mentioned above, the basic building stones are covalent molecular bonds. This does not imply Cu valencies of 2+, respectively 3+. The notation II and III indicates stereochemical configurations defined by intrinsic Cu-O bond-lengths, which are 1.93 Åfor Cu^{II} - O - Cu^{II} and 1.83 Åfor Cu^{III} - O - Cu^{III}. Because of stereochemical misfits between the CuO_2 planes and the surrounding cation-ligand complexes in the neighboring insulating planes, acting as charge reservoirs, the Cu - Cu distance of the Cu^{II} - O - Cu^{II} bonds in a square planar oxygen ligand environment

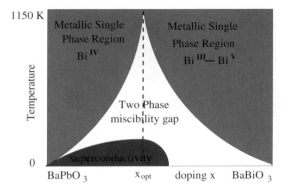

Fig. 1.   Phase diagram for the $BaBi_xPb_{1-x}O_3$ synthesization (after Ref. 6).

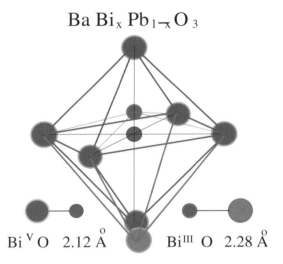

Fig. 2.   The two stable Bi^V and Bi^{III} configurations in regular and pseudo octahedral ligand environments (after Ref. 13).

are forced to reduce their lengths. They do that by a bond buckling, where the bridging oxygen of those bonds moves out of the $CDuO_2$ plane and thus respects its attributed stereochemical distance. Upon hole doping, the static bond buckling becomes dynamic, involving unbuckled linearly Cu^{III}-O-Cu^{III} bonds. The bridging oxygens then fluctuate in and out of the CuO_2 plane and thus dynamically modulate the length of the Cu-O bonds. Experimentally, this feature is manifest in a splitting of the corresponding local bond stretch mode, which sets in in low doped metallic regime and disappears upon overdoping.[16] The two stereo-chemical configurations Cu^{II} - O - Cu^{II} and Cu^{III} - O - Cu^{III}, observed in EXAFS,[17] differ by two charges and it is that which results in the dynamically fluctuating diamagnetic bonds, which are the salient features of phase fluctuation driven superfluidity of the

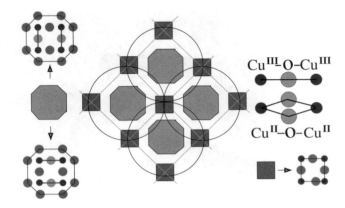

Fig. 3.   Schematic view of the local segregation of the CuO_2 planes into rotational symmetry breaking nano-size clusters (grey) (after Kohsaka *et al.* Ref. 19), containing dynamically deformable Cu-O-Cu bonds which act as pairing centers. These pairing centers are embedded in a metallic matrix of Cu_4O_4 square (brown) plaquettes, on which the charge carriers circle around the central trapping centers, indicated by black circles. The small red filled circles indicate the Cu ions and the blue filled larger ones the oxygens. The fluctuating bridging oxygens are indicated by grey filled circles.

cuprates.[18] Inspite of the huge differences in the Cu-O bond-length $\simeq 0.1$ Åof the Cu-O-Cu constituents, the cuprates do not break the overall translational symmetry caused by a charge order, but prefer to segregate into a local distribution of such bonds. They stabilize in a checkerboard structure of nano-size clusters, surrounded by small molecular Cu_4O_4 sites, such that on a local basis those systems form a bipartite structure (see an idealized form for it in Fig. 3). The dynamically fluctuating uni-directional Cu - O - Cu bonds exist in two equivalent orthogonal directions and thus break not only the local translational but also rotational symmetry.[19] This prevents any tendency to long range translational symmetry breaking.

One might have expected such a result from the outset. Superconductivity in the cuprates is destroyed by phase fluctuations of the order parameter, without that its amplitude would suffer any significant depreciation.[20,21] This can only happen when Cooper pairs are fairly local entities, such as to provide superfluidity like features with a T_c scaling with the superfluid density at zero temperature[18] and XY characteristics of the transition,[22] in strong contrast to the mean field characteristics of standard amplitude fluctuation controlled BCS superconductivity. The local nature of the Cooperons also implies that pairing correlations must develop already well above T_c. Because of the charged and local nature of the particles involved in this pairing, its has a strong effect on the local lattice deformations to which they are irrevocably coupled.

In principle one can now envisage two scenarios:
(i) if the coupling is very strong, the charge carriers form bipolarons, which in principle can condense into a superfluid phase - the Bipolaronic Superfluidity,[23] but

which are more likely to end up in an insulating state of statically disordered localized bipolarons, if not a charge ordered state of them.[24]

(ii) if the the coupling is not quite as strong, the charge carriers will be in a mixture of quasi-free itinerant states and localized states, where they are momentarily self-trapped in form of local bound pairs. Local dynamical pairing then occurs at some temperature T^*, at which the single-particle density of states develops a pseudogap, because of the electrons getting paired up on a finite time scale.[25] Upon lowering the temperature, those phase incoherent locally dynamically fluctuating pairs acquire short range phase coherence, which ultimately leads to their condensation and phase locking in a superfluid state at some temperature T_ϕ, exclusively controlled by phase fluctuations.[26] Experimental evidence for that is now well established in terms of the transient Meissner effect,[27] the Nernst effect in thermal transport,[28] the torque measurements of diamagnetism[29] and a proximity induced pseudogap of metallic films deposited on $La_{2-x}Sr_xCuO_4$.[30]

Standard low temperature BCS superconductors are controlled by amplitude fluctuations, with the feature that at T_c the expectation value of the amplitude vanishes, i.e., the number of Cooper pairs goes to zero and thus the phase of the order parameter becomes redundant. For superfluid He, the opposite is the case. The onset of superfluidity is controlled exclusively by spatial phase locking of the bosonic order parameter. The amplitude there is fixed, since the number of bosonic He atoms is conserved. The superfluid state is destroyed at a certain T_ϕ, given by the phase stiffness of the condensate. The high T_c superconductors lie between those two limiting cases. Amplitude and phase fluctuations are then strongly inter-related and have to be treated on equal footing. The onset of a finite amplitude of the order parameter happens at a temperature T^*, while that of the phase locking of the bosonic Cooperons, signaling the onset of superconductivity, occurs at T_ϕ, which can be well below T^*. Dealing with such a situation requires to disentangle the description of phase and amplitude fluctuations of the Cooperons. Cooperons are neither true bosons nor hard-core bosons, but their center of mass can be associated to bosons in real space. Using a renormalization procedure one can project out such bosons of the Cooperons, leaving us with remnants arising from kinematical interactions due to their non-bosonic statistics coming from their internal fermionic degrees of freedom. That leads to a damping of such bosons. Those bosons have an amplitude and a phase and hence can account for both: (i) their dissociation into fermion pairs at T^*, where their amplitude disappears together with the closing of the pseudogap in the fermionic single-particle subsector and (ii) the breaking up their spatial phase coherence at T_ϕ, where the bosons enter into a phase disordered boson metallic state, albeit keeping short range phase coherence in a finite temperature regime above T_ϕ. This scenario follows some old ideas in the early days of the BCS theory, when exploring the interaction between the electrons and the Cooper pairs in a perturbative way,[31] or, in a more recent study, using a phenomenological picture for strong inter-relations between the two.[32] In the approach we have been following now for several years, the inter-relation between phase and amplitude fluc-

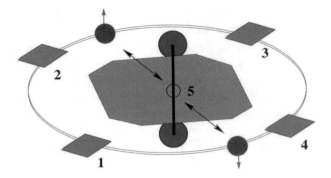

Fig. 4.   Simplified version of a local cluster of Fig. 3, composed of (i) a trapping domain Cu_4O_{10}, replaced here by a deformable dumbbell oscillator (green filled circles) at site 5, capable of trapping two electrons (red filled circles) in form of a bipolaron and (ii) four Cu_4O_4 square plaquettes, which are replaced by effective sites 1 - 4, on which the electrons circle around the central site 5.

tuations is cast into Andreev scattering of the fermions on bosons. It results in a partial trapping of the fermions in form of localized bound fermion pairs in real space, and leads to fermionic excitations, which are in a superposition of itinerant and localized states, consistent with recent STM imaging results.[19] The pairing we are considering here corresponds to a resonant pairing, induced by fluctuating molecular bonds driven by intrinsic dynamical lattice instabilities. The mechanism is similar to that of a Feshbach resonance[33] in fermionic atom gases, where fermionic atoms exist simultaneously in form of scattering electron-spin singlet states and as bound electron-spin triplet states.

3. The Model

Resonant pairing systems can most efficiently be described by a Boson Fermion Model (BFM), such as I had conjectured shortly after the proposition of the Bipolaronic Superconductivity.[23] This phenomenological model represents a mixture of itinerant fermions and localized bosons —locally bound fermion pairs— with an exchange interaction between the two. A common chemical potential is assumed such as to assure that we are dealing with a unique species of charge carriers. They must exist simultaneously in (i) itinerant states circling around the trapping centers and (ii) as localized states on the trapping centers, the moment they pair up. This picture is quite general and does not depend on any specific mechanism for pairing. It can arise from holes in a strongly correlated Hubbard system close to half filling,[34] where the exchange happens between hole singlet-pairs and itinerant uncorrelated holes[35] or, as we have been pursuing it for years, from a polaronic mechanism.

Our picture of hole-doped cuprates is based on an ensemble of effective sites, embedded in a crystalline structure, each one being composed of a deformable cation-ligand complex, surrounded by an undeformable local environment, the four-site rings, see Fig. 3. The local environments of neighboring deformable molecular clusters overlap with each other, which assures a potentially metallic phase of the charge carriers. If the coupling of electrons into bosonic bound pairs outweighs their kinetic energy, it leads to a Bose liquid, which can be either superfluid or insulating, without that any fermions are present in the background. High T_c materials do however exhibit such a background Fermi sea and at the same time a transition into a superfuid state which is governed by the dynamics of bosonic fermion-pairs rather than by the opening up of a gap in the electronic single-particle density of states. Itinerant bosons are induced upon hole doping. They form and disappear spontaneously on the deformable cation-ligand clusters and thereby break the crystal symmetry on a local level. This resonating pairing originates from the local dynamical lattice instabilities and metastability of the cuprates. It implies that those systems are intrinsically inhomogeneous, having segregated, upon doping, into effective sites of molecular clusters, partially occupied by such bosons. Resonating pairs of fermions, implies electrons fluctuating in and out of those deformable clusters, on which they self-trap themselves in form of bound pairs on a finite time scale. This situation is energetically more favorable than that of purely localized bosonic bound fermion pairs or itinerant uncorrelated fermions. It benefits from a lowering of the ionic level due to a polaron level shift and from the bound fermion pairs acquiring itinerancy.

Let us now demonstrate that on the basis of a small cluster calculations, following a previous detailed study[37] on that. The segregation of the CuO_2 plane charge distribution, upon hole doping, has been seen in STM imaging studies[19] and a sketch for such segregation is illustrated in Fig. 3. It consists of nano-size Cu_4O_{10} domains composed of deformable Cu - O - Cu bonds which are capable of trapping momentarily two itinerant holes from the surrounding Cu_4O_4 plaquettes. Those latter form the remnant back bone structure corresponding to the undoped cuprates carrying the itinerant charge carriers. The fluctuation of the charge carriers back and forth between the two subcomponents of the segregated CuO_2 planes ultimately induces a diamagnetic component among the itinerant charge carriers.

4. Resonant Pairing and Local Dynamical Lattice Instabilities

The physics elaborated above is an intrinsic property of the nano-size Cu_4O_{10} clusters, making up those materials. We illustrate that by mapping the central Cu_4O_{10} trapping domain together with its four surrounding Cu_4O_4 plaquettes, into a tractable model: a central effective site given by a deformable oscillator surrounded by four atomic sites (see Fig. 4). On the first, the itinerant charge carriers can get self trapped into localized bipolarons, corresponding to a deformation which mimics the process Cu^{III} - O - Cu^{III} $\Leftrightarrow$ Cu^{II} - O - Cu^{II} in the trapping domains of Fig. 3. Resonant pairing occurs when the energies of the bipolaronic level corresponds to the Fermi energy of the itinerant fermionic subsystem. In our toy model we shall characterize these states by two-electron eigenstates on the four-site ring.

The Hamiltonian for such a small cluster system is given by

$$H = -t \sum_{i \neq j = 1...4,\, \sigma} \left[c_{i,\sigma}^\dagger c_{j,\sigma} + h.c. \right] - t^* \sum_{i = 1...4,\, \sigma} \left[c_{i,\sigma}^\dagger c_{5,\sigma} + h.c. \right]$$

$$+ \Delta \sum_\sigma c_{5,\sigma}^\dagger c_{5,\sigma} + \hbar \omega_0 \left[a^\dagger a + \frac{1}{2} \right] - \hbar \omega_0 \alpha \sum_\sigma c_{5,\sigma}^\dagger c_{5,\sigma} \left[a + a^\dagger \right], \quad (1)$$

where $c_{i\sigma}^{(\dagger)}$ denotes the annihilation (creation) operator for an electron with spin σ on site i, and $a_5^{(\dagger)}$ the phonon annihilation (creation) operator associated with a deformable cation-ligand complex at site 5. α denotes the dimensionless electron-phonon coupling constant, ω_0 the Einstein oscillator frequency of the local dynamical deformation on the central polaronic site 5, t the intra-ring hopping integral and t' the one controlling the transfer between the ring and the central site 5. Δ denotes the bare energy of the electrons when sitting on site 5. before they couple to the local lattice deformation. This local cluster describes the competition between localized bipolaronic electron pairs on the central cation-ligand complex and itinerant electrons on the plaquette sites, when the energies of the two configurations are comparable with each other, i.e. for $2\Delta - 4\hbar \omega_0 \alpha^2 \simeq -4t$. There is a narrow resonance regime where this happens, as we can see from the strong enhancement of the efficiency rate $F_{\text{exch}}^{\text{pair}}$

$$F_{\text{exch}}^{\text{pair}} = \langle GS | c_{5,\uparrow}^\dagger c_{5,\downarrow}^\dagger c_{q=0,\downarrow} c_{q=0,\uparrow} | GS \rangle - \langle GS | c_{q=0,\uparrow}^\dagger c_{5,\uparrow} | GS \rangle \langle GS | c_{q=0,\downarrow}^\dagger c_{5,\downarrow} | GS \rangle \quad (2)$$

for transferring electron pairs back and forth between the ring sites and the central polaronic site,[5] (see Fig. 5) and where we have subtracted out the exchange due to incoherent single-electron processes. $|GS\rangle$ denotes the ground state of the cluster. This resonant scattering process induces a diamagnetic component of the electrons moving on the ring. The density n_p of such diamagnetically correlated resonantly induced pairs is given by

$$n_p = \langle n_{q=0,\uparrow} n_{q=0,\downarrow} \rangle - \langle n_{q=0,\uparrow} \rangle \langle n_{q=0,\downarrow} \rangle \quad (3)$$

The variation of n_p with α (Fig. 6) shows an equally sharp enhancement near the resonance α_c, which depends on the adiabaticity ratio ω_0/t. Its relative density is quite sizable - about 20% of the total average density of electrons on the ring at α_c. Slightly away from this resonant regime, the electrons are either pair-uncorrelated (for $\alpha \leq \alpha_c$) or not existent on the plaquette (for $\alpha \geq \alpha_c$), because of being confined to the polaronic cation-ligand complex as localized bipolarons.

The features driving these electronic exchange are local deformations of the molecular dimer, which induce a correlated dynamics of the local dimer deformations and the local charge fluctuations. Denoting the modulation in length of the dimer by $X = \sqrt{\hbar/2M\omega_0}[a^+ + a]$, the corresponding phonon Greens function is

$$D_{ph} = \frac{2M\omega_0}{\hbar} \langle\langle X; X \rangle\rangle, \quad (4)$$

where M denotes the mass of the vibrating dimer atoms (green in Fig. 4). Our previous results[37] on this local lattice dynamics are reproduced in Fig. 7, where we

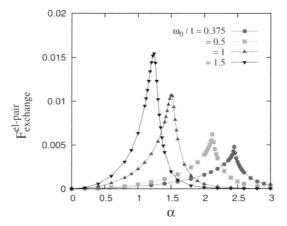

Fig. 5.   (Color on line) The efficiency factor $F_{\text{exch}}^{\text{pair}}$ as a function of α for converting a localized bipolaron on the central polaronic complex into a pair of diamagnetically correlated electrons on the ring, for several adiabaticity ratios ω_0/t (after Ref. 37).

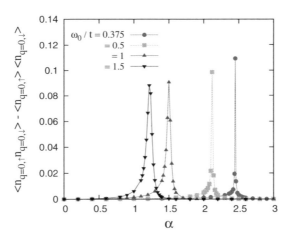

Fig. 6.   (Color on line) Density n_p of diamagneticaly correlated electron pairs on the ring as a function of α showing a strong enhancements for $\alpha = \alpha_c$ for a set of adiabaticity ratios ω_0/t (after Ref. 37).

illustrate the softening of the bare phonon mode ω_0 down to ω_0^R at the resonance. It is this frequency which determines the correlated charge-deformation fluctuations and provides us with an estimate for the effective mass of the diamagnetic pairs (see the discussion in the SUMMARY section of Ref. 37).

5. Separation of Phase and Amplitude Variables

Molecular clusters, like the one discussed in the previous section, form the building blocks of hole doped segregated CuO_2 planes and which, accordingly assembled,

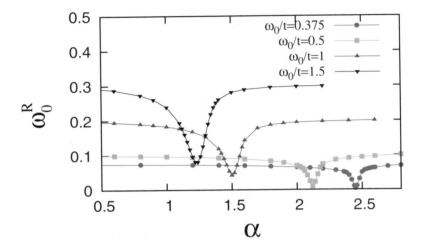

Fig. 7.   (Color on line) The phonon softening as a function of the electron-dimer coupling constant for various adiabaticity ratios $\omega_0/t = 0.375, 0.5, 1$ and 1.5 (after Ref. 37).

form the macroscopic cuprate structure: a regular bipartite lattice structure, idealized in Fig. 3. This picture is in accordance with the checker-board structure seen in STM imaging studies.[19] The local stability of such dynamically fluctuating molecular clusters is obtained at resonance, i.e., at α_c, which depends on the adiabaticity ratio but more sensibly on the exchange rate $\propto \omega_0^R$ between the localized pairs and the itinerant electrons. This latter will strongly depend on the degree of hole doping.

The microscopic mechanism for hole doping in the cuprates is a complex phenomenon and far from being understood. The undoped systems present a homogeneous lattice structure with buckled Cu^{II} - O - Cu^{II} molecular bonds. Due to strong Hubbard correlations among the electrons in such a half-filled band system, the ground state is an antiferromagnetic insulator. Upon slight hole doping it breaks down into a glassy heterogeneous structure, which we interpret in our scenario as a phase uncorrelated bipolaronic Mott insulator,[26,38] composed of Cu^{II} - O - Cu^{II} covalent molecular bonds, behaving as hardcore intersite bipolarons. Upon increasing the hole doping beyond around 0.10 per Cu ion, the system segregates in to a checkerboard structure with nano-size molecular Cu - O - Cu clusters, in a mixture of Cu^{II} - O - Cu^{II} and Cu^{III} - O - Cu^{III} molecular bonds. These nano-size clusters act as trapping centers for the itinerant electrons moving on the sublattice in which those cluster are embedded (see Fig. 3). Given the experimental findings on the effect of hole doping on the crystal structure monitored by an increase in covalency,[39] we conjecture that hole doping primarily induces hole-bipolarons (Cu^{III} - O - Cu^{III} molecular bonds) on the nano-size trapping centers. This triggers local charge fluctuations, by momentarily capturing two electrons from of the surrounding four-site ring environment, transforming Cu^{III} - O - Cu^{III} into Cu^{II} - O - Cu^{II} molecular bonds. The two initially uncorrelated electrons thereby acquire a diamagnetic component, which they keep upon returning into the itinerant subsystem via the inverse process $[Cu^{II} - O - Cu^{II}] \rightarrow [Cu^{III} - O - Cu^{III}]$.

This resonant back and forth scattering between bound and unbound electrons stabilizes the dynamical transfer of a small fraction of electrons n_F^h from the initially quasi half-filled band of itinerant charge carriers into the trapping centers. As a result, the Fermi surface shrinks, which implies a hole density $n_F^h = n_B$ of single-particle excitations. Since n_F^h varies only in a small regime, between 0.10 and 0,22, covering the superconducting phase, we choose for our resonant pairing scenario, a total number of charge carriers around $n_{tot} = n_F + n_B \simeq 1$, which will slightly decrease with hole doping. This will lead to a self-regulating boson density, such as $n_B \leq 0.08$ for the optimally doped case $n_F^h \simeq 0.16$. The electrons, being momentarily trapped into bound electron pairs on the deformable molecular clusters, can be be associated to bosonic variables $b^{(\dagger)}$ with $n_B = \langle b^\dagger b \rangle$. Our further analysis is based on the BFM which, in momentum space, is given by the Hamiltonian

$$H_{BFM} = H^0_{BFM} + H^{exch}_{BFM} \tag{5}$$

$$H^0_{BFM} = \sum_{k\sigma}(\varepsilon_k - \mu)c^\dagger_{k\sigma}c_{k\sigma} + \sum_q (E_q - 2\mu)b^\dagger_q b_q. \tag{6}$$

$$H^{exch}_{BFM} = (1/\sqrt{N})\sum_{k,q}(g_{k,q}b_q\dagger c_{q-k,\downarrow}c_{k,\uparrow} + H.c.) \tag{7}$$

$$H^{F-F}_{BFM} = \frac{1}{N}\sum_{p,k,q}U_{p,k,q}c^\dagger_{p\uparrow}c^\dagger_{k\downarrow}c_{q\downarrow}c_{p+k-q\uparrow}. \tag{8}$$

Following the basic electronic structure of the cuprates, we assume a d-wave symmetry for the boson-fermion exchange interaction $g_{k,q} = g[\cos k_x - \cos k_y]$, between (i) pairs of itinerant charge carriers $c^{(\dagger)}_{k\sigma}$ circling around the polaronic sites and (ii) the polaronicaly bound pairs of them $b^{(\dagger)}_q$, when these same charge carriers have hopped onto this site and got self trapped. The anisotropy of the itinerant charge carrier dispersion is assured by the standard expression $\varepsilon_k = -2t[\cos k_x + \cos k_y] + 4t' \cos k_x \cos k_y$ for the CuO_2 planes with $t'/t = 0.4$. The strength of the boson-fermion exchange coupling in our effective Hamiltonian can be estimated from our study of the single cluster problem and is given by F^{pair}_{exch}, which is related to the electron-lattice coupling α and the bare phonon frequency ω_0, as discussed in detail in Ref. 37. In order to get an insight into the inter-related amplitude and phase fluctuations in hole doped High T_c cuprates, we transform this Hamiltonian by a succession of unitary transformations into a block diagonal form, composed of exclusively (i) renormalized fermionic particles and (ii) renormalized bosonic particles. This gives us an access to study the characteristic single-particle spectral properties such as the pseudogap and the two-particle-properties, controlling the onset of phase locking and diamagnetic fluctuations, which govern the transport as one approaches T_c from above. Using Wegner's Flow Equation renormalization procedure[36] we eliminate the boson-fermion exchange coupling in successive steps, obtaining a Hamiltonian of the same structure as the initial one, Eqs (5–8), but with renormalized parameters $g^* = 0$, $\varepsilon_{\mathbf{k}}^*$, $E_{\mathbf{q}}^*$, $U_{\mathbf{p},\mathbf{k},\mathbf{q}}^*$, which have evolved out of the bare ones we started with, i.e., $g, \varepsilon_{\mathbf{k}}, E_{\mathbf{q}} = \Delta - \mu, U_{\mathbf{p},\mathbf{k},\mathbf{q}} = 0$. The chemical potential μ, which also evolves in the course of this renormalization procedure is determined at each step of it, such as to assure a given total density of fermionic and bosonic particles n_{tot}. From our detailed calculations[40] of this renormalized Hamiltonian and its spectral properties we see a transformation of the electron dispersion, close to the chemical potential, which changes into an S-like shape upon lowering the temperature below a certain T^*. It signals the onset of pairing and manifests itself in a corresponding opening of the pseudogap in the single-particle density of states. Simultaneously the intrinsically dispersion-less bosons acquire a q^2 like spectrum, albeit overdamped until, upon further decreasing the temperature, they finish up as well defined quasi-particle which ultimately condense into a superfluid phase below $T = T_\phi$. Their q^2 like spectrum transforms into a linear in q branch, which signals the phase locked superfluid state of those bosons.

The anisotropic d-wave structure of the boson-fermion exchange coupling implies a variation of its strength, going from zero at the nodal points $[\pm\pi/2, \pm\pi/2]$ of the Fermi surface toward its maximal value, equal to g, near the hotspots $\mathbf{k} = [0, \pm\pi]$, $[\pm\pi, 0]$. Upon moving on an arc in the Brillouin zone, corresponding to the chemical potential, i.e., $\varepsilon^*(k_x, k_y) = \mu$, one observes a well defined Fermi surface around the nodal points. But upon parting from this limited region, the Fermi surface gets hidden because of the onset of a pseudogap. This has been called "Fermi Surface Fractionation". It implies in this region of the Brillouin zone (see Fig. 8) three astonishingly related features: (i) diffusively propagating Bogoliubov branches below $\omega = 0$ for $\mathbf{k} \geq \mathbf{k}_F$, indicated by $A_{inc}^F(\mathbf{k}, \omega)$ (black lines). They are remnants of superconducting phase locking on a finite space-time scale, which have been predicted by us in 2003[41] and only very recently have been verified by ARPES in 2008,[42]; (ii) localized high energy modes above $\omega = 0$ (black) for $\mathbf{k} \geq \mathbf{k}_F$, which occur together with those diffusively propagating Bogoliubov modes. We have interpreted them as the internal degrees freedom of those two-particle collective phase modes, showing up in their high frequency response. They present locally partially trapped single-particle states making up the bipolaronic Cooperons,[40] which include; (iii) in-gap single-particle contributions, denoted by $A_{coh}^F(\mathbf{k}, \omega)$ (red delta like peaks). Their weight should be understood as the overall spectral weight for very broadened peaks, as we know from independent studies, such as self-consistent perturbative treatments[25] and DMFT procedures.[43] The characteristic three-peak structure of the single-particle spectral features for $\mathbf{k} \geq \mathbf{k}_F$ in this part of the Brillouin zone around the hidden Fermi surface is a fingerprint of resonant pairing systems, highlighting their deviations from BCS like superconductivity, for which the spectral features would invariably provide a two peak structure. The recent STM imaging studies are further proof of the non-BCS like features of the spectral properties attributed to in-gap single-particle contributions above T_c.

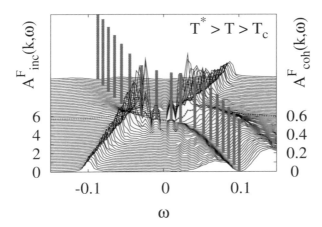

Fig. 8. The single-particle spectral function $A(\mathbf{k}, \omega) = A_{\text{inc}}(\mathbf{k}, \omega) + A_{\text{coh}}(\mathbf{k}, \omega)$ in the pseudogap phase around the hidden "Fermi surface" ($\omega = 0$) and near the anti-nodal point. The set of curves corresponds to different $\mathbf{k}$ vectors, orthogonally intersecting the hidden Fermi surface. The curve in blue indicates the spectrum exactly at this hidden Fermi surface (after Ref. 40).

6. Summary

Exploiting the intrinsic metastability of hole doped cuprates, we investigated a scenario of resonant pairing, driven by dynamical local lattice instabilities. It invokes charge carriers, which simultaneously exist as itinerant quasi-particles and as localized self-trapped bipolarons. The quantum fluctuations between the two configurations induce mutually (i) a certain degree of itinerancy of the bare localized bipolarons and (ii) localized features in the single-particle spectral function. This is evident in simultaneously featuring a low energy diffusive Bogoliubov branch below the chemical potential and single-particle excitations, accompanied by remnant bonding and anti-bonding states in the high energy response, above the chemical potential. These are finger prints of the non-BCS like physics of those cuprates, which should be interesting to test experimentally and link up with studies of anomalous lattice properties which are at the origin of the dynamical lattice instability driven segregation of the homogeneous cuprate lattice structure into checker-board structures.

Acknowledgments

I would like to thank my colleagues Tadek Domanski and Alfonso Romano for having participated in numerous theoretical studies on which this present attempt, to develop a global picture of the cuprates, is based. I thank Juergen Roehler, who kept me regularly informed about relevant experimental results as well and his critical and constructive comments, which helped me in my understanding of those materials.

References

1. H. Kamerlingh-Onnes, Leiden Comm. 120 b (1911).
2. J. G. Bednorz and K. A. Mueller, Z. Phys. B-Condensed Matter **64**, 189 (1986).
3. P. W. Anderson and B. T. Matthias, Science **144**, 373 (1964).
4. J. Bardeen, L. N. Cooper and J. R. Schrieffer, Phys. Rev, **108**, 1175 (1957).
5. J. M. Vandenberg and B. T. Matthias, Science **198**, 194 (1977).
6. A. W. Sleight, Physics Today,**44**(6), 24 (1991).
7. A. W. Sleight, J. L. Gillson and P. E. Bierstedt, Solid State Commun. **17**, 27 (1975)-
8. R. J. Cava et al., Nature (London) **332**, 814 (1988).
9. D. G. Hinks et al., Nature (London) **333**, 836 (1988).
10. I. A. Chernik and S Lykov Sov. Phys. Solid State **23**, 1724 (1981).
11. B. Ya. Moizhes and I. Drabkin Sov. Phys. Solid State **25**,1139 (1983).
12. I. Hase and T. Yanagisawa, Phys. Rev. B **76**, 174103 (2007).
13. A. Simon, Chem. Unserer Zeit **22**, 1 (1988).
14. Y. Kohsaka et al., Science **315**, 1380 (2007).
15. K. K. Gomes et al., Nature **447**, (2007).
16. D. Reznik et al., Nature **440**, 1170 (2006).
17. C. J. Zhang and H. Oyanagi, Phys. Rev. B **79**, 064521 (2009).
18. Y. J. Uemura et al., Phys. Rev. Lett., **62**, 2317 (1989).
19. Y. Kohsaka et al., Nature **454**, 1072 (2008).
20. V. J. Emery and S. A. Kivelson, Nature, **374**, 434 (1995).
21. M. Franz and A. J. Millis, Phys. Rev. B **58**, 14572 (1998).
22. M. B. Salamon et al., Phys. Rev. B **47**, 5520 (1993).
23. A. S. Alexandrov and J. Ranninger, Phys. Rev. B **23**, 1796 (1981) and Phys. Rev. B **24**, 1164 (1981).
24. B. K. Chakraverty, J. Ranninger and D. Feinberg, Phys. Rev. Lett., **81**, 433 (1998).
25. J. Ranninger, J.M. Robin, and M. Eschrig, Phys. Rev. Lett. **74**, 4027 (1995).
26. M. Cuoco and J. Ranninger, Phys. Rev. B **70**, 104509 (2004).
27. J. Corson et al., Nature (London) **398**, 221 (1999).
28. Z. A. Xu et al., Nature **406**, 486 (2000).
29. Y. Wang et al., Phys. Rev. Lett. **95**, 247002 (2005).
30. O. Yuli et al., cond-mat/0909.3963.
31. L. P. Kadanoff and P. C. Martin, Phys. Rev. **124**, 670 (1961).
32. O. Tchernyshyov, Phys. Rev. B **56**, 3372 (1997).
33. H. Feshbach, Ann. Phys. (N.Y.), **55**, 357 (1958).
34. P. W. Anderson, Science **235**, 1196 (1987).
35. E. Altman and A. Auerbach, Phys. Rev. B **65**, 104508 (2002)
36. F. Wegner, Ann. Phys. (Leipzig) **3**, 77 (1994).
37. J. Ranninger and A. Romano, Phys. Rev. B **78**, 054527 (2008).
38. T. Stauber and J. Ranninger, Phys. Rev. Lett. **99**, 045301 (2007).
39. J. Roehler, Int. J. Phys. B **19**, 255 (2005); cond-mat/ 0909.1702.
40. J. Ranninger and T. Domanski, PRB submitted.
41. T. Domanski and J. Ranninger, Phys. Rev. Lett. **91**, 255301 (2003).
42. A. Kanigel et al., Phys. Rev. Lett. **101**, 137002 (2008).
43. J.-M Robin, A. Romano and J. Ranninger, Phys. Rev. Lett. **81**, 2755 (1998).

THERMODYNAMICALLY STABLE VORTEX STATES IN SUPERCONDUCTING NANOWIRES

W. M. WU, M. B. SOBNACK* and F. V. KUSMARTSEV

*Department of Physics, Loughborough University,
Loughborough LE11 3TU, United Kingdom*
* *M.B.Sobnack@lboro.ac.uk*

Received 26 November 2009

We develop a new condensed matter theory of the formation of thermodynamically stable vortex structures in quantum nanowires. We write down the Gibbs free energy functional for the systems and we minimise the free energy to obtain the optimal position of vortices for different applied fields and temperatures. We also study the nucleation of vortices in, and their escape from, the nanostructural superconductors.

Keywords: Quantized vortices; Ginzburg-Landau theory; nanowires; surface superconductivity.

1. Introduction

In bulk systems, the transition from normal state to the superconducting state, as the temperature is lowered, takes place at the superconducting temperature T_c. In the presence of a magnetic field, superconductivity is destroyed at the thermodynamic critical field $H_c(T)$, related to the free energy difference between the normal and superconducting states, the so-called condensation energy of the superconducting state. When the transition at T_c takes place in the absence of a magnetic field, it is of second order, and the transition arising due to a change in magnetic field is of first order.

We define two characteristic lengthscales, the penetration depth λ and the Ginzburg-Landau coherence length ξ. Both λ and ξ are functions of temperature and diverge at T_c from below:

$$\lambda(T) = \frac{\lambda_0}{(1 - t^4)^{1/2}}, \quad \text{reduced temperature} \quad t = \frac{T}{T_c} \tag{1}$$

$$\xi(T) = \frac{\xi_0}{(1 - t^2)^{1/2}}. \tag{2}$$

(See, for example, Tinkham.[1]) Well below T_c, ξ approaches the Pippard coherence length ξ_0. Although both λ and ξ increase with temperature, the Ginzburg-Landau parameter κ

$$\kappa(T) = \frac{\lambda}{\xi} = \frac{\kappa_0}{(1 + t^2)^{1/2}}, \quad \kappa_0 = \frac{\lambda_0}{\xi_0}, \tag{3}$$

decreases with temperature.

The critical value $\kappa = 1/\sqrt{2}$ separates Type I and Type II superconductors. Type I superconductors in an ideal sense exclude applied magnetic field completely if the field is below the critical field H_c: the sample exhibits perfect diamagnetism and is in the Meissner state. At $H = H_c$, there is a discontinuous jump in the magnetization of the sample, and, for $H > H_c$, the applied magnetic field penetrates the sample and the sample is in the normal state. The transition at H_c arising as a result of a change in magnetic field is of first order.

For Type II superconductors, we define two critical fileds, $H_{c1} < H_c$ and H_{c2}. For $H < H_{c1}$, the sample is in the Meissner state. For $H_{c1} < H < H_{c2}$, magnetic field can enter the system as quantised magnetic flux or fluxiods – this phase is the Abrikosov mixed state – and at H_{c2} the sample becomes normal. The transition from Meissner state to normal state is of second order.

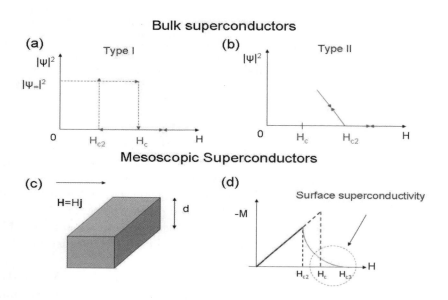

Fig. 1.   Schematics of the behaviour of the order parameter at H_{c2} in an ideal bulk Type I (Fig. 1(a)) and in Type II (Fig. 1(b)) superconductors as a function of applied magnetic field H: typical hysteretic behaviour in Type I and reversible behaviour in Type II; see, for example, reference [1]; a mesoscopic superconductor with a surface parallel to the applied field (Fig. 1(c)) and the behaviour of the magnetisation M of the sample as a function of H (Fig. 1(d))).

H_{c2} is the highest field at which superconductivity can nucleate in a decreasing magnetic field. Since Type I superconductors have $H_{c2} < H_c$, the system supercools and remains normal even at H_c (see Fig. 1(a)). At H_c the superconducting density $|\psi|^2$ undergoes a discontinuous and irreversible jump. On cooling, the system exhibits hysteresis. Of course, sample defects limit the amount of supercooling to fields higher than the theoretical limit H_{c2}. There is no such supercooling in Type II superconductors: expulsion of flux starts at H_{c2} and $|\psi|^2$ changes continuously (Fig. 1(b)).

In small samples, finite-size effects become important. For example, take a film of Type I superconductor of thickness d (see Fig. 1(c),(d)). It is known that if d is less than some critical thickness d_c, the Landau domains (the intermediate state) become stable and the system reverts to the Abrikosov mixed state similar to that in Type II superconductors – the intermediate state is replaced by a state which allows magnetic field to penetrate as quantised magnetic flux. Further, small samples have surfaces (as do real superconductors). Saint-James and de Gennes[2,3] showed that if there is a surface parallel to the applied magnetic field, superconductivity persists in a layer of the size of the coherence length up to fields $H_{c3} > H_c$. A Type I superconductor therefore is able to carry a supercurrent over the range (H_c, H_{c3}) at which there is no superconductivity in bulk. H_{c3} is a strong function of the sample shape.

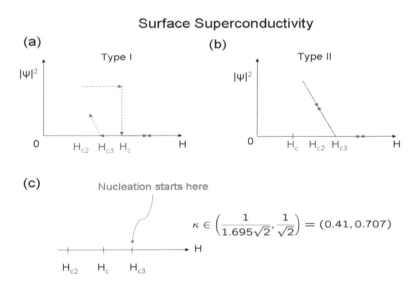

Fig. 2. Schematic behaviour of the order parameter in Type I (Fig. 2(a)) and in Type II (Fig. 2(b)) superconductors as a function of applied magnetic field H in the presence of surface superconductivity: for Type I superconductors with $\kappa \in (0.41, 0.707)$, there is no supercooling at H_c in a decreasing magnetic field (Fig. 2(c)).

An interesting consequence of surface superconductivity is that it is H_{c3}, and not H_c, that should limit the range of supercooling of a Type I superconductor (Fig. 2(a)) since the surface layer of superconductivity serves to initiate the transformation of the interior. However, for Type I superconductors with

$\kappa \in (1/(1.695\sqrt{2}), 1/\sqrt{2})$, $H_{c2} < H_c < H_{c3}$: the implication is that there is no supercooling at H_c as H is decreased from above H_c despite the fact that the volume of the sample makes a first-order transition there (Fig. 2(c)).

Recently there has been a lot of interest in superconducting nanowires. Zhang and Dai[4] experimentally studied the magnetisation of a Pb (bulk lead is a Type I superconductor) nanowire of diameter 45 nm and length 6 μm in an external magnetic field applied transversely to the nanowire. They found that the response of the nanowire depends on the temperature: below $T = 5.0°$ K, the magnetisation is irreversible and shows Type II character (H_{c1} and H_{c2} have been observed). They also found that the mean free path l of the Cooper pair is shorter than expected, and hence the measured coherent length ξ_{measured} is shorter ($1/\xi_{\text{measured}} = 1/\xi_0 + 1/l$). Michotte[5] studied Pb nanowires with diameters varying from 40 nm to 270 nm in external fields applied parallel to the axis of the nanowires. The samples showed Type II superconductor properties during field cooling (decreasing magnetic field). Zhang and Dai[4] also observed that H_c was further increased in the nanowires. Stenuit *et al.*[6] reported that the magnetisation v/s applied field results for their Pb nanowires show hysteresis and they observed Type II "mixed states." Ishii *et al.*[7] also reported that their Pb nanowires (with diameter 80 nm-100 nm) showed a wide phase transition around the critical field and hence deduced that Pb nanowires exhibited Type II structure. They also found that the coherent length ξ_0 is smaller than that expected in the clean limit, and attributed this to the fact that mean free path l is much smaller than in the bulk sample.

In the current study, we investigate theoretically the magnetisation of Pb nanowires as a function of both increasing and decreasing applied magnetic field. We also investigate the stable vortex structures in the systems. The critical temperature of bulk Pb superconductor is $T_c = 7.2°$K, with a coherent length $\xi_0 \sim 88$ nm and penetration depth $\lambda_0 \sim 38$ nm at $T = 0°$K (see, for example, references [8, 9]). However, in small samples, as discussed above, ξ_0 is smaller and λ_0 larger than these values. It must be stressed here that the current work is still ongoing and the results presented below are very preliminary.

2. Theory and Methodology

Consider a nanowire of length L and radius R, with its axis along the **z** axis, placed in an applied transverse magnetic field $\mathbf{H} = H\hat{\mathbf{y}} = \nabla \wedge \mathbf{A}_{\text{ext}}$.

The dimensionaless Gibbs free energy of the nanowire is given by

$$g_s - g_n = \int \left[-|\tilde{\Psi}|^2 + \frac{1}{2}|\tilde{\Psi}|^4 + \frac{1}{2}\left|\left(-i\tilde{\nabla} - \mathbf{a}\right)\tilde{\Psi}\right|^2 + \kappa^2(\mathbf{h} - \mathbf{b})^2 \right] dV, \quad (4)$$

where $\tilde{\Psi} = \Psi/|\Psi_\infty|$ is the normalised order parameter, $\mathbf{a} = \mathbf{A}/(c\hbar/e\xi)$ is the dimensionless vector potential (with $\nabla \wedge \mathbf{A} = \mathbf{B}$, the local magnetic field), $\mathbf{b}$ is the dimensionless local magnetic field, $\mathbf{h}$ is the dimensionless applied magnetic field and $\tilde{\nabla} = \xi\nabla = \xi(\partial_x, \partial_y, \partial_z)$ is the dimensionless gradient operator. Length scales are in units of ξ: $r = R/\xi$ and $\ell = L/\xi$. The dimensionless temperature is $t = T/T_c$.

Minimising the free energy functional (4) gives the Ginzburg-Landau pair of coupled equations

$$0 = (-i\tilde{\nabla} - \mathbf{a})^2 \tilde{\Psi} - \tilde{\Psi}(1 - |\tilde{\Psi}|^2) \tag{5}$$

$$\kappa^2 \tilde{\nabla} \wedge \tilde{\nabla} \wedge \mathbf{a} = \text{Im}(\tilde{\Psi}^* \tilde{\nabla} \tilde{\Psi}) - \mathbf{a}|\tilde{\Psi}|^2 = \mathbf{j} \tag{6}$$

($\mathbf{j}$ is the normalised superconducting current density), together with the boundary condition that on the surface $\partial\Omega$ of the nanowire, the normal component of supercurrent should vanish

$$(-i\tilde{\nabla} - \mathbf{a})\tilde{\Psi} \cdot \hat{\mathbf{n}} = 0, \tag{7}$$

and that on $\partial\Omega$ the local (dimensionless) magnetic field should satisfy

$$\mathbf{b} = \mathbf{h} = \tilde{\nabla} \wedge \mathbf{a}_{\text{ext}}. \tag{8}$$

Given the applied magnetic field $\mathbf{h}$, the two parameters $(\mathbf{a}, \tilde{\Psi})$ can be obtained by solving the two Ginzburg-Landau equations (5) and (6) self-consistently.

The dimensionless (microscopic) magnetisation $\langle\mathbf{m}\rangle$ follows from

$$\langle\mathbf{m}\rangle = \frac{1}{V} \int \frac{1}{2}\mathbf{r} \wedge \mathbf{j} \, dV, \tag{9}$$

where $\mathbf{r} = (x/\xi, y/\xi, z/\xi)$.

3. Simulation and Results

In order to solve Eqns. (5) and (6), it is necessary to input the coherent length ξ and prenetration depth λ. These two parameters can be estimated from any experimental data in the following way: for a given temperature $t = t_0$, the relation between H_{c2} and ξ is

$$\xi(t_0) = \sqrt{\frac{\phi_0}{2\pi H_{c2}(t_0)}}, \tag{10}$$

where $\phi_0 = hc/e^*$ is the flux quantum. Estimating $H_{c2}(t_0)$ from the experimental data gives $\xi(t_0)$ and $\xi(t)$ then follows from Eq. (2). Similarly the thermodynamic relation between H_c, λ and ξ is

$$\lambda(t_0) = \frac{\phi_0}{2\pi\sqrt{2}\xi(t_0)H_c(t_0)} \tag{11}$$

and $\lambda(t_0)$ can be obtained once $H_c(t_0)$ is estimated. Eq. (1) then gives the temperature dependence of λ.

The following table gives the parameters used in our simulations on nanowires of length $L = 2.1\,\mu$m and diameter 470 nm.

$T(^\circ$K)	H_c(Oe)	H_{c2}(Oe)	λ(nm)	ξ(nm)	κ
6.6	180	100	83	183	0.45
6.0	280	190	63	132	0.47
5.2	410	300	53	106	0.50
4.2	530	410	48	90	0.53

T = 6.6K

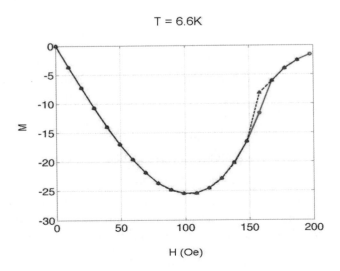

Fig. 3.   The calculated magnetisation M as a function of applied magnetic field H at $T = 6.6°$ K: the red dots and curve give the magnetisation under magnetic heating (increasing H) and the blue triangles and curve under magnetic cooling (decreasing H).

Figure 3 gives our calculated magnetisation as a function of applied magnetic field H at $T = 6.6°$ K. The curve and dots in red are the result for increasing magnetic field (from 0 Oe to above $H_c = 180$ Oe) and the blue curve and triangles that for magnetic cooling (H decreasing from above H_c to 0 Oe). $\kappa = 0.45$ at $T = 6.6°$ K. The magnetisation M and the magnetic field H follow from the non-dimensional (normalised) values through

$$M = \frac{1}{4\pi} H_{c2} m \quad \text{and} \quad H = H_{c2} h$$

respectively, with $H_{c2}(T)$ given in the table above. The magnetisation is reversible, with the dependence of M on H exactly the same for magnetic cooling and magnetic heating (except for the one numerical glitch at around $H = 160$ Oe): the wire makes a transition from Meissner state to normal state during heating and no flux penetrates the system during the transition. During magnetic cooling from above H_c, the nanowire makes a transition from normal state to Meissner state with no flux trapped, and the magnetisation has exactly the same dependence on H as for magnetic heating.

Figure 4 shows our calculated magnetisation at $T = 6.0$ K. There is clear evidence of hysteresis, with the behaviour during increasing magnetic field different from that during decreasing magnetic field. While the magnetisation shows clearly Type I behaviour in increasing field (smooth transition to the normal state), the transition from normal state to Meissner state in a decreasing field has two distinct

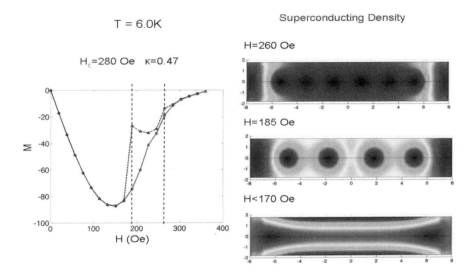

Fig. 4. Calculated magnetisation M as a function of applied magnetic field H at $T = 6.0°$ K (on the left) under increasing magnetic field (red dots and curve) and decreasing magnetic field (blue triangles and curve), showing hysteretic behaviour. There are two jumps in M during cooling, a signature that flux is being expelled from the nanowire. The figure on the right is superconducting density profile along on the nanowire at values of H on either sides of the jumps, with red representing high $|\tilde{\psi}|^2$ (pure superconducting) and dark blue color representing low $|\tilde{\psi}|^2$ (normal metal).

jumps in magnetisation at $H = H_2 = 260$ Oe and $H = H_1 = 185$ Oe. A jump in magnetisation curve is a signature of either expulsion of flux from the system (magnetic cooling) or penetration of flux (magnetic heating). The jumps separate the different vortex states of the system. The density plot on the right hand side of Fig. 4 are plots of the (normalised) superconducting density profile ($|\tilde{\Psi}|^2 = |\Psi|^2/|\Psi_\infty^2|$) along the nanowire, with deep red representing $\tilde{\Psi}^2 = 1$ and deep blue $\tilde{\Psi}^2 = 0$, at three different values of the applied field: $H > H_2$ (top profile), $H_1 < H < H_2$ (middle profile) and $H < H_1$ (bottom profile). The vortex state of the wire at H just above H_2 is one with 6 fluxoids. As H is decreased, two fluxoids are expelled at H_2 and there is a jump in the magnetisation of the nanowire. The state with four vortices then evolves with decreasing H along the magnetisation curve shown on the left and at $H = H_1$ all four fluxoids are expelled and for $H < H_1$ the system evolves to the superconducting state. The hysteresis in the magnetisation curves shows that for each values of $H \in (180, 280)$ Oe, the nanowire has at least two metastable vortex states (our calculations have identified the two shown) and different states are accessed during heating and during cooling.

At $T = 5.2°$ K ($\kappa = 0.5$), the response of the nanowire in both an increasing magnetic field (Fig.5) and in a decreasing field (Fig.6) shows typical Type II be-

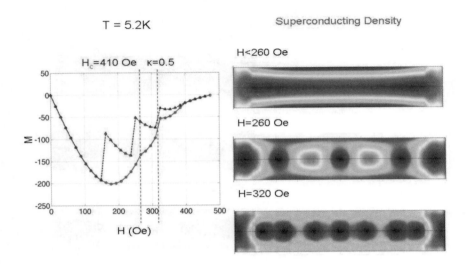

Fig. 5. $T = 5.2°$ K: magnetisation as a function of applied magnetic field (on the left). There are two jumps in M during magnetic heating (red dots and curve). The density profile plots on the right give $|\tilde{\psi}|^2$ on either sides of the jump.

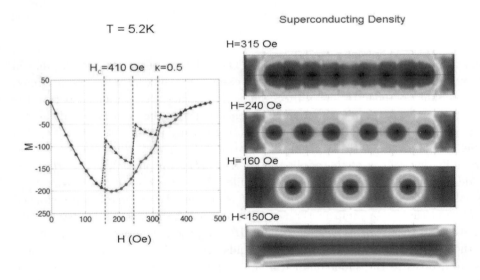

Fig. 6. $T = 5.2°$ K: as for Fig. 5, but for magnetic cooling (decreasing H). Under magnetic cooling (blue triangles and curve), the magnetisation M has three jumps, with the states of the nanowire on either sides of the jumps given in the density profile plots on the right: deep red is pure superconducting and dark blue normal metal.

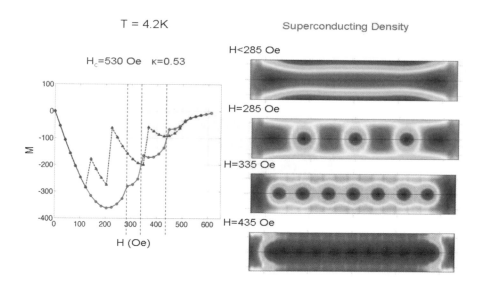

Fig. 7. Magnetisation M as a function of H (left) at $T = 4.2°$ K and density profiles (right) of the nanowire on either sides of the three jumps during magnetic heating (red dots and curve).

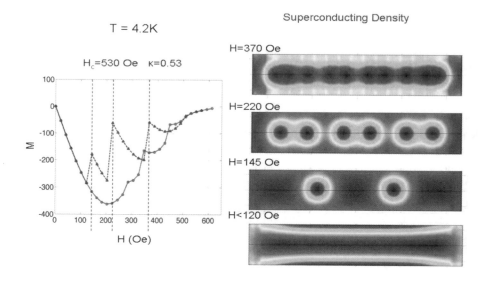

Fig. 8. $T = 4.2°$ K: as for Fig. 7, but for magnetic cooling.

Superconducting Density
T = 4.2K   H = 610 Oe

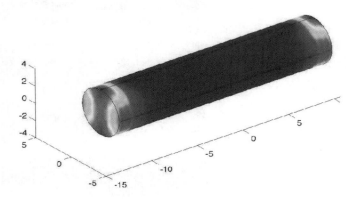

Fig. 9.   Density profile along the nanowire at $T = 4.2°$ K at an applied magnetic field of $H = 610$ Oe, showing surface superconductivity at the two ends of the nanowire parallel to the applied magnetic field. $H_{c3} \approx 610$ Oe.

haviour. As H is increased (red curve and dots), we identify two vortex states as the system evolves from the Meissner state to the normal state. The first jump in magnetisation is at $H \approx 260$ Oe and corresponds to the nucleation of three fluxoids in the wire (see the density profile plot on the right in Fig.5). For $H > 260$ Oe, this state evolves along the red curve until H reaches $H \approx 320$ Oe, when four further fluxoids penetrate and nucleate in the system. As H is increased further, the wire makes a gradual transition to the normal state.

As H is decreased from the normal state, there are three jumps in the magneti-sation curve (blue curve and triangles in Fig.5 and Fig.6). These correspond to the expulsion of fluxoids at successive threshold magnetic fields: two at $H \approx 240$ Oe, three at $H \approx 160$ Oe and the remaining three at $H \approx 150$ Oe (see density profile in Fig.6). After the last jump, the system is in the Meissner state.

Similarly at $T = 4.2°$ K ($\kappa = 0.53$), the magnetisation show Type II response in both increasing (Fig.7) and decreasing (Fig.8) magnetic fields, with jumps at $H \approx 285$ Oe, $H \approx 335$ Oe and $H \approx 435$ Oe in an increasing field and at $H \approx 220$ Oe, $H \approx 145$ Oe and $H \approx 120$ Oe as H is decreased. Each jump is a signature of flux either nucleating in the nanowire (increasing field) or expelled from it (decreasing field). The different vortex states on either sides of the jumps during magnetic heating and magnetic cooling are shown in the density profiles in Fig. 7 and Fig. 8 respectively.

4. Concluding Remarks

We have presented the results of our calculations of the response of a 20 μm long Pb nanowire of diameter 470 nm in a transverse applied magnetic field at four temperatures $T = 6.6°$ K, $T = 6.0°$ K, $T = 5.2°$ K and $T = 4.2°$ K. Although Pb is a Type II superconductor (with superconducting transition temperature $T_c = 7.2°$ K), it is only the highest temperature investigated ($T = 6.6°$ K) that the magnetisation v/s applied magnetic curve shows characteristic Type I behaviour, with the response being completely reversible (*i.e.*, symmetric under heating and cooling).

At the lower temperatures, the responses show Type II behaviour - fluxoids nucleate or are trapped in the wire (typical of the Abrikosov mixed state) and there are jumps in the magnetisation curves. These are signatures of quantised flux nucleating into (as H is increased) or being expelled from (as H is decreased) the nanowire. We have presented density profiles for the vortex states at values of H on either side of the jumps. The magnetisation curves are not reversible - they all show hysteresis; this implies that at each values of $H > H_{c1}$ there are several (at least two) metastable vortex states and different states are accessed during magnetic heating and magnetic cooling. Further, at these three lower temperatures, there is evidence of surface superconductivity in a small layer near the "ends" of the wire, *i.e.*, near the surfaces parallel to the applied magnetic field: superconductivity persists well beyond H_c at which there is no superconductivity in the bulk of the wire. This is illustrated in Fig. 9, which gives the superconducting density profile at $T = 4.2°$ K for an applied field $H = H_{c3} = 610\,\text{Oe}$ (at this temperature $H_c = 530\,\text{Oe}$).

The results presented in this manuscript are very preliminary: there are still some problems – for example, the magnetisation calculated using Eq. 9 does not coincide with that calculated using the thermodynamic definition of magnetisation, although the two results are qualitatively the same and have the same features - and work to address these is in progress. Results with more thorough calculations will be published elsewhere.

Acknowledgments

The authors are grateful to Simon Bending for very stimulating discussions. Support from the "Arrays of Quantum Dots and Josephson Junctions" Network of the ESF is acknowledged and one of the authors (M.B.S.) would like to acknowledge receipt of a Loughborough University Science Faculty "Small Grant" without which participation at CMT33 would not have been possible.

References

1. Michael Tinkham, *Introduction To Superconductivity*, Dover Publications, New York, 2004.
2. D. Saint-James and P. G. De Gennes, *Phys. Lett.* **7** 306 (1963).
3. P.G. de Gennes, *Superconductivity of Metals and Alloys*, Perseus Books Group, USA, 1999.

4. X. Y. Zhang and J. Y. Dai, *Nanotechnology* **15**, 1166 (2004).
5. S. Michotte, L. Piraux, S. Dubois; F. Pailloux, G. Stenuit and J. Govaerts, *Physica C* **377**, 267 (2002).
6. G. Stenuit, S. Michotte, J. Govaerts and L. Piraux, *Eur. Phys. J. B* **33**, 103 (2003).
7. S. Ishii, E.S. Sadki and S. Ooi, Y. Ochiai and K. Hirata, *Physcia C* **426-431**, 268 (2005).
8. Ali E. Aliev and Sergey B. Lee, *Physica C* **453**, 15 (2006).
9. R.F.Gasparovic and W.L.McLean, *Phys. Rev. B* **7**, 2519 (1970).

Part D
Quantum Information

QUANTUM INFORMATION IN OPTICAL LATTICES

ANGELA M. GUZMÁN

Physics Department, Florida Atlantic University, 777 Glades Rd.
Boca Raton, Florida 33433, USA
angela.guzman@fau.edu

MARCO A. DUEÑAS E.*

Sant' Anna School of Advanced Studies, Piazza Martiri della Libertà 33
I-56127 Pisa, Italy
m.duenasesterling@sssup.it

Received 1 December 2009

Experimental realizations of a two-qubit quantum logic gate based on cold atom collisions have been elusive mainly due to the decoherence effects introduced during the quantum gate operation, which cause transitions out of the two-qubit space and lead to a decreased gate operation fidelity. This type of decoherence effects, due to the non closeness of the interacting two-qubit system, are characteristic of the electromagnetic interaction, since the electromagnetic vacuum acts as a reservoir whose eigenmodes might become active during the gate operation. To describe the cold-atom collision we consider the quantum non-Hermitian dipole-dipole interaction instead of the less realistic s-scattering approach widely used in the literature. By adding an ancillary qubit, we take advantage of the spatial modulation of the non-Hermitian part of the interaction potential to obtain a "resonant" condition that should be satisfied to achieve lossless operation of a specific two-qubit quantum phase-gate. We demonstrate that careful engineering of the collision is required to obtain a specific truth table and to suppress the effects inherent in the openness of the system arising from the electromagnetic interaction.

Keywords: Quantum information; optical lattices; quantum logic gates.

1. Two Qubit Phase-gates

Trapped neutral atoms in optical lattices[1] provide an interesting tool for observing phenomena previously associated only with condensed matter physics, like Anderson localization and realizations of the Hubbard model and the Mott insulator. In contrast with condensed systems, neutral atoms in optical lattices constitute a system where quantum information processes like initial state preparation, quantum control, and decoherence, are usually amenable to first-principles theoretical understanding. Optical lattices offer many advantages for quantum information purposes since their geometrical structure can be engineered through external control

---

*Permanent address: Universidad Militar Nueva Granada, Carrera 11 No. 101-80, Bogotá, Colombia.

of various parameters, including laser beam geometry and intensity, laser detuning from the atomic cooling transition, and relative polarizations of the laser beams. Their magnetic character can also be determined through the use of spin-dependent optical lattice potentials.

One of the great potentialities of optical lattices would be that of providing an appropriate scenario for the realization of multiple entangled qubit states amenable to fiducial initial state preparation and collectively and individually addressable[2] to perform quantum information processing. The Quantum Information Science and Technology Experts Panel[3] considered that under the DiVincenzo criteria,[4] proposals for neutral atom quantum computing constitute a potentially viable and scalable approach, currently pursued by over twenty experimental and theoretical research groups in the US and Europe.

Quantum information processing based on collisions between cold atoms in optical lattices relies on the ability of manipulating the atoms in an efficient and controlled way, but also on an accurate theoretical description of the interactions and the type of collision leading to the expected output Early proposals for two-qubit gates[5] rely on controlled collisions of neighboring pairs of atoms that can be implemented in spin-dependent optical lattices by varying the polarization of the generating laser beams. In a largely detuned optical lattice, atoms in their internal ground state can be cooled and trapped into the lowest vibrational states of the optical potential wells. The two lowest vibrational states can then be used to define a qubit, whose memory decoherence time is of the order of $1ms$ and is mainly due to the well known mechanism of photon scattering. By varying the relative polarization of the lasers that generate the optical lattice, neighboring atoms can be brought into a collision in a controlled manner. The resulting collisional phase shifts depend on the initial two-qubit state, and can be controlled to give rise to a quantum phase gate. The speed of the gate is limited by the trap frequency, which lies in the range $10kHz - 10MHz$, or operation times of $0.1\mu s - 100\mu s$. The decoherence time to gate time ratio ($10 : 10^4$) plays a central role since the system should allow for the performance of several quantum logic operations before the system's quantum information decoheres.

The search for quantum gates and atom entanglement based on neutral atom collisions has been grounded in the idea that the atoms are weakly interacting and must be brought into close proximity, letting their wave functions overlap and undergo s-scattering, a collision that is ruled by a Fermi contact potential rather than by the more realistic long-range dipole-dipole interaction that prevails in optical lattices and acts over distances of the order of the wavelength and larger.[6] In order to increase the dipole-dipole interaction between atoms in neighboring lattice sites, alternative approaches have been suggested to induce larger dipole moments in the atoms by an auxiliary laser.[7] We have shown[8] that there is an atomic dipole induced by the laser fields that originates a long-range dipole-dipole interaction between trapped atoms and induces collision phase shifts as required to implement a phase-gate without the need of bringing the atoms into nearly complete overlap of their atomic wave functions. Upon replacing the s-scattering approach by the dipole-dipole effective interaction[6] between neighbor atoms trapped in a spin-dependent optical lattice, we found[8] a different truth table for a quantum phase gate.

One of the drawbacks of techniques like neutral atom collisions that involve moving atoms has been recognized[3] to be the inherent risk of opening up additional decoherence channels during gate operation. We demonstrated[8] that a fundamental source of decoherence is introduced by the reservoir action of the electromagnetic vacuum during the collision, the key issue being that the quantum dipole-dipole interaction takes place through the electromagnetic vacuum, and the two-atom system does not constitute a closed system. The effective dipole-dipole interaction results from tracing over the vacuum states. It is described by a non-Hermitian operator that accounts for the probability losses caused by transitions out of the two-qubit state space during gate operation, and hence for fidelity losses. But we also found that the effective two-atom relaxation rates caused by the non-Hermitian dipole-dipole effective Hamiltonian are space modulated and can turn into gain for some interatomic distances. In that regard, a quantum engineering approach to the collision can bring to light unexpected possibilities.

We consider here the effect of adding a third qubit to allow for the compensation of probability losses by quantum engineering of the collision. When two atoms are brought together, a spatially modulated loss rate decreases the final probability of the atoms returning to their original vibrational state after the collision. That is perhaps the reason why two-qubit phase gates based on atomic controlled collisions have not been demonstrated experimentally. If, on the other hand, a third atom moves away from one member of the pair, the space-modulated losses transform into an effective gain, which compensates the two-body losses and increases the probability of recovering the original three-body state after the collision. Optimal balance is obtained at a fixed minimum distance between the two atoms that perform the two-qubit operation, with the third atom serving as an equalizer of the three-body interaction with the reservoir of vacuum electromagnetic modes. We obtain a "resonance" condition to be fulfilled in order to recover high fidelity.

2. The Dipole-dipole Interaction between Trapped Neutral Atoms in a Moving Optical Lattice

We consider a 1D spin-dependent optical lattice generated by two counterpropagating laser beams with wave vectors $\vec{k}_\alpha, \alpha = 1, 2$, in a *lin* $\perp$ *lin* configuration interacting with neutral atoms in a $J = 1$ internal ground state. The optical lattice consists of two independent sublattices for atoms in the ± 1 states intertwined in a ferromagnetic structure where atoms with magnetic number $m = 1$ and $m = -1$ occupy neighboring lattice sites. When the angle β ($0 \leq \beta \leq \pi/2$) between the polarizations of the two counterpropagating beams is varied, the position and depth of the minima of the potential wells change according to the equation:

$$U_{\pm}(z,\beta) = \frac{U_0}{2} \left[2 \mp \cos\left(2k_L z \mp \beta\right) - \sin 2k_L z \sin \beta \right], \tag{1}$$

where the depth of the potential $U_0 \approx \hbar\Omega^2/(2\Delta)$ is determined by the Rabi frequency Ω and the laser detuning Δ. k_L is the wave number of the laser waves. The distance between neighbor sites of the two sublattices is given by

$$r = z_- - z_+ = \frac{1}{2k_L} \left[\pi - 2\beta - \frac{\sin 2\beta}{1 + \sin^2 \beta} \right]. \tag{2}$$

The minima of the sublattices coincide for $\beta = \pi/2$. For deep enough potential wells the hopping probability between lattice sites is negligible, and a Wannier approach to the description of the vibrational states of the trapped atoms is proper. We then assume that the vibrational states of the atoms are well described by β-dependent harmonic-oscillator eigenstates $\{|j(\beta)\rangle\}$, whose energy difference depends on β as follows:

$$\hbar\omega_{osc} = 2\sqrt{U_0 E_R (1 + \sin^2 \beta)}, \tag{3}$$

where $E_R = \hbar^2 k_L^2/(2M)$ is the atomic recoil energy acquired by an atom of mass M when absorbing a laser photon. These β-dependent harmonic oscillator eigenstates provide an adiabatic basis set, to describe the collision, whenever the collision time is long in comparison with the inverse of ω_{osc}. The matrix element of the quantum dipole interaction for a two-atom $|l, m\rangle \rightarrow |j, k\rangle$ transition between vibrational states l and j for the first atom and m and k for the second in an optical lattice is given by[6]

$$V_{jk,lm}(\beta) = \frac{V_0}{k_L^3} P \int_0^{\infty} \frac{k d\vec{k}}{k - k_L} \times$$

$$\sum_{\alpha,\beta=1}^{2} \left[A_{\alpha\beta}(\Omega_k) \chi_{\alpha,\beta}^{jklm}(\vec{r}_1, \vec{r}_2) + B_{\alpha\beta}(\Omega_k) \chi_{\alpha,\beta}^{jklm}(\vec{r}_2, \vec{r}_1) \right], \tag{4}$$

where $V_0 = 3\gamma U_0/(16\Delta)$, and γ is the spontaneous decay rate of the trapping atomic transition. P means principal part of the integral that is to be performed over all vacuum modes of wave vector $\vec{k}$. $A_{\alpha\beta}(\Omega_k)$ and $B_{\alpha\beta}(\Omega_k)$ are linear combinations of spherical harmonics of the (electromagnetic) k-space solid angle Ω_k, so that the only dependence on $\vec{r}_1$ and $\vec{r}_2$ is through the functions

$$\chi_{\alpha,\beta}^{jklm}(\vec{r}_1, \vec{r}_2) = {}_1\langle j| \exp\left[i\vec{q}_\alpha \cdot \vec{r}_1\right]|l\rangle_1 \, {}_2\langle k| \exp\left[-i\vec{q}_\beta \cdot \vec{r}_2\right]|m\rangle_2, \tag{5}$$

where $\vec{r}_{1,2}$ are the lattice site position vectors. The quantum dipole-dipole interaction can be described as a four-photon process, where an atom absorbs a photon from one of the counterpropagating laser beams and subsequently emits spontaneously a photon in a vacuum mode. The second atom absorbs the spontaneous emitted photon and emits a fourth photon into one of the laser modes. The momentum transfer $\vec{q}_\alpha = \vec{k} - \vec{k}_\alpha, \alpha = 1, 2$, appearing in Eq. (5) represents the momentum exchange between the atoms and the vacuum and applied fields.

For the purpose of a phase-gate we will consider only adiabatically controlled elastic collisions, assuming that the probability for inelastic collisions leading to vibrational excitations are low. For the first two eigen vibrational states of the harmonic oscillator $(l, l' = 0, 1)$ we obtain

$$V_{ll'} \equiv V_{ll',ll'} = V_0 e^{i\rho} F_{ll'}(\rho, \beta), \qquad (6)$$

where $\rho = k_L |\vec{r}_1 - \vec{r}_2| = k_L r$ and

$$F_{ll'} = \left[\frac{1}{\rho^3} - \frac{i}{\rho^2} - \frac{1}{\rho}\right] e^{-\xi} [\cos \rho - \sin \beta] - (l + l')\xi^2 \left[\frac{3}{\rho^4} - \frac{3i}{\rho^3} - \frac{2}{\rho^2} + \frac{i}{\rho}\right] \sin \rho. \quad (7)$$

For well confined atoms the function $\xi(\beta) = 2E_R/(\hbar\omega_{osc}) \ll 1$. The real and imag-

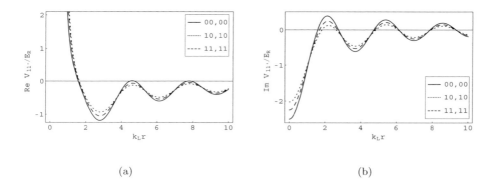

(a)   (b)

Fig. 1. (a) Real and (b)imaginary part of the two-body interaction potential $V_{ll'}, l, l' = 0, 1$ for two close neighbors in different sublattices, as a function of $\rho = k_L r$. The parameters are $\Delta = 10\gamma$ and $U_0 = 100E_R$.

inary parts of the effective dipole-dipole potential are shown in Figure 1. Note that the imaginary part tends to the finite value $-\gamma U_0/(4\Delta)$ when $\rho \to 0$. The first zero of both functions corresponds to $\rho = \pi/2$, when $\beta = 0$, and the atoms are at a distance of $\lambda/4$. When $\beta \neq 0$, two neighboring atoms approach each other and start repelling. The imaginary part of the potential becomes negative, representing the losses suffered by the two-qubit system due to open channels for transitions out of the two-qubit state through the coupling with the electromagnetic vacuum. We assume that the gate works by controlled elastic collisions, where the two atoms are brought together and back to their initial positions in their same initial vibrational state. Given the structure of the dipole-dipole interaction matrix element (see Eqs. (6) and (7)), the truth table for the gate has the general form[8]:

$$|l\rangle|l'\rangle \to |l\rangle|l'\rangle \exp\left[i(l + l')\phi\right] \qquad (8)$$

where the collisional phase ϕ depends on the specifics of the collision. Equation (8) clearly differs from the truth table predicted for collisions ruled by s-scattering.

Furthermore, the imaginary part of the potential introduces strong depletion of the two-qubit quantum state, causing probability losses as high as 80% for a π-collisional shift, and hence largely reduces the gate fidelity.

3. Quantum Engineering of Three-body Cold Atom Collisions in Optical Lattices

The imaginary part of the interaction potential shown in Fig. 1 is zero when $\beta = 0$ and $|z_- - z_+| = \lambda/4$. If the distance between the atoms diminishes or increases, the dipole interaction opens channels leading to transitions out (loss) or into (gain) the two-qubit space. We look then for the conditions that could lead to a high-fidelity gate operation, where loss and gain are compensated. This can be achieved by a three-qubit arrangement, with one atom in one sublattice flanked by two trapped atoms in neighboring sites of the other sublattice (Figure 2). Then with varying β, the atom at the center approaches one of the neighbors while moving away from the other. The atom in the center will interact via the dipole-dipole interaction with the two neighbors located at normalized distances $\rho_1(t)$ and $\rho_2(t)$,

$$\rho_1(t) = \frac{\pi}{2} - \beta(t) - \frac{\sin 2\beta(t)}{2[1 + \sin^2 \beta(t)]}$$

$$\rho_2(t) = \frac{\pi}{2} + \beta(t) + \frac{\sin 2\beta(t)}{2[1 + \sin^2 \beta(t)]}, \tag{9}$$

with $\rho_1(t) + \rho_2(t) = \pi$. The dipole-dipole interaction between the two outer atoms is identically zero. To prepare the system for phase-gate operation, we have four initial states $|l, l', l\rangle$ with the atoms in the vibrational states $l, l' = 0, 1$, with the third qubit acting as a leverage to compensate fidelity losses. Clearly any one of the outer qubits can be used as the control qubit that determines the conditional phase-shift. We have chosen the two outer atoms to be in the same vibrational state, but we could also study lossless three-qubit gates. The state of the system during a two-qubit phase operation is given by:

$$|Q_1, Q_2, Q_3\rangle = \sum_{l,l'=0}^{1} C_{ll'}(t)|l, l', l\rangle, \tag{10}$$

Considering only elastic collisions, the interaction potential can be written as a diagonal operator:

$$V_{l'l}(t)|ll'l\rangle\langle ll'l| = \{V_{ll'}[\rho_1(t)] + V_{l'l}[\rho_2(t)]\}|ll'l\rangle\langle ll'l|. \tag{11}$$

Time dependent perturbation theory results in:

$$C_{ll'}(t) = \exp\left[-\frac{i}{\hbar}\int_0^T dt V_{ll'l}(t)\right], \tag{12}$$

where T is the collision time. While the real part of $V_{ll'l}$ determines the phase shifts for the different states, its imaginary part accounts for the effect of opened channels out of the three-qubit space. The truth table for the gate operation reads:

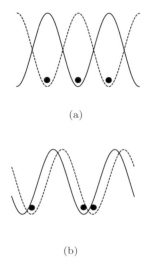

(a)

(b)

Fig. 2.   A three qubit logic gate in a spin-dependent optical lattice. The figure at the top shows
the configuration when $\beta = 0$. When the angle between the polarization of the two beams is less
than 90°, the intermediate atom gets closer to one of its neighbors and further from the other one
as shown in the diagram below. The distance between the two outer atoms remains constant.

$$|ll'l\rangle \rightarrow |ll'l\rangle \exp\left[i\varphi_{ll'}(T)\right]\exp\left[-\gamma_{ll'}(T)\right],\tag{13}$$

where

$$\varphi_{ll'}(T) = -\frac{1}{\hbar}Re\int_{0}^{T}dtV_{ll'l}(t)$$

$$\gamma_{ll'}(T) = -\frac{1}{\hbar}Im\int_{0}^{T}dtV_{ll'l}(t).\tag{14}$$

We will choose half a period of a sinusoidal temporal variation for the polarization
angle $\beta = \beta_{max}\sin \pi t/T$. The first term on the right hand side of Eq. (7) introduces
the same phase shift for all three-qubit quantum states, therefore their relative
phases will be determined only by the second term. We can also see that the first
term on the right hand side of Eq. (7) will contribute the most to the imaginary
part of the potential, and hence to the loss of fidelity. After numerical integration
of Eqs. (14), we obtain

$$\varphi_{ll'}(T) = -\alpha(l + l')\frac{\psi(\beta_{max})}{U_0/E_R}T$$

$$\gamma_{ll'}(T) = -\alpha\gamma_{col}(\beta_{max})T,\tag{15}$$

where the scaling factor

$$\alpha = \frac{3U_0\gamma}{16\hbar\pi\Delta} \tag{16}$$

contains all externally controllable parameters except for the collision time. The functions $\psi(\beta_{max})$ and $\gamma_{col}(\beta_{max})$ depend only on the distance between the atoms at their closest approach during the collision and are plotted in Figure 3 as functions of β_{max}.

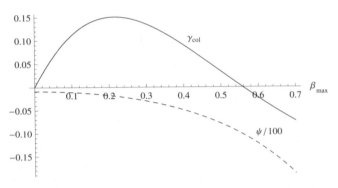

Fig. 3.   Decoherence coefficient γ_{col} (*fullline*) and the ψ phase factor divided by 100 (dashed line) as functions of β_{max}. Notice that the zero value for the decoherence factor corresponds to overall suppression of collision-induced decoherence.

With careful engineering of the collision, the overall loss can be canceled out. Note that $\gamma_{col}(\beta_{max})$ can be made zero by appropriate choice of the distance for maximum approach. We obtain a "resonance-like" condition for $\beta_{max} = 0.558753$ where $\gamma_{col} = -8.5 \times 10^{-8}$ and $\psi_{res} = -10.1779$. The polarization angle corresponds to a distance of closest approach close to $\lambda/10$. For this specific collision the contributions of the non-Hermitian part of the dipole-dipole interaction balance out due to the presence of the ancillary qubit. We recall that the resonant condition is independent of the parameters of the lasers other than their wavelength, since it reflects only the interaction with the vacuum. The truth table for the gate at resonance reads

$$|ll'l\rangle \rightarrow |ll'l\rangle \exp\left[i\phi(l+l')\right], \qquad \phi = \frac{\alpha T}{U_0/E_R}\psi_{res}. \tag{17}$$

At the distance of closest approach and for $U_0 = 100E_R$, $\Delta/\gamma = 10$, the interaction energy is smaller than the energy required for a vibrational transition, hence the assumption of elastic collision holds. While the phase shift difference relevant to the gate operation is independent of the depth of the potential, the collisional decoherence increases proportional to the depth, and the ratio between the interaction energy and the energy difference between the vibrational states increases as the square root of the depth. Therefore there seems to be no reason to use optical potentials deeper than necessary to avoid atomic heating.

The adiabatic character of the collision requires collision times larger than $\pi/(\omega_{osc}$ which for a Rb atom and $U_0 = 100E_R$ is of the order of $1ms$. The minimum time is inversely proportional to the square root of the potential depth. To obtain a $\{-\pi\}$ collisional phase shift for the state $|111\rangle$ when $\Delta/\gamma = 10$ requires a collision time of the order of $1ms$. This time could be reduced by increasing the recoil energy and/or the atomic polarizability. Two recent independent experiments have demonstrated the existence of observable effects of the long-range dipole-dipole interactions between Rydberg atoms, and the so-called "Rydberg blockade", an effect that might make long-range quantum gates between neutral atoms possible.[9,10]

4. Conclusion

The non-Hermitian character of the effective dipole-dipole interaction between neutral atoms trapped in optical lattices poses a fundamentally unsurmountable difficulty to obtain high-fidelity two-qubit gates involving vibrational states of the atoms in the lattice and operated by controlled collisions. During the collision the two colliding atoms do not constitute a closed system, and channels for transitions out of the two-qubit state become opened, leading to decreased fidelity in the gate operation. We have shown that it is possible to take advantage of the space modulated loss/gain associated with the effective dipole-dipole interaction by using a three qubit-system, where the third atom acts as a loss-gain compensator. The results show that an accurate theoretical description of the interactions taking place during the collision is required not only to properly predict the truth table for the gate but to establish what could be considered as a "resonant" condition for its lossless performance. Attempts to increase the strength of the dipole-dipole interaction by using external fields to induce large dipoles, would enhance the dissipative effects of its non-Hermitian character. A better alternative to reduce the gate operation time could be the use of trapped Rydberg atoms, which are largely polarizable. But any alternative for high-fidelity gates involving the electromagnetic interaction, due to the open nature of the interactive atom-field quantum system, would require of a similar quantum engineering approach as the one presented here.

References

1. P. S. Jessen, and I. H. Deutsch, "Optical lattices" in *Advances in Atomic, Moledular, and Optical Physics*, **137**, ed. B. Bederson and H. Walther, (Academic, San Diego, 1996), pp. 95-138.
2. I. Bloch, *Nature* **453**, 1016 (2008).
3. C. Caves, "Neutral Atom Approaches to Quantum Information Processing and Quantum Computing", in *A Quantum Information Science and Technology Roadmap*, ed. R. Hughes and T. Heinrichs (Advanced Research and Development Activity) , April 2004, http://qist.lanl.gov.
4. D. P. DiVincenzo, *Phys. Rev. A*, **51**, 1015 (1995).
5. G. K. Brennen, C. M. Caves, P. S. Jessen, and I. H. Deutsch, *Phys. Rev. Lett.* **82**, 1060-1063 (1999).

6. A. M. Guzmán, P. Meystre, *Phys. Rev. A* **57**, 1139 (1998).
7. D. Jaksch, H.-J. Briegel, J. I. Cirac, C. W. Gardiner, and P. Zoller. *Phys. Rev. Lett.* **82**,1975 (1999).
8. M.A. Dueñas, A. M. Guzmán, *Proc. SPIE* **5622**, p. 342-347, (2004).
9. E. Urban et al., *Nature Phys.* **5**, 110 (2009).
10. A. Gaetan et al., *Nature Phys.* **5**, 115 (2009).

Part E
Theory and Applications of Molecular Dynamics
and Density Functional Theory

EXCHANGE-CORRELATION FUNCTIONALS FROM THE IDENTICAL-PARTICLE ORNSTEIN-ZERNIKE EQUATION: BASIC FORMULATION AND NUMERICAL ALGORITHMS

ROGELIO CUEVAS-SAAVEDRA

Department of Chemistry; McMaster University
Hamilton, Ontario, Canada L8S 4M1
cuevasr@mcmaster.ca

PAUL W. AYERS

Department of Chemistry; McMaster University
Hamilton, Ontario, Canada L8S 4M1
ayers@mcmaster.ca

Received 24 December 2009

We adapt the classical Ornstein-Zernike equation for the direct correlation function of classical theory of liquids in order to obtain a model for the exchange-correlation hole based on the electronic direct correlation function. Because we explicitly account for the identical-particle nature of electrons, our result recovers the normalization of the exchange-correlation hole. In addition, the modified direct correlation function is shorted-ranged compared to the classical formula. Functionals based on hole models require six-dimensional integration of a singular integrand to evaluate the exchange-correlation energy, and we present several strategies for efficiently evaluating the exchange-correlation integral in a numerically stable way.

Keywords: Exchange-correlation hole; Ornstein-Zernike equation; DFT.

1. Introduction

Density-functional theory (DFT)[1-3] is an enticing approach to the electronic structure problem, but it is not without its challenges. First, one needs formulas for computing physical observables from the electron density. (That is, one must develop functionals of the electron density for every interesting physical observable.) A second, related, challenge is accounting for the effects of Pauli repulsion and electron correlation in a theory that depends only on a single spatial coordinate. In order to compute the ground-state energy and ground-state density of system of N-electrons bound by the external electrostatic potential $\nu(r)$ (usually the potential due to atomic nuclei), one uses the variational principle for the electron density[4]

$$E_{gs}[\nu; N] = \min_{\{\rho(\mathbf{r}) \geq 0; N = \int \rho(\mathbf{r})d(\mathbf{r})\}} \left\{ F[\rho] + \int \rho(\mathbf{r})\nu(\mathbf{r})d(\mathbf{r}) \right\} \tag{1}$$

The Hohenberg-Kohn functional, $F[\rho]$, is defined as[4,5]

$$F[\rho] = T_s[\rho] + J[\rho] + E_{xc}[\rho] \tag{2}$$

where $T_s[\rho]$ is the kinetic energy for a system of noninteracting electrons with the electron density $\rho(r)$, $J[\rho]$ is the classical self-repulsion energy of the electron density with itself, and $E_{xc}[\rho]$ is called the exchange- correlation energy. As is standard in chemistry,[6,7] we denote the electron density as

$$\rho(\mathbf{r}) = \langle \Psi | \sum_{i=1}^{N} \delta(\mathbf{r} - \mathbf{r}_i) | \Psi \rangle, \tag{3}$$

instead of $n(r)$.

In the approach of Kohn and Sham, $T_s[\rho]$ and $J[\rho]$ are computed exactly.[5] The only unknown term is $E_{xc}[\rho]$, which is defined as "whatever has to be added to $T_s[\rho] + J[\rho]$ so that Eq. (1) is correct." In order to impart a physical interpretation to $E_{xc}[\rho]$, it is computed in terms of the exchange-correlation hole, which can be computed from the two-electron distribution function by

$$h_{xc}(\mathbf{r}, \mathbf{r}') = \frac{\rho_2(\mathbf{r}, \mathbf{r}') - \rho(\mathbf{r})\rho(\mathbf{r}')}{\rho(\mathbf{r})\rho(\mathbf{r}')} \tag{4}$$

That is, $\rho(\mathbf{r})\rho(\mathbf{r}')h_{xc}(\mathbf{r}, \mathbf{r}')$ is the cumulant of the 2-density: it represents the correction to the assumption that electrons are statistically independent. All of the exchange-correlation holes in this paper are averaged over the adiabatic connection between the noninteracting system of electrons (in which the electron- electron repulsion term is omitted from the Hamiltonian) and the physical system.[8-10] We do not explicitly indicate the adiabatic-connection averaging, however. The exchange-correlation energy is

$$E_{xc}[\rho] = \frac{1}{2} \int \int \frac{\rho(\mathbf{r})\rho(\mathbf{r}')h_{xc}(\mathbf{r}, \mathbf{r}')}{|\mathbf{r} - \mathbf{r}'|} d\mathbf{r} d\mathbf{r}' \tag{5}$$

The exchange-correlation hole is thus revealed as the key element in DFT. Unfortunately computationally convenient formulae for the exchange-correlation hole are unknown, so it has to be approximated.

When developing approximations for the exchange-correlation energy through the exchange- correlation hole, it is important that certain properties of the exact exchange-correlation hole are imposed on its approximant forms. Some of these restrictions are, in perceived order of importance: (a) the normalization of the hole (avoiding self-interaction error),[9,11] (b) the holes depth (the probability of finding two electrons with specified spins at the same position; this is also called the on-top value),[12-20] (c) the holes near-coalescence behavior (the probability of finding two electrons with specified spins very close together, but not quite at the same place; this is where the local kinetic energy becomes important),[18-24] (d) the holes asymptotics (related to the probability of finding two electrons far apart; this is important for dispersion forces),[2,25-28] and (e) reproduction of the uniform electron gas limit.[29,30]

Most common approximations for the exchange-correlation energy functional do not explicitly depend on the exchange correlation hole; partly this is because the six-dimensional singular integrand in Eq. (5) cannot be integrated analytically, and numerical integration is difficult due to the dimensionality and the singularity. Instead, most approaches are based on the exchange-correlation energy density per particle,

$$E_{xc}[\rho] = \int \rho(\mathbf{r}) e_{xc}[\rho; \mathbf{r}] d\mathbf{r} \tag{6}$$

Approximations using Eq. (6) are commonly used in many applications, from atoms and small molecules up to biological macromolecules and complex materials. In most cases, DFT calculations give highly accurate results, suitable for experimental comparison and interpretation. Specifically, for systems where the electron hole is localized, and for properties whose values are dominated by the short-distance portion of the hole, existing exchange-correlation energy functionals are almost always sufficient. By contrast, for systems where the exchange-correlation hole is delocalized[6,7,31–35] (e.g., systems with "nonclassical" chemical bonding, transition states of chemical reactions, and antiferromagnetic open-shell systems like Mott insulators) or where the long-range properties of the exchange-correlation hole is decisive (e.g., dispersion interactions),[26,35–42] existing exchange-correlation functionals fail catastrophically. For example, commonly used exchange-correlation energy functionals often overestimate chemical reaction rates by orders of magnitude and give a qualitatively incorrect description of dispersion.

The goal of this project is to construct exchange-correlation holes that model long-range exchange and correlation effects. This is a very challenging task because the long-range portion of the exchange- correlation hole is highly structured and strongly system-dependent. By contrast, the short-range portion of the exchange-correlation hole has a relatively simple structure and a relatively straightforward dependence on the density, density gradient, kinetic energy density, etc. This is why "conventional" DFT approximations work well in cases where long-range form of $h_{xc}(\mathbf{r}, \mathbf{r}')$ may be modeled by simply "extending" the short- range form.

2. The Ornstein-Zernike Equation for the Exchange-correlation Hole

Recently, March and his coworkers have developed approximations for the exchange-correlation hole based on the classical direct correlation function (DCF) for the homogeneous electron liquid.[43–45] Their approach is based on using the classical Ornstein-Zernike equation for classical liquids to approximate the exchange-correlation hole.[46–48] I.e.

$$h_{xc}(\mathbf{r}, \mathbf{r}') = c_{xc}(\mathbf{r}, \mathbf{r}') + \int c_{xc}(\mathbf{r}, \mathbf{x}) \rho(\mathbf{x}) h_{xc}(\mathbf{x}, \mathbf{r}') d\mathbf{x} \tag{7}$$

In classical liquids, the hole-correlation function (usually called the "total correlation function") has long- range stucture arising from steric ("packing") and electrostatic interactions, leading to a series of peaks and valleys in $h_{xc}(\mathbf{r}, \mathbf{r}')$. So $h_{xc}(\mathbf{r}, \mathbf{r}')$ is very sensitive to the interaction potential. (In fact, reproduction of the total correlation function is a common test used to assess approximate interaction potentials.) For this reason, the total correlation function is rarely constructed directly. Instead an auxiliary function, called the direct correlation function (DCF), is introduced. The DCF usually has a shorter range and simpler form than the total correlation function. The DCF also seems to have a more "universal" form than the total correlation function; it is less sensitive to the details of the interparticle interactions. The relationship between the total correlation function and the DCF is defined by the Ornstein-Zernike (OZ) equation (7).[46–48]

Our hypothesis is that a suitably defined electronic DCF will have a simple form that we can approximate computationally, and that this form will be "universal" enough to reproduce the complicated, highly-system dependent exchange correlation hole.

Unfortunately, the classical OZ equation, Eq. (7), proves to be unacceptable for atoms and molecules. To see why, multiply both sides of the OZ equation by $\rho(\mathbf{r}')$ and integrate with respect to $\mathbf{r}'$, obtaining,

$$\int \rho(\mathbf{r}')h_{xc}(\mathbf{r}, \mathbf{r}')d\mathbf{r}' = \int \rho(\mathbf{r}')c_{xc}(\mathbf{r}, \mathbf{r}')d\mathbf{r}' + \int \int \rho(\mathbf{r}')c_{xc}(\mathbf{r}, \mathbf{x})\rho(\mathbf{x})h_{xc}(\mathbf{x}, \mathbf{r}')d\mathbf{x}d\mathbf{r}'$$

(8)

Assuming that the integrands are sufficiently well-behaved, we can interchange the integration with respect to $\mathbf{r}'\mathbf{r}$ and $\mathbf{x}$ in the latter integral. Then, using the normalization condition on the exchange-correlation hole

$$\int \rho(\mathbf{r})h_{xc}(\mathbf{r}, \mathbf{r}')d\mathbf{r} = -1$$

(9)

we obtain the contradiction

$$-1 = \int \rho(\mathbf{r}')h_{xc}(\mathbf{r}, \mathbf{r}')d\mathbf{r}' - \int \rho(\mathbf{x})c_{xc}(\mathbf{r}, \mathbf{x})d\mathbf{x}$$

(10)

This means that the assumptionthat the integrands were well-behavedis invalid. In particular, the normalization integral for the direct correlation function (cf. Eq. (10)) must diverge. This means that every well-behaved direct correlation function will give an exchange correlation hole with incorrect normalization. This error is not that important in the thermodynamic limit, because an error in the normalization of the pair distribution function of order one is not important compared to Avogadros number. But in the context of molecular electronic structure theory, where the number of electrons is twenty orders of magnitude smaller, this error is devastating.

3. The Ornstein-Zernike Equation for Indistinguishable Particles

To motivate our generalization of the Ornstein-Zernike equation for the electronic structure problem, let us first interpret the OZ equation for classical liquids. Repeatedly substituting the OZ equation for $h_{xc}(\mathbf{r}, \mathbf{r}')$ into the right-hand-side of Eq.

(7) gives

$$h_{xc}(\mathbf{r}, \mathbf{r}') = c_{xc}(\mathbf{r}, \mathbf{r}') + \int c_{xc}(\mathbf{r}, \mathbf{x_1})\rho(\mathbf{x_1})c_{xc}(\mathbf{x_1}, \mathbf{r}')d\mathbf{x_1}$$

$$+ \int \int c_{xc}(\mathbf{r}, \mathbf{x_1})\rho(\mathbf{x_1})c_{xc}(\mathbf{x_1}, \mathbf{x_2})\rho(\mathbf{x_2})c_{xc}(\mathbf{x_2}, \mathbf{r}')d\mathbf{x_1}d\mathbf{x_2}$$

$$+ \int \int \int c_{xc}(\mathbf{r}, \mathbf{x_1})\rho(\mathbf{x_1})c_{xc}(\mathbf{x_1}, \mathbf{x_2})\rho(\mathbf{x_2})c_{xc}(\mathbf{x_2}, \mathbf{x_3})\rho(\mathbf{x_3})c_{xc}(\mathbf{x_3}, \mathbf{r}')d\mathbf{x_1}d\mathbf{x_2}d\mathbf{x_3}$$

$$(11)$$

Thus, the total correlation function is the sum of the DCF (the direct interaction between two particles) and the sum of all possible "chains" of direct interactions including one or more of the remaining particles in the system. This captures the fact that a particle at $\mathbf{r}$ can interact directly with a particle at $\mathbf{r}'$. Alternatively, the particle at $\mathbf{r}$ can interact with a particle at $\mathbf{x_1}$ that, in turn, is interacting with the particle at $\mathbf{r}'$. The probability of this type of interaction is modulated by the probability of observing a particle at $\mathbf{x_1}$, $\rho(\mathbf{r})$. Particles at $\mathbf{r}$ and $\mathbf{r}'$ can also interact through two, three, four, etc. "intermediate" particles. Since the series does not truncate, the OZ equation implicitly assumes that the number of particles in the system (ergo the number of possible "intermediate" particles) is infinite.

To facilitate a probabilistic interpretation of the OZ equation, we first multiply both sides of Eq. (11) by $\rho(\mathbf{r})\,\rho(\mathbf{r}')$. We then modify the coefficients of each term in the series to account for the facts that (a) electrons are indistinguishable particles and (b) the number of electrons in a molecule is finite number, N. Thus, since there are two ways of assigning $\mathbf{r}$ and $\mathbf{r}'$ to the "reference electrons" that we are studying the correlation between, we divide the left-hand-side of the equation by 2. Since (a) the electron density is normalized to the number of electrons, N, and (b) the number of different ways to select a subset of k electrons is $\binom{N}{k} = N(N-1)(N-2)(N-k+1)/k!$, we multiply the terms on the right-hand side of Eq. (11) by $N^{-k}\binom{N}{k}$. The final expression is then:

$$\underbrace{\frac{\rho(\mathbf{r})\rho(\mathbf{r}')}{2}h_{xc}(\mathbf{r}, \mathbf{r}')}_{(a)} = \underbrace{\frac{1}{N^2}\binom{N}{2}\rho(\mathbf{r})\rho(\mathbf{r}')\,c_{xc}(\mathbf{r}, \mathbf{r}')}_{(b)}$$

$$+ \int \underbrace{\frac{1}{N^3}\binom{N}{3}\rho(\mathbf{r})\rho(\mathbf{r}')\rho(\mathbf{x_1})\,c_{xc}(\mathbf{r}, \mathbf{x_1})c_{xc}(\mathbf{x_1}, \mathbf{r}')d\mathbf{x_1}}_{(c)}$$

$$+ \int \int \underbrace{\frac{1}{N^4}\binom{N}{4}\rho(\mathbf{r})\rho(\mathbf{r}')\rho(\mathbf{x_1})\rho(\mathbf{x_2})\,c_{xc}(\mathbf{r}, \mathbf{x_1})c_{xc}(\mathbf{x_1}, \mathbf{x_2})c_{xc}(\mathbf{x_2}, \mathbf{r}')d\mathbf{x_1}d\mathbf{x_2}}_{(d)} + \dots$$

$$(12)$$

where: (a) is the probability deficit when electrons are at $\mathbf{r}$ and $\mathbf{r}'$; (b) is the probability of observing independent particles at $\mathbf{r}$ and $\mathbf{r}'$; (c) is the probability of observing independent particles at $\mathbf{r}$, $\mathbf{r}'$ and $\mathbf{x_1}$; and, (d) is the probability of observing independent particles at $\mathbf{r}$, $\mathbf{r}'$, $\mathbf{x_1}$ and $\mathbf{x_2}$.

The exchange-correlation energy functional built from Eq. (12) is not size consistent. For two systems that are far apart, the density cumulant must be separable,

$$\rho^{(AB)}(\mathbf{r})\rho^{(AB)}(\mathbf{r}')h^{(AB)}(\mathbf{r},\mathbf{r}') \xrightarrow{\ |\mathbf{R}^A-\mathbf{R}^B|\to\infty\ } \rho^{(A)}(\mathbf{r})\rho^{(A)}(\mathbf{r}')h^{(A)}(\mathbf{r},\mathbf{r}')$$
$$+ \rho^{(B)}(\mathbf{r})\rho^{(B)}(\mathbf{r}')h^{(B)}(\mathbf{r},\mathbf{r}') \quad (13)$$

Eq. (12) is not exactly separable because the coefficients of the terms on the right-hand-side are modified by the addition of a subsystem infinitely far away through the change in the overall number of electrons (e.g., $N^{(AB)}$ becomes $N^{(A)} + N^{(B)}$). To reimpose size consistency, we take the limit as the number of electrons goes to infinity, obtaining,

$$\frac{1}{2!}h_{xc}(\mathbf{r},\mathbf{r}') = \frac{1}{2!}c_{xc}(\mathbf{r},\mathbf{r}') + \frac{1}{3!}\int c_{xc}(\mathbf{r},\mathbf{x_1})\rho(\mathbf{x_1})h_{xc}(\mathbf{x},\mathbf{r}')d\mathbf{x}$$
$$+ \frac{1}{(k+2)!}\int\int \cdots \int c_{xc}(\mathbf{r},\mathbf{x_1})\rho(\mathbf{x_1})c_{xc}(\mathbf{x_1},\mathbf{x_2})\rho(\mathbf{x_2})$$
$$\times \cdots \rho(\mathbf{x_k})c_{xc}(\mathbf{x_k},\mathbf{r}')d\mathbf{x_1}\cdots d\mathbf{x_k} \quad (14)$$

Equation (14) is precisely the "classical" OZ equations, corrected by factorial coefficients to account for the indistinguishability of identical quantum particles. Because we have hypothesized that the electronic DCF has a simple, universal form that is nearly transferable between different systems, we can parameterize the DCF using the uniform electron gas. The result is shown in Figure 1. Notice that the DCF obtained using Eq. (14) is more localized than the one obtained by Amovilli and March using Eq. (7). In fact, the asymptotic decay of our DCF is the same as that of the exchange-correlation hole, while that of Amovilli and March is significantly slower. The OZ equation for indistinguishable particles, Eq. (14), resolves the issue with the normalization of the exchange-correlation hole. As long as the DCF is appropriately normalized,

$$\int c_{xc}(\mathbf{r},\mathbf{r}')\rho(\mathbf{r}')d\mathbf{r}' = -1.5936 \quad (15)$$

the exchange-correlation hole will be normalized also.

4. Challenges in Numerical Implementation

As written, Eq. (14) requires evaluating an infinite series, with each term in the series requiring integration of a function with increasing dimension. A practical method for evaluating this integral can be achieved b

$$\mathbf{H}(\mathbf{r},\mathbf{r}') = \sqrt{\rho(\mathbf{r})}h_{xc}(\mathbf{r},\mathbf{r}')\sqrt{\rho(\mathbf{r}')}$$
$$\mathbf{C}(\mathbf{r},\mathbf{r}') = \sqrt{\rho(\mathbf{r})}c_{xc}(\mathbf{r},\mathbf{r}')\sqrt{\rho(\mathbf{r}')} \quad (16)$$

In terms of these matrices, Eq. (14) can be rewritten as

$$\mathbf{H} \cdot \mathbf{C} = 2\left[e^{\mathbf{C}} - \mathbf{C} - \mathbf{I}\right] \quad (17)$$

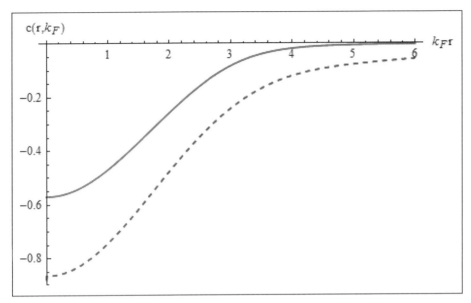

Fig. 1. Electronic direct correlation functions for only-exchange. The solid curve corresponds to the N model direct correlation function. The dashed curve corresponds to Amovilli and March direct correlation function.

where $\mathbf{I}$ is the identity matrix and the product of continuous matrices is defined as

$$(\mathbf{A} \cdot \mathbf{B})(\mathbf{r}, \mathbf{r}') = \int \mathbf{A}(\mathbf{r}, \mathbf{x}) \mathbf{B}(\mathbf{x}, \mathbf{r}') d\mathbf{x} \qquad (18)$$

Using Eq.(17), the exchange-correlation hole can be evaluated quickly using standard linear algebra methods. By exploiting the nearsightedness of the DCF, Eq. (17) can be evaluated with computational cost growing linearly with the size of the electronic system. After the exchange-correlation hole has been obtained, the exchange-correlation energy can be evaluated using Eq. (5). The integral in Eq. (5) is challenging for two reasons. First of all, because the integral is six-dimensional, unlike the usual three-dimensional integrals in DFT, it is computationally demanding to compute. Efficient numerical integration techniques are required. Second, it is challenging to evaluate the integral numerically because of the singularity inherent to the Coulomb potential. We have considered several possible approaches for evaluating Eq. (5), which we now briefly summarize.

4.1. Short-range/Long-range decomposition

Andreas Savin has developed a general approach that resolves the Coulomb singularity known as the short-range/long-range decomposition.[49-54] The idea is to express the potential as

$$\frac{1}{r} = \nu_{ee}^{\mu}(r) + \bar{\nu}_{ee}^{\mu}(r) \qquad (19)$$

where $\nu_{ee}^{\mu}(r)$ is a long-range interaction and $\bar{\nu}_{ee}^{\mu}(r)$ is the complementary short-range interaction. This separation is controlled by the parameter μ. Many different forms for $\nu_{ee}^{\mu}(r)$ have been proposed.[50,53,55,56] We used the erfgau separation,[53]

$$\nu_{ee,erfgau}^{\mu}(r) = \frac{erf(\mu r)}{r} - \frac{2\mu}{\sqrt{\pi}}e^{-\frac{1}{3}\mu^2 r^2} \tag{20}$$

The long-range portion of the exchange-correlation energy is then a non-singular six-dimensional integral,

$$E_{xc}^{\mu}[\rho] = \int\int \rho(\mathbf{r})\rho(\mathbf{r}')h_{xc}(\mathbf{r},\mathbf{r}')\nu_{ee}^{\mu}(|\mathbf{r}-\mathbf{r}'|)d\mathbf{r}d\mathbf{r}' \tag{21}$$

We approximated the remaining component of the exchange-correlation energy

$$\bar{E}_{xc}^{\mu}[\rho] = \int\int \rho(\mathbf{r})\rho(\mathbf{r}')h_{xc}(\mathbf{r},\mathbf{r}')\bar{\nu}_{ee}^{\mu}(|\mathbf{r}-\mathbf{r}'|)d\mathbf{r}d\mathbf{r}' \tag{22}$$

using the appropriate short-ranged local density approximation (LDA) for the integral. Another approach would be to approximate the short-range contribution by substituting the small $|\mathbf{r}-\mathbf{r}'|$ expansion of $h_{xc}(\mathbf{r},\mathbf{r}')$ into Eq. (5), and then perform the short-range integral analytically. In both cases, because we can only approximate the short-ranged integral, we have to extrapolate to infinite μ,

$$E_{xc}[\rho] = \underbrace{\lim}_{\mu\to\infty} \left(E_{xc}^{\mu}[\rho] + \bar{E}_{xc}^{\mu}[\rho]\right) \tag{23}$$

Performing this procedure for the Argon atom leads to en exchange energy of -28.01 Hartrees, compared to -27.86 Hartrees for the LDA and the exact value of -30.98 Hartrees.

4.2. Center-of-mass transformation of coordinates

If one uses center-of-mass coordinates,

$$\mathbf{R} = (\mathbf{r}+\mathbf{r}')/2$$
$$\mathbf{u} = (\mathbf{r}'-\mathbf{r})/2 \tag{24}$$

he exchange-correlation energy expression in Eq. (5) becomes

$$E_{xc}[\rho] = \frac{1}{4}\int d\mathbf{R}\int d\mathbf{u}\rho(\mathbf{R}+\mathbf{u})\rho(\mathbf{R}-\mathbf{u})h_{xc}(\mathbf{R},\mathbf{u})\frac{1}{u} \tag{25}$$

Focusing on the inner integral we have

$$\int d\mathbf{u}\rho(\mathbf{R}+\mathbf{u})\rho(\mathbf{R}-\mathbf{u})h_{xc}(\mathbf{R},\mathbf{u})\frac{1}{u} \tag{26}$$

$$= \int_0^{2\pi}d\phi_{\mu}\int_0^{\pi}d\theta_{\mu}\int_0^{\infty}du\left[\rho(\mathbf{R}+\mathbf{u})\rho(\mathbf{R}-\mathbf{u})h_{xc}(\mathbf{R},\mathbf{u})u^2 sin\theta_{\mu}\left(\frac{1}{u}\right)\right]$$

Observe that the Coulomb singularity is cancelled out by the volume element in the center-of-mass coordinate system. By repeatedly integrating by parts with respect to u, we eventually obtain the expression,

$$E_{xc}[\rho] = \frac{1}{2} \int \int \left\{ \frac{d^4}{d(|\mathbf{r} - \mathbf{r}'|)^4} \left[|\mathbf{r} - \mathbf{r}'|^2 \rho(\mathbf{r}) \rho(\mathbf{r}') h_{xc}(\mathbf{r}, \mathbf{r}') \right] \right\}$$
$$\times \left\{ \frac{|\mathbf{r} - \mathbf{r}'|}{6} \left(\ln |\mathbf{r} - \mathbf{r}'| - \frac{11}{6} \right) \right\} d\mathbf{r} d\mathbf{r}' \qquad (27)$$

Notice that none of the above presented integrals are infinite and the problematic term (first term in {}) has been replaced by a term that is zero at $|\mathbf{r} - \mathbf{r}'| \to 0$. The problem with this approach is the presence of the fourth derivative, which is ill-conditioned numerically and extremely complicated to evaluate analytically.

4.3. *Greens theorem*

In the preceding methods, one still needs to evaluate a six-dimensional integral. To derive an alternative, notice that applying Greens theorem,

$$\int \phi(\mathbf{R}, \mathbf{u}) \nabla^2 \psi(\mathbf{u}) d\mathbf{u} = \int \left(\nabla^2 \phi(\mathbf{R}, \mathbf{u}) \right) \psi(\mathbf{u}) d\mathbf{u}$$
$$+ \underbrace{\lim_{\mu \to \infty}} \oint \left(\phi(\mathbf{R}, \mathbf{u}) \nabla \psi(\mathbf{u}) \cdot \psi(\mathbf{u}) \nabla \phi(\mathbf{R}, \mathbf{u}) \right) \cdot \mathbf{n} da \quad (28)$$

to

$$\psi(\mathbf{u}) = \frac{1}{u} \qquad (29)$$

$$\nabla_{\mathbf{u}}^2 \phi(\mathbf{R}, \mathbf{u}) = \rho(\mathbf{R} + \mathbf{u}) \rho(\mathbf{R} - \mathbf{u}) h(\mathbf{R}, \mathbf{u}) \qquad (30)$$

gives

$$-4\pi \int \phi(\mathbf{R}, \mathbf{u}) \delta(\mathbf{u}) d\mathbf{u} = \int \rho(\mathbf{R} + \mathbf{u}) \rho(\mathbf{R} - \mathbf{u}) h_{xc}(\mathbf{R}, \mathbf{u}) \frac{1}{u} d\mathbf{u}$$
$$-4\pi \phi(\mathbf{R}, 0) = \int \rho(\mathbf{R} + \mathbf{u}) \rho(\mathbf{R} - \mathbf{u}) h_{xc}(\mathbf{R}, \mathbf{u}) \frac{1}{u} d\mathbf{u} \qquad (31)$$

The surface terms in Eq. (28) are zero because of the exponentially fast asymptotic decay of the electron density in any finite system. The exchange-correlation energy can then be obtained by solving the (three-dimensional) Poisson-like equation, Eq. (30), near $\mathbf{u} = 0$ to obtain $\phi(\mathbf{R}, 0)$. Three-dimensional integration then gives the exchange-correlation energy.

$$E_{xc}[\rho] = -\pi \int \phi(\mathbf{R}, 0) d\mathbf{R} \qquad (32)$$

An alternative to this two-step procedure is to solve the three-dimensional Poisson equation,

$$\nabla_{\mathbf{u}}^2 \Phi(\mathbf{u}) = P(\mathbf{u}) \tag{33}$$

where

$$\Phi(\mathbf{u}) = \int \phi(\mathbf{R}, \mathbf{u}) d\mathbf{R}$$

$$P(\mathbf{u}) = \int \rho(\mathbf{R} + \mathbf{u})\rho(\mathbf{R} - \mathbf{u})h(\mathbf{R}, \mathbf{u}) d\mathbf{R} \tag{34}$$

The exchange-correlation energy is then

$$E_{xc}[\rho] = -\pi \Phi(0) \tag{35}$$

These approaches only require performing 3-dimensional integrals and solving-three dimensional differential equations. Unfortunately, one must either solve a three-dimensional differential equation at many different points in space (Eq. (30)) or evaluate a three-dimensional integral at many different points in space (Eq. (35).)

4.4. Eigenvector decomposition

Consider the eigenvector decomposition of the exchange-correlation hole,

$$h_{xc}(\mathbf{r}, \mathbf{r}') = \sum_i \tau_i \eta_i(\mathbf{r}) \eta_i(\mathbf{r}') \tag{36}$$

where τ_i and $\eta_i(\mathbf{r})$ are the eigenvalues and eigenvectors, respectively. The exchange-correlation energy can then be evaluated as

$$E_{xc}[\rho] = \frac{1}{2} \sum_i \tau_i \int \rho(\mathbf{r})\eta_i(\mathbf{r}) \left(\int \frac{\eta_i(\mathbf{r}')\rho(\mathbf{r}')}{|\mathbf{r} - \mathbf{r}'|} d\mathbf{r}' \right) d\mathbf{r} \tag{37}$$

The inner integral is the electrostatic potential due to the charge distribution $\eta_i(\mathbf{r}')\rho(\mathbf{r}'$ and it can be evaluated by solving the three-dimensional Poisson equation.

$$\nabla_{\mathbf{r}}^2 \Phi(\mathbf{r}) = -4\pi \eta_i(\mathbf{r})\rho(\mathbf{r}) \tag{38}$$

Then a simple three-dimensional integral suffices to determine the exchange-correlation energy,

$$E_{xc}[\rho] = \frac{1}{2} \int \rho(\mathbf{r}) \sum_i \tau_i \eta_i(\mathbf{r}) \Phi_i(\mathbf{r}) d\mathbf{r} \tag{39}$$

The most difficult step here is solving the Poisson equation. We have observed that many of the eigenvalues are nearly zero, so one only has to solve for $\Phi_i(\mathbf{r})$ for a few of the eigenvectors. We could solve the Poisson equation using purely grid-based methods, but it seems more efficient to use collocation. We expand the potential in a complete basis set,

$$\Phi_i(\mathbf{r}) = \sum_{j=1} k_{ij} \chi_j(\mathbf{r}) \tag{40}$$

and then equation (40) turns into

$$\sum_{j=1} k_{ij} \nabla^2 \chi_j(\mathbf{r}_a) = -4\pi \eta_i(\mathbf{r}_a)\rho(\mathbf{r}_a) \tag{41}$$

where $\{\mathbf{r}_a\}$ are the points on the numerical integration grid. The number of grid points is always much greater than the number of basis functions, so the error in the linear equation (41) is minimized in the least-squares sense. The cost of this method grows quadratically with the number of basis functions, so it is important to develop a good basis set. At first, we considered products of spherical harmonics with Laguerre polynomials and products of spherical harmonics with spherical Bessel functions. Currently we are exploring the Gaussian potentials

$$\Phi_{lm}^{(\alpha)}(\mathbf{R}) = \int \frac{r^l Y_l^m(\theta, \phi)e^{-\alpha r^2}}{|\mathbf{r} - \mathbf{R}|} d\mathbf{r} \tag{42}$$

and the Slater potentials,

$$\Phi_{lm}^{(\alpha)}(\mathbf{R}) = \int \frac{r^l Y_l^m(\theta, \phi)e^{-\alpha r}}{|\mathbf{r} - \mathbf{R}|} d\mathbf{r} \tag{43}$$

These basis sets seem more appropriate for atomic and molecular basis calculations. We are also combining these methods with our previous work on computational methods for solving the Poisson equation[57] and for numerical integration.[58-60]

5. Summary

March and his coworkers have recent explored models for the exchange-correlation hole in density-functional theory based on the Ornstein-Zernike (OZ) equation for classical liquids. Unfortunately, the OZ equation for classical liquids is inconsistent with the normalization of the exchange-correlation hole in the sense that for any well-behaved direct correlation function, the resulting exchange-correlation hole is not normalized. To rectify this, we propose a revised Ornstein-Zernike-like equation that is appropriate for indistinguishable particles, Eq. (14).

Because the OZ-based approach to the exchange-correlation energy is based on a truly nonlocal, fully six-dimensional, exchange-correlation hole, this approach can address long-range correlations like those associated with left-right correlation, antiferromagnetic coupling, nonclassical bonding, and dispersion interactions. For the same reason, however, this approach is also significantly more computationally demanding than conventional approaches based on the exchange-correlation energy density. We present several efficient algorithms for evaluating the exchange-correlation hole and the exchange-correlation energy. Of these, the eigenvector decomposition technique (section 4.4) seems to be the most efficient possibility to us, but other approaches may be more useful in certain implementations. Approaches based on Greens theorem (section 4.3) might be particular useful in plane-wave DFT codes. Any of these approaches would be preferable to a "brute strength" implementation of the sorts of integrals that occur in kernel-based nonlocal density functionals.

Acknowledgments

We thank NSERC, Sharcnet, and the Canada Research Chairs for funding. RCS thanks CONACYT for a graduate fellowship as well as ITESM and DGRI-SEP for complementary financial support.

References

1. R. G. Parr and W. Yang, Density-Functional Theory of Atoms and Molecules. (Oxford UP, New York, 1989).
2. R. M. Dreizler and E. K. U. Gross, Density Functional Theory: An Approach to the Quantum Many- Body Problem. (Springer-Verlag, Berlin, 1990).
3. W. Kohn, Reviews of Modern Physics 71, 1253 (1999).
4. P. Hohenberg and W. Kohn, Phys.Rev. 136, B864 (1964).
5. W. Kohn and L. J. Sham, Phys.Rev. 140, A1133 (1965).
6. E. J. Baerends and O. V. Gritsenko, J. Phys. Chem. A 101, 5383 (1997).
7. P. W. Ayers, W. Yang, P. Bultinck, H. de Winter, W. Langenaeker, and J. P. Tollenaere, in Computational Medicinal Chemistry for Drug Discovery (Dekker, New York, 2003), pp. 571.
8. J. Harris and R. O. Jones, J.Phys.F 4, 1170 (1974).
9. O. Gunnarsson and B. I. Lundqvist, Phys. Rev. B 13, 4274 (1976).
10. D. C. Langreth and J. P. Perdew, Phys. Rev. B 15, 2884 (1977).
11. J. P. Perdew and A. Zunger, Phys. Rev. B 23, 5048 (1981).
12. J. P. Perdew, A. Savin, and K. Burke, Phys. Rev. A 51, 4531 (1995).
13. J. P. Perdew, M. Ernzerhof, K. Burke, and A. Savin, Int. J. Quantum Chem. 61, 197 (1997).
14. H. Yasuhara and Y. Kawazoe, Physica A 85, 416 (1976).
15. A. W. Overhauser, Can. J. Phys. 73, 683 (1995).
16. K. Burke, J. P. Perdew, and M. Ernzerhof, J. Chem. Phys. 109, 3760 (1998).
17. V. A. Rassolov, J. A. Pople, and M. A. Ratner, Phys. Rev. B 59, 15625 (1999).
18. J. C. Kimball, Journal of Physics A 8, 1513 (1975).
19. M. Corona, P. Gori-Giorgi, and J. P. Perdew, Phys. Rev. B 69, 045108 (2004).
20. P. Gori-Giorgi and J. P. Perdew, Phys. Rev. B 6415, 155102 (2001).
21. V. A. Rassolov and D. M. Chipman, J. Chem. Phys. 104, 9908 (1996).
22. J. C. Kimball, Phys. Rev. A 7, 1648 (1973).
23. T. Kato, Commun.Pure Appl.Math. 10, 151 (1957).
24. R. T. Pack and W. B. Brown, J. Chem. Phys. 45, 556 (1966).
25. O. Gunnarsson and R. O. Jones, Phys. Scr. 21 (3-4), 394 (1980).
26. M. Dion, H. Rydberg, E. Schroder, D. C. Langreth, and B. I. Lundqvist, Phys. Rev. Lett. 92, 246401 (2004).
27. P. Gori-Giorgi and P. Ziesche, Phys. Rev. B 66, 235116 (2002).
28. M. H. Cohen and M. V. Ganduglia-Pirovano, J. Chem. Phys. 101, 8988 (1994).
29. P. Gori-Giorgi and J. P. Perdew, Phys. Rev. B 69, 041103 (2004).
30. P. Gori-Giorgi and J. P. Perdew, Phys. Rev. B 66, 165118 (2002).
31. Y. Zhang and W. Yang, J. Chem. Phys. 109, 2604 (1998).
32. O. V. Gritsenko, B. Ensing, P. R. T. Schipper, and E. J. Baerends, J. Phys. Chem. A 104, 8558 (2000).
33. A. J. Cohen, P. Mori-Sanchez, and W. T. Yang, Science 321, 792 (2008).
34. A. J. Cohen, P. Mori-Sanchez, and W. T. Yang, J. Chem. Phys. 129, 121104 (2008).
35. P. Mori-Sanchez, A. J. Cohen, and W. T. Yang, Phys. Rev. Lett. 102, 066403 (2009).

36. A. D. Becke and E. R. Johnson, J. Chem. Phys. 123 (15), 154101 (2005).
37. A. D. Becke and E. R. Johnson, J. Chem. Phys. 122 (15), 154104 (2005).
38. E. R. Johnson and A. D. Becke, J. Chem. Phys. 123 (2), 024101 (2005).
39. A. D. Becke and E. R. Johnson, J. Chem. Phys. 124 (1), 014104 (2006).
40. A. D. Becke and E. R. Johnson, J. Chem. Phys. 127, 154108 (2007).
41. J. Angyan, J. Chem. Phys. 127, 024108 (2007).
42. P. W. Ayers, J. Math. Chem. 46, 86 (2009).
43. C. Amovilli and N. H. March, Phys. Rev. B 76, 195104 (2007).
44. N. H. March, Phys. Chem. Liq. 46, 465 (2008).
45. C. Amovilli, N. H. March, and A. Nagy, Phys. Chem. Liq. 47, 5 (2009).
46. L. S. Ornstein and F. Zernike, Proc. Akad. Sci. (Amsterdam) 17, 793 (1914).
47. N. H. March, Electron correlation in molecules and condensed phases. (Plenum, Oxford, 1996).
48. J. K. Percus, H. L. Frisch, and J. L. Lebowitz, in The Equilibrium Theory of Classical Fluids (W. A. Benjamin, New York, 1964).
49. B. Miehlich, H. Stoll, and A. Savin, Mol. Phys. 91, 527 (1997).
50. T. Leininger, H. Stoll, H. J. Werner, and A. Savin, Chem. Phys. Lett. 275, 151 (1997).
51. J. Toulouse, F. Colonna, and A. Savin, J. Chem. Phys. 122 (1) (2005).
52. J. Toulouse, F. Colonna, and A. Savin, Phys. Rev. A 70 (6) (2004).
53. J. Toulouse, A. Savin, and H. J. Flad, Int. J. Quantum Chem. 100 (6), 1047 (2004).
54. W. Yang, J. Chem. Phys. 109, 10107 (1998).
55. A. Savin and H. J. Flad, Int. J. Quantum Chem. 56, 327 (1995).
56. R. Baer and D. Neuhauser, Phys. Rev. Lett. 94, 043002 (2005).
57. D. C. Thompson and P. W. Ayers, Int. J. Quantum Chem. 106, 787 (2006).
58. J. S. M. Anderson, J. I. Rodriguez, D. C. Thompson, and P. W. Ayers, in Quantum Chemistry Research Trends (Nova, Hauppauge, NY, 2007).
59. J. I. Rodriguez, D. C. Thompson, P. W. Ayers, and A. M. Koster, J. Chem. Phys. 128, 224103 (2008).
60. J. I. Rodriguez, D. C. Thompson, J. S. M. Anderson, J. W. Thomson, and P. W. Ayers, Journal of Physics A 41, 365202 (2008).

FEATURES AND CATALYTIC PROPERTIES OF RhCu: A REVIEW

SILVIA GONZALEZ

Instituto de Química Aplicada, Universidad Técnica Particular de Loja,
San Cayetano Alto s/n, Ap. P. 1101608 Loja, Ecuador
sgonzalez@utpl.edu.ec

CARMEN SOUSA

Departament de Química Física and Institut de Química Teorica i Computacional (IQTCUB),
Universitat de Barcelona, C/ Martíi Franques 1, E-08028 Barcelona, Spain
c.sousa@ub.edu

FRANCESC ILLAS

Departament de Química Física and Institut de Química Teorica i Computacional (IQTCUB),
Universitat de Barcelona, C/ Martíi Franques 1, E-08028 Barcelona, Spain
francesc.illas@ub.edu

Received 28 November 2009

The study of bimetallic catalysts has scientific and technologic importance because of special catalytic activity towards several reactions. $RhCu$ is an interesting bimetallic system due to combination of the very different catalytic activities of Rh and Cu. The catalytic activity of this bimetallic does not result from simple interpolation of the constituents. In fact, at low Cu content, the catalytic activity of $RhCu$ is superior to that of Rh but when the Cu content is higher the activity decays. This is a curious trend which theoretical works had attempted to explain. This paper reports an overview of the most recent research works about this bimetallic system with emphasis in its especial characteristics.

Keywords: RhCu; bimetallic; catalysis; DFT.

1. Introduction

Since many years ago and until nowadays, bimetallic systems had been extensively used as catalysts in a many heterogeneously catalyzed reactions.[1-4] This is because the activity and/or selectivity of this class of materials are superior of those observed in both original metal constituents. The enhanced properties of bimetallic systems are exploited in many industrial reactions, as oxidation, reforming and reduction, among others.[1,3] Consequently, the experimental and theoretical studies of these catalysts had become an interesting field of fundamental and applied research. To explain the behavior of these materials it is usual to invoke the ensemble, ligand and strain effects provoked by the second metal in the monometallic catalyst.[1,3,5-9] Ensemble effects correspond to the structural changes introduced by the second metal; these changes can induce or remove certain active centers. Ligand effects are a consequence of the heterometallic bonding that may induces changes in the electronic density of a metal catalyst. Finally, strain effects include the deformations of the original metallic lattice induced by a different metal.

The interest in $RhCu$ as a particular bimetallic catalyst started in 1970[11] and it has been evolving throughout the years,[12–15] for instance for hydrocarbons hydrogenation,[18] CO oxidation[12] and more recently for nitrate and nitrite reduction in water.[14,15] The interest in this bimetallic system arises from the fact its catalytic properties are different from those of the pure parent metals and they cannot be simply interpolated from the bimetallic composition. Indeed, the catalytic behavior of $RhCu$ systems does not follow a monotonic trend in relation to its metallic composition. For example, at low Cu content H_2 adsorption is higher than that of a Rh catalyst until a maximum is reached while at higher Cu contents the adsorption ability of the catalyst towards H_2 decreases.[19] Similar trends have been reported for the NO reduction by CO.[20]

This review provides a brief compilation of experimental and theoretical works dedicated to the study of $RhCu$ and its particularities, since 1995 until currently, and it is aimed to complement the complete works of Rodríguez[3] and Ponec[1] about bimetallic systems in general with emphasis on the theoretical treatment. This paper is organized in three main sections. In the first one we comment on the studies focusing on the $RhCu$ structure and its electronic properties. Section two describes the $RhCu$ adsorption related quantities whereas in the third and last section we provide information about several reactions where $RhCu$ acts as a catalyst.

2. RhCu: Structure and Other Properties

The alloy phase diagram of this bimetallic system indicates a very low miscibility between Rh and Cu resulting in either Rh- or Cu-rich compositions in the $RhCu$ alloys.[21,22] In recent years, this bimetallic has attracted interest from both experiment and theory. Different structures such as clusters, monolayers or single crystals either isolated or interacting with an extended variety of supports have been considered.

2.1. *Structure*

Bimetallic clusters with different metallic composition of $RhCu$ had been studied using EXAFS (Extended x-Ray Absorption Fine Structure). The results indicate that the average diameter size of these particles is of 14 with a $Rh - Cu$ distance of 2.64 , a $Rh - Rh$ distance of 2.68 and a $Cu - Cu$ one of 2.63. At a higher Cu content, this metal is predominantly in the particles surface.[23] Infrared data showed that the interaction of Cu with the support prevents the re-dispersion or oxidation of Rh crystallites.[19] In different $RhCu$ samples, using Transmission Electron Microscopy (TEM) and hydrogen chemisorptions, the results indicated a larger dispersion to smaller particles size.[24] The influence of small amounts of Rh on the structure, and the reducibility of Cu-based catalysts deposited on different supports has been studied by X-Ray Diffraction (XRD) and Temperature Programmed Reduction (TPR) techniques. The analysis of obtained results suggested that addition of Rh does not affect the reduction mechanism of Cu, and Rh addition to Cu favors the formation of small particles.[25]

The results of the characterization of $RhCu/Al_2O_3$ catalysts with low metal-contents using TPR, Diffuse Reflectance Spectroscopy (DRS), X-ray Photoelectron spectroscopy (XPS), XRD and Fourier Transformed Infrared spectroscopy (FT-IR) techniques; indicate that the catalysts display an enrichment of Rh at the surface, and the presence of Cu allows the restructuring of the surface complexes.[19] Infrared spectroscopy results of adsorbed CO and ydrogenolysis of methylcyclopentane have shown that in $RhCu/Al_2O_3$ catalysts, Cu is selectively deposited on Rh surface atoms with a low coordination number, because significant changes in the spectra as well as in the reactivity were observed upon addition of Cu to Rh/Al_2O_3.[16] X-Ray Absorption Near-Edge Spectroscopy (XANES) and infrared spectroscopy detected initially only the presence of Cu aluminates and Rh oxides in $RhCu/Al_2O_3$ catalysts; increasing the quantity of both metals, bimetallic particles are formed with a Rh-rich core and a surface enriched in Cu.[26] Chou *et al.* according to results provided by Nuclear Magnetic Resonance (NMR) and H_2 chemisorption estimated a surface enrichment of Cu on supported $Cu-Rh$ crystallites supported in SiO_2. They explained their results by the formation of two alloy phases, i.e. a Rh-rich phase [(Rh), X-Cu approximate to 0.05] and a Cu-rich phase, [(Cu), X-Cu approximate to 0.8], on the surface of bimetallic crystallites.[27]

Anderson *et al.* with IR spectra of CO adsorbed on silica-supported Rh, Cu, and $RhCu$ catalysts reported that the surface of bimetallic particles was enriched with Cu, but all surfaces contained alloy phases.[28] A study employing TEM indicated that addition of Cu to Rh/SiO_2 catalysts increases the Rh dispersion and modifies the Rh clusters, either by alloy formation and/or by electron transfer from Cu to Rh metal clusters in intimate contact.[18]

Films of $RhCu$ on MgO were characterized by using different spectroscopic techniques, a perfect parallel epitaxial layer was observed[29] and enrichment in Cu in the surface.[30] The difference observed in the infrared spectra of adsorbed CO at 303 K and 513 K in the presence and absence of reactants shows that CO induces surface reconstruction at 513 K.[31]

2.2. *Electronic distribution*

IR spectral and methylcyclopentane conversion results in supported Rh and $RhCu$ indicated that charge transfer could take place from Cu to partially oxidized Rh.[16] Results of Electron Paramagnetic Resonance (EPR) detected that Rh was electron-ically perturbed in the binary particles with respect to a monometallic reference and showed that Rh is positively charged.[26] Results of XPS of monolayers of Rh on $Cu(100)$ were interpreted in terms of that Rh centers have a noticeable positive charge.[32]

2.3. *Other studied properties*

The magnetic properties of *RhCu* clusters as well as its electronic heat capacity, have been studied using NMR.[33,34] The results of the last study indicate that the *Cu* conduction electrons initially fill the *d*-band of *Rh*, and once this is filled the rest of electron population of the *Cu* conduction band goes into the *s*-band. Characterization of surface area of different *RhCu* catalysts using hydrogen chemisorption exhibits an inverse behavior with the isosteric heat of adsorption.

3. Adsorption Properties

Diverse studies had been realized for measure the effects of a second metal (*Cu* or *Rh*) on the adsorption properties of a metallic catalyst (*Rh* or *Cu*), in order to determine if its presence improves the catalytic properties of the original metal. In this review, results about the changes in the adsorption of H_2, *CO* and sulfur molecules are presented; additionally we included several results of theoretical studies of molecular adsorption on *Rh*, *Cu* and the *RhCu* bimetallic.

3.1. *Hydrogen adsorption*

Chou *et al.* studied the irreversible hydrogen uptake of *RhCu* crystallites supported samples. They found that the adsorption increases with a small amount of *Cu* in $RhCu/Al_2O_3$ but decreases with a larger quantity of *Cu*. For explain the enhanced activity in the *Rh*-rich catalyst, the authors suggested that new active sites are formed by the interaction of the two metals and surface contains some *Rh* atoms surrounded by *Cu* thus leading to a high ability for hydrogen chemisorption. Upon increasing *Cu* content, *Cu* atoms tend to occupy more surface sites and eventually segregate *Cu* islands which finally cover the whole surface and prevent the H_2 adsorption.[27] Similar behavior was observed by others authors and according their IR results, they deduced that the presence of *Cu* decreases top site and increases bridge one.[50]

3.2. *CO adsorption*

IR studies of *CO* adsorption in *Rh* and *Cu* supported in Al_2O_3 detected as the preferred adsorption site, atop, and in *Rh−Cu* particles no changes were observed.[19] In another study using *in-situ* IR reveals that *Cu* blocks the formation of bridged *CO* and that addition of *Cu* causes little variation in carbon monoxide insertion activity. The authors attributed the change to a high hydrogenation activity of *Cu* and a possible electronic interaction between both metals leading to a reduced *Rh* surface that is less active for carbon monoxide insertion than the oxidized *Rh*.[31]

Using IR Anderson *et al.* observed that addition of *Cu* to *Rh* supported on SiO_2 blocked sites for the formation of $Rh(CO)_2$ and decreased the proportion of atop sites and favored bridge-bonded sites. On the other hand, *Cu* largely poisoned the catalytic activity of Rh.[28]

3.3. *Sulfur adsorption*

Rodríguez *et al.* observed that the rate of sulfurization in a Cu adlayer on $Rh(111)$ significantly enhances with respect to a clean Rh surface, and explained this behavior by direct and indirect interactions between $S - Cu - Rh$. Another findings about the presence of Cu is an increase in the thermal stability of S on Rh and the presence of sulfured species different to the formed in Rh surface.[51]

3.4. *Theoretical studies*

Theoretical studies of $RhCu$ crystals,[35–37] monolayers, clusters and nanotubes[38–46] had been carried out and the electronic structure has been obtained.[47,48] More recently the site-specific average atomic concentrations had been computed.[49] With this study the authors predicted an abnormal heat-capacity, which had been attributed to atomic interactions.

DFT theoretical calculations of H_2 adsorption on $RhCu$ cluster models, compared with both monometallic ones had shown that the presence of a Cu atom in a Rh rich environment is enough to eliminate the presence of the energy minimum associated to adsorbed molecular hydrogen, and to convert H_2 dissociation in an activated process. Besides, if the Cu content increases the effect is a change in the active site (from 3-fold site to bridge one), in agreement with explanations arising from the interpretation of experimental data. To explain the enhanced molecular adsorption, the authors suggested a combination of structural and electronic effects. They speculated about the possibility of that a Rh-rich alloy would have regions with a composition where bridge sites are preferred, thus favoring an increase in the hydrogen uptake, whereas an enhancement of Cu content would lead the existence of Rh atoms completely coordinated to Cu with a concominant decrease in the reactivity as found in experiments.[39]

CO adsorption properties were studied using Rh, Cu and $RhCu$ cluster models and DFT calculations. The authors concluded that the structural properties of CO adsorbed on bimetallic $RhCu$ surfaces can be used to test the preferred adsorption site but it is not possible to extract information about the composition, because the calculated properties indicate that CO chemisorption has a local character and it is very sensitive to the metal atom directly interacting with the molecule.[40] The last conclusions were confirmed by periodic calculations using slab models which also seem to suggest that the interaction of the CO molecule with the surface is dominated by the atomic site composition with little influence of the remaining metallic environment, hence the presence of a second metal in a metallic catalyst is not easily predictable, and the resulting properties of a bimetallic catalyst are not always a result of interpolation of the properties of the parent metals.[41]

In a theoretical study of $RhCu$ bimetallic monolayers supported on $Ru(0001)$ surface model was found a small increment in the $CO - Rh$ bond strength due to the presence of Cu, which can be attributed to the enhancement of electronic density of Rh caused again by charge transfer from Cu. Conversely this charge transfer disfavors the $CO - Cu$ interaction. The results indicate that the nature of the atom directly interacting with CO is the key factor controlling the surface adsorption properties.[43]

4. Reactions

Several experimental studies indicate that Cu in Rh affects the properties of Rh, for instance its activity and selectivity. In this section, three reactions are reviewed focusing the effects by the metallic composition of the catalyst.

4.1. *Hydrocarbons*

$RhCu$ has been employed to many reactions involving hydrocarbons, in particular for hydrogenation and dehydrogenation. Many of the authors coincide that these effects depend on Cu:Rh ratio, and the catalytic propertiesstrongly depend on the alloy composition.[50,52,53] In the following discussion we focus in some particular in more detail.

(i) The activity for dehydrogenation of cyclohexanol decreases when Cu is added to Rh. The selectivity of the Rh catalyst toward cyclohexene formation prevails, and the selectivity of the bimetallic catalysts is toward dehydrogenation. The authors concluded that the changes can be explained by ligand effects. Furthermore the second metal affects the surface sites enhancing the metallic or bimetallic ones for dehydrogenation, indicating additionally ensemble effects.[54]

(ii) A small addition of Cu to Rh increases the selectivity of conversion of 2,2,3,3-tetramethylbutane (TeMB) to isobutane (iC_4). A further addition decreases the selectivity to iC_4 and increases the selectivity to isopentane (iC_7). This last reaction takes place in well-dispersed catalysts, hence the authors concluded that the selectivity of TeMB hydrogenolysis is controlled by the coordination of the active sites. The initial shift of the catalytic properties lets the authors propose the topological segregation of Cu at the edge and corner sites on the surfaces of small particles.[55]

(iii) Cu in Rh supported in SiO_2 provokes a poison effect on the activity of the coinage metal for 1,3 butadiene hydrogenation. Cu improves the selectivity of Rh for butadiene hydrogenation to 1-butene and 2-butene -*trans* and -*cis*. Nevertheless the addition of a large amount of Cu inhibits the reaction. The authors explained these results by the formation of Cu-Rh alloy which alters the catalytic properties of Rh irreversibly.[56,57]

(iv) n-hexane conversion on $RhCu$ supported in silica showed a selectivity of 25% for pentane and hexane (C_5+C_6) cyclic products whereas selectivity of both monometallic catalysts for cyclization is small or negligible. Reaction of n-pentane and of neohexane on $RhCu$ clusters was predominantly by hydrogenolysis with relatively minor amounts of isomerization. The hydrogenolysis of neohexane (to neopentane and methane) was not decreased on Cu incorporation.[58,59]

(v) The activity of a Cu in Rh/SiO_2 catalyst for carbon monoxide hydrogenation and the rate of formation of methane, C_2 and C_3+ hydrocarbons decreases with respect to that of the monometallic phase. The selectivity for C_2 oxygenate formation shows a slight variation, and exhibits a high initial selectivity for methanol hydrogenation, but decays with time. Finally, the selectivity for propionaldehyde on $RhCu$ is higher than that for the Rh or Cu catalysts studied.[31]

(vi) A study of reforming of CH_4 with CO_2 on Al_2O_3-supported Rh catalysts, indicated that the methane activation takes place on the Rh phase while carbon dioxide is activated on the support surface. The addition of Cu promotes methane activation. The presence of Cu and the support do not cause any effect on the initial activity per surface exposed site. It indicates that the dry reforming of methane is not a structure-sensitive reaction and that catalytic activity is dependent on the number of surface exposed Rh centers.[60]

(vii) Reyes et al. using XPS in several $RhCu$ supported in Al_2O_3 concluded that Cu(I) species remain trapped in the support matrix, changing the metal-support interaction. A significant enhancement in both activity and selectivity for crotonaldehyde hydrogenation to crotyl alcohol was observed depending on these interactions. Selectivity to the unsaturated alcohol higher than 60% was obtained for $RhCu$ catalysts having a Rh atom ratio close to 35%.[18]

4.2. NO reduction and CO oxidation

Research in the field of atmospheric pollution abatement is a hot topic nowadays, due to the importance to reduce pollutant gases that attack the quality of life of all people around the world. The use of automotive exhaust catalysis was a great progress in the field and the research is growing to further improve it. Nowadays, it is evident than the noble metals are the most effective ones as catalysts of the oxido-reduction of NO and CO. Among these metals the most active ones are Rh, Pd and Pt and the used catalyst actually is a combination of these three metals,[10,61] because the performance of a isolated metal is not enough for the complete reaction. Nevertheless, one must note that all these metals are fairly expensive. A cheaper proposal is the use of a different bimetallic, for example, $RhCu$ has been studied as catalyst of this reaction. Under laboratory conditions, its behavior had been promising.[20]

(i) The activity of several $RhCu$ catalysts supported on ceria and on zeolites had been tested for CO oxidation and NO reduction and compared with industrial Rh containing catalysts for exhaust gas treatment. Two catalysts with both metals in their compositions showed, under laboratory conditions, better performance than the monometallic catalysts.[20]

(ii) $RhCu$ catalysts deposited on three supports with different oxygen storage capacity (OSC) were tested under three-way catalytic cycling conditions. Both Cu and Rh metals favor the reduction of supports at low temperature. For monometallic Cu catalysts, Cu participates in the regulation of the oxidant/reducer ratio and is determinant if the OSC of the support is insufficient. Additionally, the interaction between Cu and the mobile oxygen of the support greatly favors the CO oxidation at low temperature, whereas it has less influence on C_3H_6 oxidation and disfavors the NO reduction at low temperature. In this case, the activity is ruled by the metal or the association metal-support, which is the most active in each temperature range. The association Cu-support exhibiting mobile oxygen is the most active for CO conversion. NO reduction depends mainly on the Rh content, especially at low temperature, and C_3H_6 conversion is a little improved by Rh addition to Cu catalyst.[62]

(iii) The three-way catalytic activity of a Cu/Al_2O_3 catalyst modified by addition of small amounts of Rh, at two different temperatures. No synergetic effect in the catalytic properties has been observed between Cu and Rh. Moreover, CO, NO and C_3H_6 conversions at low temperature increase with Rh concentration. At high temperature, the use of Cu which exhibits oxygen storage properties improves the CO and NO conversion.[63]

(iv) A microwave irradiation method has been developed to prepare a wide variety of pure metallic and bimetallic alloy nanoparticles with controlled size and shape. High activity for CO oxidation and thermal stability have been observed for the nanoalloys according to the order $CuPd > CuRh > AuPd > AuRh > PtRh > PdRh > AuPt$.[12]

4.3. *Reactions of compounds with nitrogen*

Since a few years ago, $RhCu$ bimetallic has been studied as catalyst for reactions as reduction of nitrates and nitrites in water,[13] in similar ways to $PdCu$, the most studied catalyst for these reactions.[64–66] In particular we highlight to importance cases:

(i) The activities and selectivities of numerous Rh, Cu and bimetallic catalysts supported on activated carbon were studied in order to optimize the metallic composition for the reduction of nitrate in water with hydrogen. The activity of the catalysts is quite different depending on the Cu content. The $RhCu$ system was the most active among the bimetallic catalysts tested; however, significant amounts of ammonium are obtained.[14,15]

(ii) Several bimetallic catalyst supported in NaY zeolite have been tested for the gas phase hydrogenation of acetonitrile and butyronitrile. Addition of several noble metals, in particular Rh to Cu enhances the reduction of Cu_2+ and lowers the activity for acetonitrile hydrogenation. All catalysts including Cu display high selectivity toward the formation of a secondary amine.[67]

4.4. *Theoretical studies*

A theoretical study proposed an explanation about the promotion of catalytic activity of Rh by Cu in NO reduction by CO, using Rh, Cu and bimetallic slabs models. The results suggested that Cu in Rh at low content promotes Rh catalytic activity by two main reasons: (i) NO adsorption on bimetallic sites is weaker than on clean Rh thus decreasing the reaction barrier for NO dissociation (the limiting step of this reaction[10,68]) and (ii) because in bimetallic sites, Cu favors the apparition of a new transition state, a tilted precursor NO species. The precursor was identified as promoter for NO breaking. At higher contents of Cu this reaction is blocked.[44] The NO dissociation on sites at Rh and $RhCu$ stepped surface models indicated that they are more favored than those on perfect surfaces, and no noticeable effects were determined by the presence of Cu on Rh sites.[46] Furthermore, a slighter effect in CO oxidation in $RhCu$ models compared with Rh was predicted from the calculations using similar approximations. The only effect was a minor decrease in the barrier energy of the reaction in the case of CO adsorbed on top Rh site and O atomic in a bimetallic site.[45]

5. Conclusion

Bimetallic systems generally provide an exceptional kind of catalysts, with new properties and capabilities; the case of $RhCu$ is among the most exceptional. Its catalytic behavior can not be anticipated from that of both parent metals and its catalytic properties do not follow a monotonic trend with respect its metallic composition. Surely subsequent studies will help us further elucidate new features of this interesting system.

In the 30 years elapsed since the study of catalytic activity of $RhCu$ was started, initiated, the uses and requirements for catalyst have substantially changed due to new regulations and social needs. From this short review one can predict that the trend will be continued for $RhCu$ and other systems and new applications for $RhCu$ will emerge.

References

1. V. Ponec, G. C. Bond, *Catalysis by metals and alloys* (Elsevier, Amsterdam, 1995).
2. V. Ponec, *Appl. Catal. A* **222**, 31 (2001).
3. J. A. Rodríguez, *Surf. Sci. Rep.* **24**, 223 (1996).
4. J. H. Sinfelt, *Surf. Sci.* **500**, 923 (2002).
5. W. M. H. Sachtler, *Faraday Discuss. Chem. Soc.* **72**, 7 (1981).
6. J. A. Rodríguez, D. W. Goodman, *Science* **257**, 897 (1992).
7. C. T. Campbell, *Annu. Rev. Phys. Chem.* **41**, 775 (1990).
8. J. R. Kitchin, J. K. Norskov, M. A. Barteu, J.G. Chen, *J. Chem. Phys.* **120**, 10240 (2004).
9. J. R. Kitchin, J. K. Norskov, M. A. Barteu, J.G. Chen *Phys. Rev. Lett.* **93**, 156801 (2004).
10. G. Ertl, H. Knözinger, J. Weitkamp, *Handbook of Heterogeneous Catalysis, Vol. 3-5* (Wiley-VCH, Munich, 1997).
11. J. K. A. Clarke, *Chem. Rev.* **75**, 291 (1975).
12. V. Abdelsayed, A. Aljarash, M. S. El-Shall, Z. A. Al Othman, A. H. Alghamdi, *Chem. Materials* **21**, 2825 (2009).
13. I. Witonska, S. Karski, J. Goluchowska, *Kinetics and Catal.* **48**, 823 (2007).
14. O.S.G.P. Soares, J. M. Orfao, M. F. R. Pereira, *Catal. Lett.* **126**, 253 (2008).
15. O.S.G.P. Soares, J. M. Orfao, M. F. R. Pereira, *Appl. Catal. B* **91**, 441 (2009).
16. J. M. Dumas, C. Geron, H. Hadrane, P. Marecot, J. Barbier, *J. Mol. Catal.* **77**, 87 (1992).

17. J. Sanyi, D. W. Goodman, *J. Catal.* **145**, 508 (1994).
18. P. Reyes, G. Pecchi, J. L. G. Fierro, *Langmuir* **17**, 522 (2001).
19. F. M. T. Mendes, M. Schmal, *Appl. Catal. A* **151**, 393 (1997).
20. L. Petrov, J. Soria, L. Dimitrov, R. Cataluna, L. Spasov, P. Dimitrov, *Appl. Catal. B* **8**, 9 (1996).
21. T. B. Massalsiki, H. Okamoto, P. R. Subramanian, L. Kacprezak, *Binary alloy phase diagrams*, 2nd ed. (ASM, Materials Parks, Ohio, 1992).
22. L. Irons, S. Mini, B. E. Brower, *Mater. Sci. Eng.* **98**, 309 (1998).
23. G. Meitzneir, G. H. Via, F. W. Lytle, J. H. Sinfelt, *J. Chem. Phys.* **78**, 882 (1983).
24. M. N. Zauwen, A. Crucq, L. Degols, G. Lienard, A. Frennet, N. Mikhalenko, P. Grange, *Catalysis Today* **5**, 237 (1989).
25. X. Courtois, V. Perichon, M. Primet, G. Bergeret, *Stud. Surf. Sci. Catal.* **130**, 1031 (2000).
26. M. Fernández-García, A. Martínez-Arias, I. Rodríguez-Ramos, P. Ferreira-Aparicio, A. Guerrero-Ruiz, *Langmuir* **15**, 5295 (1999).
27. S. C. Chou, C. T. Yeh, T. H. Chang, *J. Phys. Chem. B* **101**, 5828 (1997).
28. J. A. Anderson, C. H. Rochester, Z. J. Wang, *J. Mol. Catal. A* **139**, 285 (1999).
29. F. Reniers, M. P. Delplancke, A. Asskali, V. Rooryck, O. Van Sinay, *Appl. Surf. Sci.* **92**, 35 (1996).
30. F. Reniers, M. P. Delplancke, A. Asskali, M. Jardinier-Offergeld, F. Bouillon, *Appl. Surf. Sci.* **81**, 151 (1994).
31. R. Krishnamurthy, S. S. C. Chuang, K. Ghosal, *Appl. Catal. A* **114**, 109 (1994).
32. J. A. Rodríguez, C. T. Campbell, D. W. Goodman, *J. Phys. Chem.* **94** (1990) 6936.
33. J. Abart, J. Voitlnder, *Solid State Comm.* **40**, 277 (1981).
34. R. Viswanathan, H. R. Khan, Ch. J. Raub, *J. Phys. Chem. Solids* **37**, 431 (1976).
35. Z. W. Lu, S. -H. Wei, Alex Zunger, L. G. Ferreira, *Solid State Comm.* **78**, 583 (1991).
36. J. Kudrnovsk, V. Drchal, *Solid State Comm.* **65**, 613 (1988).
37. P. Mohn, K. Schwarz, *Solid State Comm.* **57**, 103 (1986).
38. W. Q. Deng, X. Xu, W. A. Goddard, *Nano Lett.* **4**, 2331 (2004).
39. S. González, C. Sousa, M. Fernández-García, V. Bertin, F. Illas, *J. Phys. Chem. B* **106**, 7839 (2002).
40. S. González, C. Sousa, F. Illas, *Surf. Sci.* **531**, 39 (2003).
41. S. González, C. Sousa, F. Illas, *J. Phys. Chem. B* **109**, 4654 (2005).
42. J. R. B. Gomes, S. González, D. Torres, F. Illas, *Russ. J. Phys. Chem. B* **1**, 292 (2007).
43. S. González, F. Illas, *Surf. Sci.* **598**, 144 (2005).
44. S. González, C. Sousa, F. Illas, *J. Catal.* **239**, 431 (2006).
45. , S. González, C. Sousa, F. Illas, *Phys. Chem. Chem. Phys.* **9**, 2877 (2007).
46. S. González, D. Loffreda, F. Illas, P. Sautet, *J. Phys. Chem. C* **111**, 11376 (2007).
47. R. Kuentzler, H. Ebert, J. Voitlander, H. Winter, J. Abart, *Solid State Comm.* **62**, 145 (1987).
48. J. Maek, J. Kudrnovsk, *Solid State Comm.* **54**, 981 (1985).
49. M. Polak, L. Rubinovich L, *Phys. Rev. B* **71**, 125426 (2005).
50. A. Guerrero-Ruiz, B. Bachiller-Baeza, P. Ferreira-Aparicio, I. Rodríguez-Ramos, *J. Catal.* **171**, 374 (1997).
51. J. A. Rodríguez, S. Chaturvedi, M. Kuhn, *J. Chem. Phys.* **108**, 3064 (1998).
52. B. H. Davis, G. A. Westfall, J. Watkins, J. Pezzanite, *J. Catal.* **42**, 247 (1976).
53. H. C. de Jongste, V. Ponec, *J. Catal.* **63**, 389 (1980).
54. F. M. T. Mendes, M. Schmal, *Appl. Catal. A* **163**, 153 (1993).
55. B. Coq, R. Dutartre, F. Figueras, A. Rouco, *J. Phys. Chem.* **93**, 4904 (1989).
56. Z. Ksibi, A. Ghorbel, *J. Chim. Phys. Phys. Chim. Biolog.* **92**, 1418 (1995).

57. Z. Ksibi, A. Ghorbel, B. Bellamy *J. Chim. Phys. Chim. Biolog.* **94**, 1938 (1997).
58. J. K. A. Clarke, K. M. G. Rooney, T. Baird, *J. Catal.* **111**, 1988 (1988).
59. J. K. A. Clarke, I. Manninger, T. Baird, *J. Catal.* **54**, 230 (1988).
60. P. Ferreira-Aparicio, M. Fernández-García, A. Guerrero-Ruiz, I. Rodríguez-Ramos, *J. Catal.* **190**, 296 (2000).
61. H. S. Gandhi, G. W. Graham, R. W. McCabe, *J. Catal.* **216**, 433 (2003).
62. X. Courtois, V. Perrichon, M. Primet *Comptes Rendus de lAcadmie des Sciences - Series IIC - Chemistry* **3**, 429 (2000).
63. X. Courtois, V. Perichon, *Appl. Catal. B* **57**, 63 (2005).
64. F. J. J. G. Janssen, R. A. Santen, *Enviromental Catalysis* (Imperial College Press, London, 1999).
65. A. Edelmann, W. Schieber, H. Vinek, A. Jentys, *Catal. Lett.* **69**, 11 (2000).
66. B. P. Chaplin, E. Roundy, K. A. Guy, J. R. Shapley, C. J. Werth, *Environ. Sci. Technol.* **40**, 3075 (2006).
67. Y. Y. Huang, W. M. H. Sachtler, *J. Catal.* **188**, 215 (1999).
68. V. P. Zhdanov, B. Kasemo, *Surf. Sci. Rep.* **29**, 31 (1997).

KINETIC ENERGY FUNCTIONALS: EXACT ONES FROM ANALYTIC MODEL WAVE FUNCTIONS AND APPROXIMATE ONES IN ORBITAL-FREE MOLECULAR DYNAMICS

VALENTIN V. KARASIEV

Centro de Química, Instituto Venezolano de Investigaciones Científicas, IVIC,
Apartado 21827, Caracas 1020-A, Venezuela
vkarasev@gmail.com

XABIER LOPEZ

Kimika Fakultatea, Euskal Herriko Unibertsitatea, Posta Kutxa 1072,
20080 Donostia, Euskadi, Spain

JESUS M. UGALDE

Kimika Fakultatea, Euskal Herriko Unibertsitatea, Posta Kutxa 1072,
20080 Donostia, Euskadi, Spain
jesus.ugalde@ehu.es

EDUARDO V. LUDEÑA

Centro de Química, Instituto Venezolano de Investigaciones Científicas, IVIC,
Apartado 21827, Caracas 1020-A, Venezuela

Received 18 December 2009

We consider the problem of constructing kinetic energy functionals in density functional theory. We first discuss the functional generated through the application of local-scaling transformations to the exact analytic wavefunctions for the Hookean model of ^4He (a finite mass three-particle system), and contrast this result with a previous one for $^\infty He$ (infinite mass system). It is shown that an exact non-Born-Oppenheimer treatment not only leads to mass-correction terms in the kinetic energy, but to a basically different functional expression. In addition, we report and comment on some recently advanced approximate kinetic energy functionals generated in the context of the constraint-based approach to orbital-free molecular dynamics. The positivity and non-singularity of a new family of kinetic energy functionals specifically designed for orbital-free molecular dynamics is discussed. Finally, we present some conclusions.

Keywords: Kinetic energy functionals; Hookean systems; Orbital-free Molecular Dynamics.

1. Introduction

One of the outstanding problems in density functional theory as well as in molecular dynamics is the construction of sufficiently accurate kinetic energy functionals[1–4] so that the evaluation of the kinetic energy through the use of orbitals (as is done in the Kohn-Sham version of DFT[5] and in the Carr-Parinello version of molecular dynamics[6]) is bypassed.

The reason why this is such a difficult task stems from the Coulomb virial theorem which states the equivalence between the kinetic and the total energies. Thus, to obtain an explicit and accurate expression for total electronic kinetic energy density functional $T[\rho] = \langle \Psi | \hat{T} | \Psi \rangle$ for a many-electron system corresponding to the state $|\Psi\rangle$ with electron number density ρ is as difficult as obtaining the energy density functional for the total energy, $E[\rho] = \langle \Psi | \hat{H} | \Psi \rangle$ where $\hat{H}$ is the electronic Hamiltonian. This situation contrasts markedly with the usual one in DFT where the task is just to approximate the exchange and correlation energy functionals, where the former is commonly less than 10% and the latter of the order of 1% of the total energy.

The search for a kinetic energy functional expressed solely in terms of the one-particle density ρ started shortly after the inception of quantum mechanics. The Thomas-Fermi[7,8] and and von Weizsäcker[9] functionals are examples of these early attempts. Using these functionals, new ones have been proposed as combinations of these two functionals.[4] Moreover, throughout the years, there have been quite a number of efforts directed at constructing these kinetic energy functionals. A comprehensive review of these efforts was given by Ludeña and Karasiev[10] in 2002. For another important review, see the work of Carter *et al.*, of 2000,[11] and for more recent developments, Refs. 12–17 and Karasiev *et al.*, in Refs. 18 and 19. Also, some recent work on the construction of kinetic energy functionals using local-scaling transformations is given by Ludeña *et al.* in Refs. 20–25.

Some of the work on the application of local-scaling transformations for the purpose of devising kinetic energy functionals is based on exact analytic wavefunctions obtained for Hookean models of few-particle systems.[26–29] A recent review on the development and application of these Hookean models is given in Ref. 30.

In Section 2 we present translationally invariant wavefunctions for the ground states of $^\infty He$ and ^4He. These wave functions are expressed in terms of the inter-particle coordinates. In a previous works,[24,25] we had discussed the construction of the kinetic energy functional arising from application of local-scaling transformations. In the present work, we extend these results to the case of the finite mass ^4He system. As in these model systems the results are truly non-Born-Oppenheimer we are able to discuss the effect of a finite mass on the form of the kinetic energy functional.

In Section 3 we report on some recent developments of approximate functionals designed within the constraint-based approach.[31] Some new trends are discussed, although no definitive solution is presented to the kinetic energy functional construction problem.

2. Exact non-Born-Oppenheimer Kinetic Energy Functionals for Hookean Model Systems

2.1. *Hookean model systems and their exact analytic solutions*

The Hookean atoms or $^{\infty}He$ is a two-electron system defined by the Hamiltonian

$$\widehat{H} = -\frac{h^2}{8\pi^2 m_e}\nabla^2_{\mathbf{r}_1} - \frac{h^2}{8\pi^2 m_e}\nabla^2_{\mathbf{r}_2} + K\frac{e^2}{a_o^3 4\pi\epsilon_0}(r_1^2 + r_2^2) + \frac{e^2}{4\pi\epsilon_0|\mathbf{r}_1 - \mathbf{r}_2|} \tag{1}$$

The Hamiltonian for ^4He (a three-particle system) is:[28]

$$\widehat{H} = -\frac{h^2}{8\pi^2 m_e}\nabla^2_{\mathbf{r}_1} - \frac{h^2}{8\pi^2 m_e}\nabla^2_{\mathbf{r}_2} - \frac{h^2}{8\pi^2 m_h}\nabla^2_{\mathbf{r}_3} + K_{13}\frac{e^2}{a_o^3 4\pi\epsilon_0}(\mathbf{r}_1 - \mathbf{r}_3)^2$$
$$+ K_{23}\frac{e^2}{a_o^3 4\pi\epsilon_0}(\mathbf{r}_2 - \mathbf{r}_3)^2 + \frac{e^2}{4\pi\epsilon_0|\mathbf{r}_1 - \mathbf{r}_2|} \tag{2}$$

Actually, letting the mass m_h (the mass of the He nucleus) go to infinity, this Hamiltonian yields that of Eq. (1).

2.2. *Analytic solutions in terms of interparticle coordinates*

The wavefunction expressed in terms of interparticle coordinates for $^{\infty}He$ is:

$$\Psi_{(\mathbf{r}_1, \mathbf{r}_2)} = N(1 + \frac{1}{2}r_{12})\exp(-\frac{1}{4}r_1^2)\exp(-\frac{1}{4}r_2^2) \tag{3}$$

Similarly, for the finite mass ^4He system, the wave function is:[28]

$$\Psi_{(\mathbf{r}_1, \mathbf{r}_2, \mathbf{r}_3)} = N\exp-\frac{1}{4}(1 - \sqrt{m_h/(m_h + 2m_e)}r_{12}^2)$$
$$\times (1 + \frac{1}{2}r_{12})\exp(-\frac{1}{4}\sqrt{m_h/(m_h + 2m_e)}r_{13}^2)$$
$$\times \exp(-\frac{1}{4}\sqrt{m_h/(m_h + 2m_e)}r_{23}^2) \tag{4}$$

The Hookean model for the system H^- is easily obtained by replacing m_h by the proton mass m_p. Also, a particular realization of this wave function arises when the He nucleus is replaced a positron; in this case we have the Ps^- system, whose wavefunction is:

$$\Psi_{(\mathbf{r}_1, \mathbf{r}_2, \mathbf{r}_3)} = N\exp-\frac{1}{4}(1 - \sqrt{1/3}r_{12}^2)(1 + \frac{1}{2}r_{12})$$
$$\times \exp(-\frac{1}{4}\sqrt{1/3}r_{13}^2)$$
$$\times \exp(-\frac{1}{4}\sqrt{1/3}r_{23}^2) \tag{5}$$

In Fig. 1 we sketch the densities of these systems. Although the wavefunctions giving rise to the finite and infinite mass He kinetic energies are analytically different, the fact that the exponential term for r_{23} is very small due to the mass differences between the He nuclei mass and that of the electron , the graphic display of their

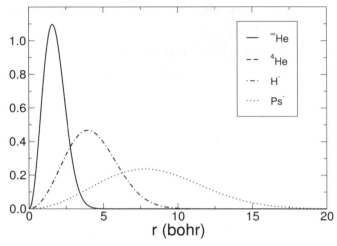

Fig. 1.   Electron-pair probability distribution functions for the $^\infty He$, ^4He, H^- and Ps^- two-electron systems.

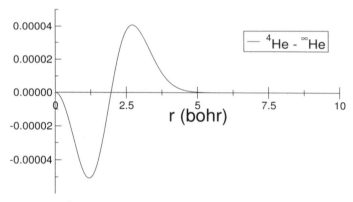

Fig. 2.   Difference (^4He-$^\infty He$) of the electron-pair probability distribution functions.

respective electron-pair probability distribution functions shows that they are undistinguishable at the scale presented in Fig. 1. In Fig. 2 we graph their differences with a scale large enough to allow us to clearly assess this mass effect.

2.3. *Generation of kinetic energy functionals through local-scaling transformations*

The local-scaling transformation version of density functional theory, LS-DFT,[4,32,33] is an "explicit constructive procedure", which allows us – by means of the introduction of density-dependent coordinates $r^T \equiv r^T([\rho]; \mathbf{r})$ – to transform an ordinary wavefunction into a density-dependent wavefunction. For example, for an N-electron system, described by $\Psi_0(\mathbf{r}_1, \ldots, \mathbf{r}_N)$ we obtain the transformed density-dependent wave function

$$\Psi_0([\rho]; \mathbf{r}_1, \ldots, \mathbf{r}_N) = \sqrt{\frac{\rho(r_1)}{\rho_\Psi(r_1^T)}} \times \cdots \times \sqrt{\frac{\rho(r_N)}{\rho_\Psi(r_N^T)}} \Psi_0(\mathbf{r}_1^T, \ldots, \mathbf{r}_N^T) \qquad (6)$$

The density-dependent coordinate r^T is obtained through the application of locally-scaled transformations to the coordinate r. Explicitly, the transformed coordinate is given by

$$r^T = \sqrt{(x^T)^2 + (y^T)^2 + (z^T)^2} \qquad (7)$$

where the Cartesian locally-scaled components are:

$$x^T = \lambda(\mathbf{r})x, \qquad y^T = \lambda(\mathbf{r})y, \qquad \text{and} \qquad z^T = \lambda(\mathbf{r})z \qquad (8)$$

Here, $\lambda(\mathbf{r})$ is the local-scaling transformation function. This function is determined by solving the following first-order differential equation

$$\lambda(\mathbf{r}) = \left(\frac{\rho(\mathbf{r})}{\rho_\Psi(\mathbf{r}^T)}\right)^{1/3} L(\mathbf{r})^{-1/3} \qquad (9)$$

which relates an initial density $\rho_\Psi(\mathbf{r})$ coming from the known function $\Psi_0(\mathbf{r}_1, \ldots, \mathbf{r}_N)$ to a final (known and fixed) density $\rho(\mathbf{r})$. In Eq. (9) we have defined $L(\mathbf{r})$ as:

$$L(\mathbf{r}) = (1 + \mathbf{r} \cdot \nabla_\mathbf{r} \ln \lambda(\mathbf{r})) \qquad (10)$$

As in the present case we have at our disposal exact analytic wave functions for ground-states of Hookean models of few-particles physical systems, then, through the application of local-scaling transformations, we can generate the corresponding exact density-dependent wavefunctions $\Psi_0([\rho])$ which, in turn, may be used to generate the kinetic energy functionals.

$$T[\rho] = < \Psi_0([\rho]) | \widehat{T} | \Psi_0([\rho]) > \qquad (11)$$

The local-scaling transformation procedure may also be used to generate the non-interacting kinetic energy functional $T_s[\rho]$. This is accomplished by letting $\Phi([\rho] : \mathbf{r}_1, \ldots, \mathbf{r}_N)$ denote a single Slater determinant. As an example, for atoms, we may construct this Slater determinants from a set of three-dimensional density-dependent orbitals comprising transformed radial functions times spherical harmonics: $\{\phi_i([\rho]; \mathbf{r}) = R_i([\rho]; r) Y_{l_i, m_{l_i}}(\theta, \phi)\}_{i=1}^N$. The non-interacting kinetic energy functional is then given by $T_s[\rho] = \langle \Phi[\rho] | \widehat{T} | \Phi[\rho] \rangle$. In this context, we can actually **derive** the following form for $T_s[\rho]$[10]

$$T_s[\rho] = T_W[\rho] + \frac{1}{2} \int_0^\infty dr r^2 \rho^{5/3}(r) A_N([\rho]; r) \qquad (12)$$

where $T_W[\rho]$ is the Weizsäcker term and the second term is the non-local correction.[10] This second term is also known as the Pauli term, and is usually expressed as:

$$T_\theta[\rho] = \int_0^\infty dr r^2 \rho^{5/3}(r) F_\theta([\rho]; r) \qquad (13)$$

where $F_\theta([\rho], \mathbf{r})$ is called the "enhancement" factor and is related to A_N by $2F_\theta([\rho], \mathbf{r}) \equiv A_N$. Explicit analytical expressions for A_N for the case when we use specific sets of radial orbitals $\{R_i([\rho]; r)\}$ generated by local-scaling transformations have been obtained in previous works.[20,33] Let us just mention in passing that the $\rho^{5/3}$ dependence in Eq. (13) comes about as a result of an *exact derivation*, where no assumption other than the existence of an initial orbital set has been introduced. It is interesting, to observe, however, that in this exact treatment, the enhancement factor, represented in the present case by A_N (or equivalently by $F_\theta([\rho], \mathbf{r})$) depends upon the number N of electrons. Hence, this term is not universal.

The details of the application of local-scaling transformations for the generation of the kinetic energy functional in the case of the infinite mass He system $^\infty He$ are given in Ref. 24. Note that $^\infty He$ is a two-particle system, whereas ^4He is a three-particle one. The explicit form of the kinetic energy functional for $^\infty He$ is:

$$T[\rho, \Psi] = T_W[\rho] + \int_0^\infty dr\, r^2 \rho^{5/3}(r) F_{5/3}([\rho]; f(r)) \tag{14}$$

where $f(r) \equiv \lambda(r)r$. The explicit expression for the enhancement factor is

$$F_{5/3} = 4\pi L(r)^{4/3} \left[\frac{b(f)}{2\sqrt{2}(16\pi + 10\pi^{3/2})(\rho_g(f))^{5/3}} \right.$$
$$\left. - \frac{f^2}{8(\rho_g(f))^{2/3}} - \frac{f\frac{d\rho(f)}{df}}{4(\rho_g(f))^{5/3}} - \frac{\left(\frac{d\rho(f)}{df}\right)^2}{8(\rho_g(f))^{8/3}} \right] \tag{15}$$

where we have defined $b(f) = e^{-1/2f^2}$ and $L(r) = \left(1 + r\frac{d}{dr}\ln\lambda(r)\right)$.

Also, applying this same procedure we can also generate the kinetic energy expression for the wave function for the finite-mass system ^4He. The result is:

$$T[\rho, \Psi] = T_W[\rho] + \int_0^\infty dr\, r^2 \rho^{5/3}(r_1) F_{5/3}([\rho]; f(r))$$
$$+ \int_0^\infty dr\, r^2 \rho(r_1) F_{mass}([\rho]; f(r)) \tag{16}$$

where the enhancement factor $F_{5/3}([\rho]; f(r))$ is the same as in Eq. (15). The mass-dependent enhancement factor $F_{mass}([\rho]; f(r))$ is given by

$$F_{mass} = \frac{e^{-b_1 f^2}}{\rho_g(f)} \left[X(a_1, a_2, a_3; b_1, b_2, b_3; f) \right.$$
$$+ X(0, a_8, a_9; b_1, b_2, b_3; f)f^2$$
$$\left. - X(0, 0, a_{10}; b_1, b_2, b_3; f)f^4 \right] + a_4 \frac{e^{-b_4 f^2}}{\rho_g(f)f}$$
$$+ \frac{f}{\rho_g(f)} \left[a_5 e^{-b_4 f^2} + [a_0 e^{-b_5 f^2} - [a_7 e^{-b_1 f^2} \right] \tag{17}$$

where $X(a, b, c; d, g, h; f) = a - bErf(gf) - 2cErf(df) - 2hErf(hf)$. The rather complicated form of this enhancement factor clearly evinces the non-trivial influence of the non-Born-Oppenheimer finite mass effect on the infinite mass kinetic energy expression (the latter strictly obtained within the Born-Oppenheimer approximation).

3. Constraint-based Kinetic Energy Functionals for Orbital-free Molecular Dynamics

Highly accurate kinetic energy functionals are also important in several domains of physics and material science, where the simulation of the properties of complex materials presents a very difficult and demanding challenge. In effect, for systems away from equilibrium and under the influence of solvents or mechanical stress successful application of molecular dynamics requires the calculation of accurate Born-Oppenheimer forces in particular regions of the material such as the reactive zones. Hence, in these regions, the application of a high level quantum mechanical treatment for determining the forces between atoms is mandatory. As other regions may be treated by means of potentials or classical approximations, it quite usual to operate under a multi-graded regime leading to multi-scale simulation.

Obviously, for material-science simulations, the most demanding step from a computational point of view is the application of quantum mechanical-based methods to the calculation of forces in the reactive zones. In the last years, density functional theory (DFT) has acquired a very prominent status in electronic structure calculations, particularly in applications that rely on the Kohn-Sham (KS) equations. But in spite of the progress achieved in DFT through the incorporation of pseudo-potentials and exchange and correlation functionals that allow a linear dependence of the energy on the number of electrons (an N-order dependence), the very fact that one must solve the Kohn-Sham equations for N orbitals makes the problem quite untractable for very large N.

The several approaches that have been advanced in order to treat this problem, may be combined, as has been proposed in the Graded-Sequence of Approximations scheme[34] to form a ladder of increasing difficulty involving classical potentials, simple reactive (charge re-distribution potential,[35-37] orbital-free (OF) DFT, Quasi-spin density DFT,[38] and fully spin-polarized DFT (this last level is necessary for an adequate treatment of bond-breaking).

Clearly, at the level of OF-MD we require very accurate kinetic-energy (KE) density functionals. These are demanded for the calculation of quantum-mechanical forces in multi-scale molecular dynamics simulations.[18] To be computationally acceptable, these kinetic energy functionals besides leading to accurate forces on the nuclei must be simple to operate (i.e., they must contain few parameters).[19] Following an earlier proposal based on modifications of GGA-type functionals constrained to give positive-definite Pauli potentials, we report here on a new family of constraint-based kinetic energy functionals.[31] The advantage of this new family of functionals is that they do not present singularities (this was a problem in the modified GGA-type ones).

3.1. *Behavior of GGA-type functionals near the nucleus*

The OF-DFT energy functional is $E^{\mathrm{OF\text{-}DFT}}[n] = T_s[n] + E_{\mathrm{Ne}}[n] + E_{\mathrm{H}}[n] + E_{\mathrm{xc}}[n] + E_{\mathrm{NN}}$ where $T_s[\rho]$ is the non-interacting kinetic energy functional given by Eq. (12), and where $E_{\mathrm{Ne}}[\rho]$, $E_{\mathrm{H}}[\rho]$, $E_{\mathrm{xc}}[\rho]$ are the electron-nuclear, Hartree (electron-electron repulsion) and exchange-correlation functionals and where E_{NN} is the internuclear repulsion.

Using the above OF-DFT energy functional one obtains the force on nucleus I at $\mathbf{R}_I$:

$$\mathbf{F}_I = -\nabla_{\mathbf{R}_I} E^{\mathrm{OF\text{-}DFT}}$$

$$= -\nabla_{\mathbf{R}_I} E_{\mathrm{NN}} - \int \rho(\mathbf{r})\, \nabla_{\mathbf{R}_I} v_{\mathrm{Ne}} d^3\mathbf{r}$$

$$- \int \left[\frac{\delta T_s[\rho]}{\delta \rho(\mathbf{r})} + v_{\mathrm{KS}}([\rho];\mathbf{r}) \right] \nabla_{\mathbf{R}_I} \rho(\mathbf{r})\, d^3\mathbf{r}. \tag{18}$$

where $v_{\mathrm{KS}} = \delta(E_{\mathrm{Ne}} + E_{\mathrm{H}} + E_{\mathrm{xc}})/\delta\rho$

In Eq. (18) the largest error in the force comes from the functional derivative of the non-interacting kinetic energy with respect to the density. The reason is that the contribution to the energy of $T_s[\rho]$ is an order of magnitude larger than $E_{\mathrm{xc}}[\rho]$.

In terms of the parameter

$$s = \frac{|\nabla\rho|}{2(3\pi^2)^{1/3}\rho^{4/3}} \tag{19}$$

one has the general form of the GGA non-interacting kinetic energy functional

$$T_s^{\mathrm{GGA}}[\rho] = T_{\mathrm{W}}[\rho] + T_\theta[\rho] \tag{20}$$

where the Pauli kinetic energy term is defined as

$$T_\theta[\rho] = c_0 \int d^3\mathbf{r}\, \rho^{5/3}(\mathbf{r}) F_\theta(s(\mathbf{r})) \tag{21}$$

It seems that a general feature of kinetic energy functionals generated within the GGA framework, is that they are not positive-definite near the nucleus. In Fig. 3, as a typical example of this behavior, we plot twice the enhancement factor $2F_\theta([\rho],\mathbf{r}) \equiv A_N$ for three GGA functionals. These are the Perdew-Wang (1991), PW91,[39] the Becke (1986) A, B86A,[40] and the DePristo and Kress (1987), DPK,[41] functionals. The correct behavior is given by the exact result (the bell-shaped curve depicted by the heavy black continuous line), which shows a monotonically decreasing and positive-definite curve that goes to zero at the origin and at infinity In contrast, all three GGA functionals plotted here (which have almost identical behavior) become negative near the nucleus. It seems that this same behavior is general for GGA kinetic energy functionals (at least it is so for those analyzed in Ref. 10). For completeness, we graph in Fig. 3 also the density variable $\tilde{p}$ defined by

$$\tilde{p} = \Delta\rho/2\rho^{5/3} \tag{22}$$

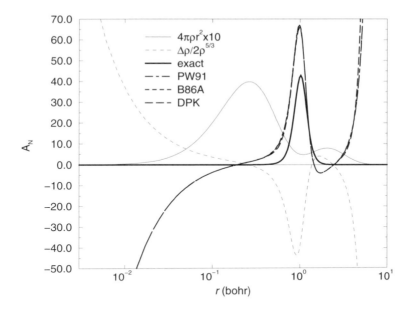

Fig. 3.   Graph of the kinetic energy enhancement factor $2F_\theta([\rho], \mathbf{r}) \equiv A_N$ for the GGA functionals PW91, B86A and DPK for the Beryllium atom. The exact enhancement factor, radial density and $\tilde{p} = \Delta\rho/2\rho^{5/3}$ are also depicted.

(the dashed line that starts at the left at plus infinity and ends at the right at minus infinity) and the radial density multiplied times 10 (the positive double-humped curve). It is also interesting to observe that the GGA functionals depicted here have the wrong behavior at infinity (they go to plus infinity instead of to zero).

Two important constraints that must be satisfied by GGA functionals are the positive-definite behavior of both the kinetic energy enhancement factor $F_\theta(s(\mathbf{r})) \geq 0$ and of the Pauli potential:

$$v_\theta([\rho]; \mathbf{r}) = \delta T_\theta[\rho]/\delta\rho(\mathbf{r}) \geq 0 , \quad \forall\, \mathbf{r} . \tag{23}$$

The explicit form of the Pauli potential for any GGA functional is:[31]

$$v_\theta^{\text{GGA}} = \frac{5}{3} c_0 n^{2/3} F_\theta +$$
$$c_0 n^{5/3} \frac{\partial F_\theta}{\partial s} \left[\frac{\partial s}{\partial n} - \frac{5}{3} \frac{\nabla n}{n} \cdot \frac{\partial s}{\partial \nabla n} - \nabla \cdot \frac{\partial s}{\partial \nabla n} \right]$$
$$- c_0 n^{5/3} \frac{\partial^2 F_\theta}{\partial s^2} \left(\nabla s \cdot \frac{\partial s}{\partial \nabla n} \right) . \tag{24}$$

In terms of the variables s [Eq. (19)], $p = \tilde{p}/2(3\pi^2)^{2/3}$ where $\tilde{p}$ is given by Eq. (22) and q, defined by

$$q \equiv \frac{\nabla n \cdot (\nabla\nabla n) \cdot \nabla n}{(2k_F)^4 n^3} = \frac{\nabla n \cdot (\nabla\nabla n) \cdot \nabla n}{16(3\pi^2)^{4/3} n^{13/3}} \tag{25}$$

it is possible to rewrite the Pauli potential as:[31]

$$v_\theta^{\text{GGA}}(s^2) = c_0 n^{2/3} \left\{ \frac{5}{3} F_\theta(s^2) - \left(\frac{2}{3} s^2 + 2p \right) \frac{\partial F_\theta}{\partial(s^2)} \right.$$
$$\left. + \left(\frac{16}{3} s^4 - 4q \right) \frac{\partial^2 F_\theta}{\partial(s^2)^2} \right\}. \tag{26}$$

Since near the nucleus, the density ρ must satisfy Kato's cusp condition, it follows that in this region we may represent ρ as hydrogen-like 1s density:

$$n_H(\mathbf{r}) \sim \exp(-2Z|\mathbf{r}|) \tag{27}$$

In addition, since at $r = 0$, s and q remain finite while $p \to -4Z/(2k_F)^2 r$, we have that Eq. (26) goes into

$$v_\theta^{\text{GGA}}(r \to 0) = \frac{3Z}{5r} \frac{\partial F_\theta^{\text{GGA}}}{\partial(s^2)} + \text{nonsingular terms}. \tag{28}$$

Moreover, if for sufficiently small s^2 we take $F_\theta^{\text{GGA}} \approx 1 + as^2$ Eq. (28) becomes simplifies to

$$v_\theta^{\text{GGA}}(r \to 0) = \frac{3aZ}{5r} + \text{nonsingular terms}, \tag{29}$$

It follows, therefore, v_θ^{GGA} goes to infinity at the nuclei (with the same sign as the GGA parameter a). Since the present analysis is made for small values of s^2 it turns out that this is a general property of all GGA functionals. We may, thus, conclude that GGA Pauli potentials have singularities in the vicinity of nuclear sites, a behavior that is at odds with the exact Kohn-Sham case.

3.2. *Ways to remedy the singularity problem*

In Ref. 31 we have proposed a possible procedure to eliminate the singularities. This requires, however, that we go beyond the GGA approximation. Applying a gradient expansion to the Pauli term, we may rewrite it as:

$$T_\theta[n] = \int d^3\mathbf{r} \left[t_\theta^{(0)}([n]; \mathbf{r}) + t_\theta^{(2)}([n]; \mathbf{r}) + t_\theta^{(4)}([n]; \mathbf{r}) + ... \right] \tag{30}$$

where

$$t_\theta^{(2i)}([n]; \mathbf{r}) = t_0([n]; \mathbf{r}) F_\theta^{(2i)}(s, p, ...), \tag{31}$$

As this is clearly similar to the GGA form, we may identify

$$F_\theta^{(0)} + F_\theta^{(2)} = 1 + a_2 s^2, \tag{32}$$

with $a_2 = -40/27$ and

$$F_\theta^{(4)} = a_4 s^4 + b_2 p^2 + c_{21} s^2 p, \tag{33}$$

with coefficients $a_4 = 8/243$, $b_2 = 8/81$, and $c_{21} = -1/9$.

3.3. *A new family of orbital-free molecular dynamics kinetic energy functionals*

To go beyond the GGA approximation we treat the values of a_2, a_4, b_2, c_{21} as parameters and seek values or relationships among them which would yield a non-singular v_θ through a given order. The details of this procedure are discussed in Ref. 31 In this way, a second- and fourth-order reduced density derivatives (RDD) are found:

$$\kappa_2 = s^2 + b_1 p \,, \tag{34}$$

$$\kappa_4 = s^4 + \frac{18}{13}p^2 - \frac{30}{13}s^2 p \,. \tag{35}$$

In terms of this new variable, for example, the fourth-order enhancement factor is given by

$$F_\theta^{(4)}(\kappa_4) = a_4 \kappa_4 \,, \tag{36}$$

In general, therefore, we may introduce the following Reduced Derivative Approximation (RDA) kinetic energy functionals:

$$T_s^{\mathrm{RDA}}[n] \equiv T_W[n] + \int d^3\mathbf{r}\, t_0([n];\mathbf{r}) F_\theta(\kappa_2(\mathbf{r}), \kappa_4(\mathbf{r})) \tag{37}$$

It is important to remark that the cancellation of singularities leading to the definition of κ_4 are based on a particular form of the density near the nucleus given by Eq.(27). For a more general form of the density expansion such as

$$\rho(r) \sim (1 + C_1 r + C_2 r^2 + C_3 r^3). \tag{38}$$

it can be shown that the singularity is eliminated if the following variable is introduced

$$\tilde{\kappa}_4 = \sqrt{s^4 + b_2 p^2}, \quad b_2 > 0 \,. \tag{39}$$

Thus, a more general for for RDA kinetic energy functionals is:

$$T_s^{\mathrm{RDA}}[n] \equiv T_W[n] + \int d^3\mathbf{r}\, t_0([n];\mathbf{r}) F_\theta(\kappa_2(\mathbf{r}), \tilde{\kappa}_4(\mathbf{r})) \tag{40}$$

The performance is these RDA functionals is currently under examination. Some preliminary results have been presented in Ref. 31.

4. Conclusion

Even though the availability of highly accurate kinetic energy functionals depending on the one-particle density is a very desirable goal, the actual construction of these functionals still remains as an unsolved problem. In this work we have examined two approaches for reaching this goal. The first is based on exact solutions of Hookean model system (which, unfortunately, must be restricted to few-particles) from which, in turn, by the application of local-scaling transformations we generate exact expressions for these functionals. As in this approach nuclei and electrons are treated under the same footing, we obtain the analytic exact non-Born-Oppenheimer kinetic energy functional for ^4He. The complicated form of the mass-dependent enhancement factor indicates that the inclusion of nuclear motion in a strict non-BO regime, affects in a non-trivial manner the behavior of electrons.

In the second approach, we have examined the construction af approximate functionals designed for OF-MD. The shortcomings of GGA functionals have been made explicit and a remedy to the singularity problem has been advanced through the introduction of RDD variables. However, as there are many possible ways of defining these singularity-free variables, there is not a single and fixed procedure for generating the RDA functionals. In this sense, lacking a systematic way to ensemble these functionals, we must still rely on trial and error constructions, or perhaps, in the future, on a systematically enforcement of necessary constraints.

References

1. P. Hohenberg and W. Kohn, Phys. Rev. B **136**, 864 (1964).
2. R.G. Parr and W. Yang, *Density Functional Theory of Atoms and Molecules* (Oxford, New York, 1989).
3. R.M. Dreizler and E.K.U. Gross, *Density Functional Theory* (Springer-Verlag, Berlin, 1990).
4. E.S. Kryachko and E.V. Ludeña, *Energy Density Functional Theory of Many-Electron Systems* (Kluwer, Dordrecht, 1990).
5. W. Kohn and L.J. Sham, Phys. Rev. **140**, A1133 (1965).
6. R. Car and M. Parrinello, Phys. Rev. Lett. **55**, 2471 (1985).
7. L.H. Thomas, Proc. Cambridge Phil. Soc. **23**, 542 (1927).
8. E. Fermi, Atti Accad. Nazl. Lincei **6**, 602 (1927).
9. C.F. von Weizsäcker, Z. Phys. **96**, 431 (1935).
10. E.V. Ludeña and V.V. Karasiev, in *Reviews of Modern Quantum Chemistry: a Celebration of the Contributions of Robert Parr*, edited by K.D. Sen (World Scientific, Singapore, 2002) pp. 612–665.
11. Y. A. Wang and E. A. Carter, "Orbital-free Kinetic-energy Density Functional Theory", Chap. 5 in *Theoretical Methods in Condensed Phase Chemistry*, edited by S. D. Schwartz (Kluwer, NY, 2000) pp. 117–184.
12. B.-J. Zhou and Y.A. Wang, J. Chem. Phys. **124**, 081107 (2006).
13. D. García-Aldea and J.E. Alvarellos, Phys. Rev. A **77**, 022502 (2008); J. Chem. Phys. **127**, 144109 (2007) and references in both.
14. J. P. Perdew and L.A. Constantin, Phys. Rev. B **75**, 155109 (2007).
15. C.J. Garcia-Cervera, Commun. Computat. Phys. **3**, 968 (2008).
16. L.M. Ghiringhelli and L. Delle Site, Phys. Rev. B **77**, 073104 (2008).
17. W. Eek and S. Nordholm, Theoret. Chem. Accounts **115**, 266 (2006).
18. V.V. Karasiev, S.B. Trickey, and F.E. Harris, J. Comp.-Aided Mater. Des. **13**, 111 (2006).
19. V.V. Karasiev, R.S. Jones, S.B. Trickey, and F.E. Harris, in *New Developments in Quantum Chemistry*, J.L. Paz and A.J. Hernández eds. (Transworld Research Network, Trivandum, Kerala, India, 2009), p. 25.
20. V.V. Karasiev, E.V. Ludeña and A.N. Artemyev, Phys. Rev. A **A 62**, 062510 (2000)

21. A. Artemyev, E.V. Ludeña, and V. Karasiev, J. Mol. Struct. (Theochem) **580**, 47 (2002)
22. E.V. Ludeña, V. Karasiev, and L. Echevarría Int. J. Quantum Chem. **91**, 94 (2003)
23. E.V. Ludeña, V. Karasiev and P. Nieto (2003) Theor. Chem. Acc. **110**, 395 (2003)
24. E.V. Ludeña, D. Gómez, V. Karasiev and P. Nieto, Inter. J. Quantum Chem. **99**, 297 (2004)
25. D. Gómez, E.V. Ludeña, V.V. Karasiev and P. Nieto Theor. Chem. Acc. **116**, 608 (2006)
26. X. Lopez, J.M. Ugalde and E.V. Ludeña Chem. Phys. Lett. **412**, 381 (2005).
27. E.V. Ludeña, X. Lopez and J.M. Ugalde, J. Chem. Phys. **123**, 024102 (2005).
28. X. Lopez, J.M. Ugalde, L. Echevarria and E.V. Ludeña, Phys. Rev. A **74**, 042504 (2006).
29. X. Lopez, J.M. Ugalde and E.V. Ludeña (2006) Eur. Phys. J. D **37**, 351 (2006).
30. X. Lopez, J.M. Ugalde, L. Echevarría, E.V. Ludeña, in: *New developments in Quantum Chemistry*. J.L. Paz and A.J. Hernández eds. (Transworld Research Network, Trivandum, Kerala, India, 2009), p. 303.
31. V.V. Karasiev, R.S. Jones, S.B. Trickey, and F.E. Harris, Phys. Rev. B **80** (in press) (2009).
32. E.V. Ludeña and R. López-Boada, Topics Current Chem. **180**, 169 (1996).
33. E.V. Ludeña, V. Karasiev, R. López-Boada E. Valderrama, and J. Maldonado, J. Comp. Chem. **20**, 55 (1999)
34. D.E. Taylor, V.V. Karasiev, K. Runge, S.B. Trickey, and F.E. Harris, Comp. Mater. Sci. **39**, 705 (2007).
35. F.H. Streitz and J.W. Mintmire, Phys. Rev. B **50**, 11996 (1994).
36. M.J. Buehler, A.C. van Duin, and W.A. Goddard III, Phys. Rev. Lett. **96**, 095505 (2006).
37. A.K. Rappe and W.A. Goddard III, J. Phys. Chem. **95**, 3358 (1991).
38. V.V. Karasiev, S.B. Trickey, and F.E. Harris, Chem. Phys. **330**, 216 (2006).
39. J.P. Perdew, J.A. Chevary, S.H. Vosko, K.A. Jackson, M.R. Pederson, D.J. Singh, and C. Fiolhais, Phys. Rev. B **46**, 6671 (1992).
40. A.D. Becke, J. Chem. Phys. **85**, 7184 (1986).
41. A.E. DePristo and J.D. Kress, Phys. Rev. A **35**, 438 (1987).

NUMERICAL ANALYSIS OF HYDROGEN STORAGE
IN CARBON NANOPORES

CARLOS WEXLER,* RAINA OLSEN and PETER PFEIFER

Department of Physics and Astronomy, University of Missouri
Columbia, Missouri 65211, USA
* wexlerc@missouri.edu

BOGDAN KUCTHA

Labortoire Chimie Provence, University of Aix-Marseille 1, Marseille, France

LUCYNA FIRLEJ

LCVN, University of Montpellier 2, Montpellier, France

SZTEPAN ROSZAK

Institut of Physical and Theoretical Chemistry,
Wroclaw University of Technology, Poland

Received 2 December 2009

Carbon-based materials, due to their low cost and weight, have long been considered as suitable physisorption substrates for the reversible storage of hydrogen. Nanoporous carbons can be engineered to achieve exceptional storage capacities: gravimetric excess adsorption of 0.073 ± 0.003 kg H_2/kg carbon, gravimetric storage capacity of 0.106 ± 0.005 kg H_2/kg carbon, and volumetric storage capacity of 0.040 ± 0.002 kg H_2/liter carbon, at 80 K and 50 bar. The nanopores generage high storage capacity by having a very high surface are, by generating a high H_2-wall interaction potential, and by allowing multi-layer adsorption of H_2 (at cryogenic temperatures). In this paper we show how the experimental adsorption isotherms can be understood from basic theoretical considerations and computational simulations of the adsorption in a bimodal distribution of narrow and wide pore spaces. We also analyze the possibility of multi-layer adsorption, and the effects of hypothetical larger adsorption energies. Finally, we present the results of coupled *ab initio* calculations and Monte Carlo simulations showing that partial substitution of carbon atoms in nanoporous matrix with boron results in significant increases of the adsorption energy and storage capacity.

Keywords: Hydrogen Storage; Nanoporous Carbon; Ab Initio, Alternative Fuels.

---

*Corresponding author.

1. Introduction

Practical implementation of hydrogen fuel for automotive applications is still impeded by the lack of a storage medium capable of delivering a driving range of hundreds of kilometers without compromising safety, cost, or cargo capacity.[1,2] The main problem of hydrogen is that, being a gas at ambient temperatures, its volumetric energy density is quite low requiring either very bulky storage tanks, liquefaction at very low temperatures or extremely high compression (5,000–10,000 psi is common in current test vehicles). These extreme conditions are not only complicated from a practical point of view, but also pose materials challenges (e.g., steel becomes brittle when exposed to high-pressure hydrogen) and significant additional energy costs (compression or liquefaction requires 30–50% of the chemical energy stored). In recent years, therefore, significant efforts have been made to find alternative ways to store hydrogen. For these new methods, the U.S. Department of Energy (DOE) has set targets of 0.045 kg H_2/kg system, 28 kg H_2/m^3 system at room temperature for 2010.[2]

Early proposals focused on the formation of hydrides, and storage in zeolites and metal organic frameworks,[1,3] but modest performance and relatively high cost lead to the search of alternative methods and materials. Carbon-based adsorbants (activated carbon and carbon nanotubes) are amongst the most promising materials for reversible storage of hydrogen by physisorption. However, overly optimistic early reports of 20–60 wt.% storage at room temperature in carbon nanofibers and alkali-doped nanotubes[4,5] have been significantly scaled down in later studies.

The problem with carbon is that the hydrogen-carbon interaction is relatively weak: the binding energy of a single hydrogen molecule on a single graphitic surface is 4–6 kJ/mol,[6–11] which is too small for practical storage.[12,13] However, recent "nanopore engineering" in activated carbons[10] resulted in materials very high surface area materials ($\sim$3,000 m^2/g[10,14]). These materials perform exceedingly well at cryogenic temperatures ($\sim$0.1 kg H_2/kg carbon at $P = 100$ bar, $T = 77$–80 K[10,14]). The high performance can be qualitatively explained by the presence of a large number of sub-nm scale pores in the carbon. Such nanopores generate the observed high surface areas (i.e., many adsorption sites) and host adsorption potentials that are deeper by overlapping the adsorbant-adsorbate potentials from both sides of the pore (resulting in observed binding energies of 5–9 kJ/mol).[10,11,15] Also, surface functionalization of the adsorbant may result in substantial increases in the interaction energies.

In this paper we analyze numerically recent hydrogen adsorption experiments in nanoporous carbon. We show that grand canonical Monte Carlo (GCMC) simulations using standard hydrogen-carbon[6–8] and hydrogen-hydrogen[8] interaction potentials, with Feynman-Hibbs quantum corrections[16–18] accounts reasonably well for the experimentally observed hydrogen adsorption at cryogenic temperatures. The experimental results are explained in terms of a bimodal distribution of slit-shaped pores of 7 Å (with a binding energy of 9 kJ/mol) and 20 Å (5 kJ/mol).

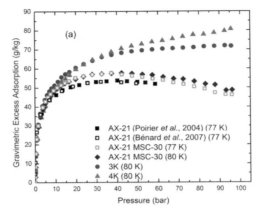

Fig. 1.   Gravimetric excess adsorption isotherms of H_2 on carbon samples 3K, 4K, AX-21 MSC-30, and AX-21 (see text), all measured on a Hiden HTP1 high-pressure volumetric analyzer (U. Missouri) at indicated temperatures (data from the literature for AX-21 is also shown for comparison). Experimental uncertainties are less than 5%. Data courtesy of Pfeifer et al.[25]

We also discuss theoretical limits for the storage of hydrogen in carbon-based systems based on hypothetical increased values of the adsorbate-adsorbant interaction, and analyze possible multi-layer adsorption in these cases. We show that if the hydrogen binding energy can be raised by a factor of two, practical storage at room temperature and moderate pressures is possible. Based on these results, we discuss the possibility of raising the adsorption energy by substituting some of the carbon atoms by boron. Boron being a p-type dopant to carbon introduces holes in the graphite electronic structure, lowering the chemical potential for electrons and increasing the surface polarizability.[19–21] We show from first principles that this chemical substitution increases the hydrogen binding energy, and how this results in enhanced storage capacity that has the potential to reach DOE goals.

2. Experimental Background and Analysis

Various activated carbon samples were produced by the *Alliance for Collaborative Research in Alternative Fuel Technology* (ALL-CRAFT),[22] at the University of Missouri. Figure 1 shows gravimetric excess adsorption isotherms of H_2 on two ALL-CRAFT carbon samples "3K" and "4K,"[22] and reference/callibration samples AX-21 MSC-30[23] and AX-21.[13,24] Samples 3K and 3K have record-breaking gravimetric excess adsorption of 0.073 ± 0.003 kg H_2/kg carbon, gravimetric storage capacity of 0.106 ± 0.005 kg H_2/kg carbon (including gas in pore space[10]), and volumetric storage capacity of 0.040 ± 0.002 kg H_2/liter carbon, at 80 K and 50 bar.[25] Samples surface areas and porosities are 3K ($\Sigma = 2{,}500$ m^2/g, $\phi = 0.78$), 4K ($\Sigma = 2{,}600$ m^2/g, $\phi = 0.81$) and AX-21 MSC-30 ($\Sigma = 3{,}400$ m^2/g, $\phi = 0.81$).[25]

Samples AX-21 MSC-30 and AX-21 exhibit a local maximum at $P \sim 40$–50 bar, characteristic of high-surface-area, AX-21-type carbons. (On the high-pressure

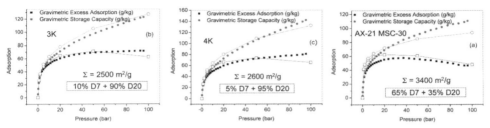

Fig. 2.   Comparison of experimental (full symbols)[25] and GCMC simulations (open symbols) of H_2 adsorption isotherms for samples 3K and 4K (80K) and AX-21 MSC-30 (77K). GCMC simulations were performed with a bi-modal distribution of slit-shaped pores of 7 Å (binding energy 9 kJ/mol) and 20 Å (5 kJ/mol) in the proportion indicated in each figure. Discrepancy of the excess adsorption at high pressures are due to low binding-energy sites not included in the simulations (absent in AX-21 MSC-30). The discrepancy of the storage capacity for this sample is likely due to macropores also ignored in the simulations.

side of the maximum, the density of the nonadsorbed gas grows more rapidly than the density of the adsorbed film.) By contrast, samples 3K and 4K show no such maximum in the pressure range investigated, and their excess adsorption exceeds that of AX-21 and AX-21 MSC-30. Samples 3K and 4K are members of a series of carbons, made from corncob, in which the ratio of potassium hydroxide (KOH) to carbon during chemical activation (treatment at 790 C) was varied from 2:1 ("2K") to 6:1 ("6K").[25] The absence of a local maximum in samples 3K and 4K is due to variations in binding energies, and possibly vibrational frequencies of adsorbed H_2 in the adsorption potential of sub/supra-nm pores.[10]

Grand Canonical Monte Carlo (GCMC) simulations of adsorbed H_2 in bimodal distributions of sub/supra-nm pores show good agreement with 3K, 4K, and AX-21 MSC-30 isotherms up to $\sim$ 30 bar, by modeling systems with binding energies of $\sim$ 9 kJ/mol and $\sim$ 5 kJ/mol from pores of width $\sim$ 7 Å and $\sim$ 20 Å, respectively (best fits in the 0–30 bar region are 3K: 10% D7+90% D20, 4K: 5% D7+95% D20, AX-21 MSC-30: 65% D7+35% D20, see Fig. 2). Above 30 bar, however, experimental isotherms were found to exceed the simulated ones. This led to the conclusion that the absence of a local maximum in samples 3K and 4K is due to a significant presence of binding energies *less* than 5 kJ/mol, presumably located at edge sites of fragments of graphitic sheets. The significant fraction of supra-nm pores in 3K and 4K is consistent with a significant fraction of such edge sites. The mechanism by which supra-nm pores can give rise to higher excess adsorption than sub-nm pores (Figs. 1 and 2), even though binding energies may be low, is multilayer adsorption by attractive H_2-H_2 interactions, effective at low temperature and in wide pores,[11] but ineffective in sub-nm pores, too narrow to hold multilayers (see Section 3).

3. Numerical Estimation of Hydrogen Storage

Numerical studies of different adsorption processes have shown that the optimal pore width for slit-shaped pores in carbon (center-to-center distance between the center of the carbon atoms) corresponds to the accommodation of ca. two adsorption layers. For a Lennard-Jones (LJ) (6-12) model, this corresponds to $D/\sigma_{fs} \simeq 2.9$–3.0, where σ_{fs} is the fluid-solid LJ interaction length.[12] For our calculations we considered a solid-solid (carbon-carbon) $\sigma_{ss} = 0.34$ nm,[7] whereas for hydrogen we use the value of $\sigma_{ff} = 0.296$ nm.[8] Employing the Lorentz-Berthelot mixing rules, the binary parameter is $\sigma_{fs} = 0.318$ nm, resulting in an optimal slit-pore with $D \simeq 0.92$–0.95 for hydrogen adsorption. Furthermore, if one assumes that σ_{ff} corresponds to the size of a hydrogen molecule, and considers a two-dimensional hexagonal close packing of LJ spheres with spacing $2^{1/6}\sigma_{fs}$, the effective area per H_2 is $S_{ml} = 0.0956$ nm^2.[12] Given that for an ideal graphene sheet the surface area (both sides) is 2,630 m^2/g, the maximum hydrogen capacity of an idealized carbon slit-pore material (two layers of hydrogen per pore) is of 9.0–9.5 wt.%. Larger pores could result in higher storage capacities, if additional layers can be stabilized.

In this Section, we analyze how the storage capacity of slit-shaped pores depends on the energy of adsorption (between 4.5 and 15 kJ/mol), temperature (77–298 K), slit-pore size (5–20 Å) and pressure (0–100 bar). The purpose of the calculations is to obtain estimates of the maximum theoretical storage capacity that could be achieved in the adsorbent. For each combination of these parameters we have determined the total amount adsorbed in pores. Various models of the H_2-H_2 interaction have been exhaustively described in the literature.[8] Here, we consider the H_2 molecules as structureless superatoms interacting via LJ (6-12) potential. The H_2-graphene interaction we conside here is the analytical form proposed by Steele.[7] Since the corrugation potential is rather small, we consider only the *laterally averaged* potential $U_0(z)$.[7,26]

We estimate the hydrogen uptake in homogeneous slit pores of various width using four different potential models that differ in the energy of adsorption on the free surface. In addition to typical H_2-carbon energies widely used in literature and coming from: (1) Steele average ($E_a \simeq 4.5$ kJ/ mol),[6–11] and (2) *ab initio* estimations ($E_a \simeq 6.5$ kJ/mol),[13,27] we also considered two *hypothetical* models with (3) $E_a \simeq 10$ kJ/mol, and (4) $E_a \simeq 15$ kJ/mol. As explained before, for slit-pores the actual energy of adsorption is higher because of the cummulative effect of the two surfaces (this becomes important for sub-nm pores). Feynman-Hibbs quantum corrections[16–18] were used for all cases. In general, this correction makes the energy of adsorption slightly weaker and the molecule effectively larger (larger effective σ's). At room temperature these variations are negligible. At 77 K, however, the corrections are on the order of 10%. This difference is sufficiently important to significantly modify the estimation of the adsorbed amount. Figure 3 (left) compares both classical and quantum models of inter-molecular H_2-H_2 interaction at $T = 77$ K. The most important difference appears at short distances, comparable to the effective size of the hydrogen molecules. When the density of the adsorbed molecules increases the quantum size of molecules should be taken into account trying to maximize the amount adsorbed.

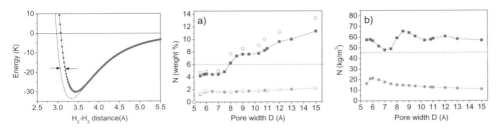

Fig. 3. Left panel: The hydrogen-hydrogen interaction profile in classical (solid) and quantum corrected (circles) models at $T = 77$ K. The most important effect is the larger effective size of the molecules, which limits the compressibility of the adsorbed phase at high pressures. Center and right panels: total gravimetric (in wt.%, center), and volumetric (in kg/m^3, right) quantity of H$_2$ molecules adsorbed at 100 bars in slit graphene pores (Matera model) at $T = 298$ K (circles) and 77 K (squares). For comparison, the results at 77 K calculated without quantum correction are also shown (open symbols).

Figures 3 (center and right) and 4 present the main results for the most realistic model of pure carbon adsorbent ($E_a = 4.5$ kJ/mol, free graphene surface). It is noteworthy that at room temperatures, the amount of H$_2$ stored is quite small, it seems unlikely that the DOE goals can be achieved as-is (the goals are readily achievable in cryogenic conditions, but this creates significant complications for a practical storage tank). Experimentally, our recent results (Section 2) are in excellent agreement with the theoretical predictions. The difference between classical and quantum corrected results at 77 K increases with the size of the slit. At low temperature the quantum correction increases the effective size of molecules due to the zero point motion (Fig. 3). As a result, a smaller number of molecules can fit into the pore volume. This results in a decrease of ca. 1 wt.% in the stored amount for pore of $D = 0.9$ nm and more than 2 wt.% for $D = 1.5$ nm. This difference originates both from the quantum correction of the H$_2$-wall interaction and the quantum correction of the H$_2$-H$_2$ interactions. The non-negligible role of the intermolecular energy can be deduced from the density profiles shown in Fig. 4. If the temperature is low enough, a middle (third) layer can be stabilized even in pores larger than ~ 1.15nm. At high temperature the thermal fluctuations destabilize it, and in the middle of the pore there is essentially a gas, which does not contribute to the excess adsorption.

For the various H$_2$-wall interaction models discussed above, we present a summary of results in Fig. 5: gravimetric and volumetric storage capacities at 298 and 77 K. It is noteworthy that at room temperatures and for narrow pores ($D \leq 1$ nm), the efficiency of the storage material depends strongly on the adsorption energy, but saturates quickly for $E_a \geq 10$ kJ/mol. In larger pores, however, larger E_a results in the stabilization of a third (middle) layer, whose density depends crucially on the pressure, temperature and H$_2$-H$_2$ interaction potentials. At room temperature, however, thermal fluctuations prevent the system from stabilizing this structure.

At $T = 298$ K the volumetric capacity decreases when the pores become larger

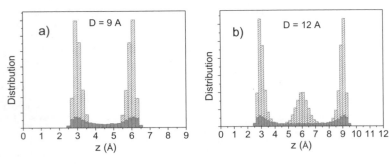

Fig. 4.   Density profiles of adsorbed molecules at 100 bar, for two slit sizes: (a) $D = 9$ Å and (b) $D = 12$ Å, at $T = 298$ K (full) and 77 K (shadowed). The profiles represent the total amount adsorbed.

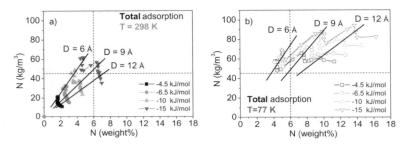

Fig. 5.   Total volumetric versus gravimetric hydrogen storage capacities at $P = 100$ bar, for $T = 298$ K (closed symbols, a) and 77 K (open symbols, b) for four different H_2-wall interaction models (see text). The points correspond to the pore sizes between 5.5 and 15 Å. The solid straight lines are guides for the eye, traced through the points corresponding to three sizes of slit pores: 6, 9 and 12 Å.

due to the reduction in the overlap of wall potentials. It seems evident that optimal pore sizes will correspond to $D \sim 0.8$–1.1 nm. It is also evident that to achieve substantial storage at high temperatures, the "standard" binding energy (~ 5 kJ/mol) is unlikely to be sufficient: 10–15 kJ/mol is required. See Section 4.

4. Enhanced Hydrogen-Substrate Interaction in Boron Substituted Carbon

One way to achieve a higher hydrogen-substrate binding energy, and therefore a high room-temperature hydrogen uptake is the modification of the carbon material by substitution of the carbon with other atoms. Substitution with boron atoms has been long speculated to be a "silver bullet" for hydrogen storage through "strong physorption." Prior *ab initio* analyses have shown that the strongly localized empty p_z orbital of boron is important and enhances the interaction via coupling with the occupied σ orbital of H_2,[19,20] perhaps by producing a partial charge transfer between the adsorbant and the adsorbate. In C36 Fullerenes, the calculated enhancement lead to adsorption energies (calculated) as high as 35 kJ/mol.[19] How-

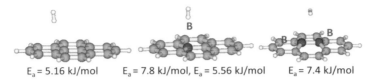

$E_a = 5.16$ kJ/mol $E_a = 7.8$ kJ/mol, $E_a = 5.56$ kJ/mol $E_a = 7.4$ kJ/mol

Fig. 6.   Minimum energy configurations for H_2 on pyrene and boron-doped pyrene, calculated at the second order Møller-Plesset theory and restricted open Hartree-Fock wavefunctions (see text).

ever, so far both experimental and computational evidence is scant. It is also not clear whether a single boron atom substituted into a graphene-like structure can create high binding energies not just at the boron sites but also on its periphery, and how this relates to the adsorption characteristics of the system with a (random) distribution of boron dopants, given the significant heterogeneity introduced.[15]

Initially *ab intio* calculations were performed to get the energy of interactions between hydrogen molecules and a pyrene molecule ($C_{16}H_{10}$) and a variant where one of the "central" carbons was substituted by boron (Fig. 6). The external carbon frame was kept frozen to better approximate the actual structure of a larger graphene-like material. All calculations were performed applying in GAUSSIAN 03[28] at the second order Møller-Plesset level of theory utilizing restricted open Hartree-Fock wavefunctions.[29] The effective core potential SBKJC VDZ basis[30] set was extended to include polarization functions.[31] The H_2 molecule forms two perpendicular complexes with the pyrene surface. The minimum energy structures were determined for hydrogen interacting with B-center ($\Delta E = 7.8$ kJ/mol, R(B-H_2) = 3.12 Å) and C-center ($\Delta E = 5.56$ kJ/mol, R(C-H_2) = 3.24 Å), see Fig. 6. For comparison, the energy for H_2 on pyrene (no boron) resulting from our calculations is $\Delta E = 5.16$ kJ/mol, comparing favorably with accepted values,[6–8,10] thus validating the method. It is worth mentioning that in our calculations no deviations from planarity of the pyrene molecule were observed vs. some deviations suggested in Ref. 32.

Once the interaction potentials of H_2 and the boron-substituted pyrene were calculated, we used the results to generate a potential map for a randomly doped graphene sheet. For simplicity we assumed the H_2 molecules to be structureless superatoms interacting via the Lennard-Jones (LJ) (6-12) potential. We further assumed an additive LJ-type potential between the hydrogen superatoms and the pore wall with a cut-off distance of 15 Å. The energy and radius parameters of the LJ interaction were chosen so that the energy experienced by the hydrogen over the surface agrees to that of the *ab initio* results for pyrene and its boron-doped variant. This required (see Fig. 6) modifications to the interaction to the nearest and next-nearest neighboring carbons to the boron. For calculations, Feynman-Hibbs quantum corrections were employed in the usual way.[16–18]

To explore the adsorption of hydrogen in these systems, boron-substituted graphene sheets were prepared with random distributions of boron impurities (we imposed no nearest- or next-nearest neighbor pairing of boron atoms to insure that

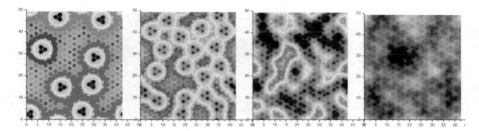

Fig. 7.   Adsorption potentials for B-substituted graphene containing 1%, 2.5%, 5% and 10% boron (left to right). The lightest/darkest regions correspond to -580 K/-1,500 K potentials.

the "additivity" of potentials from the pyrene calculations was adequate) at 1%, 2.5%, 5% and 10% boron concentration. Figure 7 shows the potential profiles for these cases. For low substitution ratio the strong adsorption regions around the boron atoms remains well separated (Fig. 7, leftmost panel). However, even at a modest substitution ratio of 5% these regions start to overlap and the average adsorption energy increases with respect to pure-carbon graphite (from $\sim$ 600 K to $\sim$ 900 K); in fact, the energy of the strongest adsorption site even doubles (Fig. 7, 3$^{\rm rd}$ panel). At a substitution ratio of 10% the strongest adsorption energy reaches $\sim$ 1,500 K (Fig. 7, rightmost panel). GCMC simulations were then performed for various boron concentrations and several slit-pore widths to generate H_2 adsorption isotherms at liquid nitrogen (77 K) and room temperatures (298 K).

Figure 8 (left panel) shows adsorption isotherms (shown: total amount stored[10]) for slit-shaped pores of 12 Å width. This size is close to optimal (see Section 3) as these pores can accommodate up to three layers of hydrogen (one next to each pore surface, plus one between the layers). Under cryogenic conditions the addition of boron produces a dramatic increase in the total amount of gas stored; however, for the higher concentrations of boron, the adsorption-desorption cycle shows that the binding energy may already be a bit too large for H_2 *delivery* under 1–100 bar pressure sweeps (see discussion on Refs. 10–12). More importantly, the improvements at room temperature are quite significant: the amount stored in the 10% boron/carbon sample is approximately double that of the pure carbon system. The end-result is a material that is potentially capable of achieving the 2010 DOE target at moderate pressures. Furthermore, the complete adsorption-desorption cycle is highly efficient, achieving ca. 97% delivery. The right panel of Fig. 8 shows how the volumetric and gravimetric storage capacities depend on both boron/carbon ratio and pore sizes. It is evident that most activated carbons will have a substantial heterogeneity of pore sizes, but it is clear that there is a sufficiently large "sweet spot" of pore sizes in the vicinity of 1 nm width, a region of widths that is reasonably common in activated carbon.[10,14]

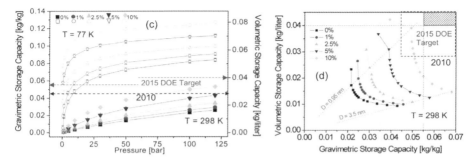

Fig. 8. Left: Simulated gravimetric and volumetric storage capacities for H_2 on B-doped carbon with slit-shaped pores of 12 Å width for 0%, 1%, 2.5%, 5% and 10% boron/carbon concentration at liquid nitrogen and room temperatures. Right: Volumetric and gravimetric storage capacities at room temperature as function of boron concentration and pore width. DOE *system* target goals for 2010 and 2015 are presented for comparison (note that the calculated numbers are for *material* only, not including tanks, pumps, etc.).

5. Summary

Nanoporous carbons have shown promise as a potential storage medium for hydrogen. Under cryogenic conditions (liquid nitrogen temperatures) the H_2-carbon binding energy is close to ideal, resulting in reversible storage that can be operated with good delivery efficiency by performing pressure sweeps. Unfortunately, at room temperature the low H_2-carbon binding energy results in an insufficient amount of H_2 being stored. Our *ab initio* calculations of the interaction potentials of H_2 with boron-substituted carbon and subsequent GCMC simulations of the system show that this chemical modification of activated carbons may be able to achieve high storage capacities with a high delivery rate. We are hopeful that these results will be experimentally confirmed shortly, as this would result in a material with the potential to store hydrogen for vehicular applications.

Acknowledgments

This material is based upon work supported in part by the U.S. Department of Energy under Award Nos. DE-FG02- 07ER46411 and DE-FG36-08GO18142. We acknowledge the University of Missouri Bioinformatics Consortium for the use of their computational facilities. The authors acknowledge fruitful discussions with G. Suppes, J. Burress, M. Beckner, M. Kraus and M.F. Hawthorne.

References

1. S. Satyapal, J. Petrovic, C. Read, G. Thomas, and G. Ordaz, *Catalysis Today* **120**, 246 (2007).
2. The U.S. D.O.E. targets for 2010, 2015 and "ultimate fleet" are published at http://www1.eere.energy.gov/hydrogenandfuelcells/mypp/pdfs/storage.pdf.
3. See L. Schlapbach and A. Zuttel, *Nature* (London) **414**, 353 (2001), and references therein.

4. A. Chambers, C. Park, R.T.K. Baker, and N.M. Rodriguez, *J. Phys. Chem. B* **102**, 4253 (1998).
5. P. Chen, X. Wu, J. Lin, and K.L. Tan, *Science* **285**, 91 (1999).
6. L. Mattera, F. Rosatelli, C. Salvo, F. Tommasini, U. Valbusa, and G. Vidali, *Surf. Sci.* **93**, 515 (1980).
7. W.A. Steele, *The interaction of gases with solid surfaces*, Pergamon Press (New York, 1974).
8. V. Buch, *J. Chem. Phys.* **100**, 7610 (1994).
9. D.Y. Sun, J.W. Liu, X.G. Gong, and Z.-F. Liu, *Phys Rev B* **75**, 075424 (2007).
10. J. Burress, M. Kraus, M. Beckner, R. Cepel, G. Suppes, C. Wexler, and P. Pfeifer, *Nanotechnology* **20**, 204026 (2009).
11. B. Kuchta, L. Firlej, P. Pfeifer, and C. Wexler, *Carbon* **48**, 223 (2010).
12. S. K. Bhatia and A. L. Myers, *Langmuir* **22**, 1688 (2006).
13. P. Benard, and R. Chahine, *Scripta Materialia* **56**, 803 (2007).
14. P. Pfeifer, *et al.*, *Mater. Res. Soc. Symp. Proc.* **1041**, R02-02 (2008).
15. L. Firlej, B. Kuchta, C. Wexler and P. Pfeifer, *Adsorption* **15**, 312 (2009).
16. Q. Wang and J. K. Johnson, *J. Phys. Chem. B* **103**, 277 (1999).
17. P. Kowalczyk, H. Tanaka, R. Hoyst, K. Kaneko, T. Ohmori, and J. Miyamoto, *J. Phys. Chem. B* **109**, 17174 (2005).
18. V. A. Kumar, H. Jobic, and S. K. Bhatia, *Adsorption* **13**, 501 (2007).
19. Y.-H. Kim, Y. Zhao, A. Williamson, M. J. Heben, and S. B. Zhang, *Phys. Rev. Lett.* **96**, 016102 (2006).
20. Y. Ferro, F. Marinelli, A. Jelea, and A. Allouche, *J. Chem. Phys.* **120**, 11882 (2004).
21. L. Firlej, Sz. Roszak, B. Kuchta, P. Pfeifer, and C. Wexler, *J. Chem. Phys.* **131**, 164702 (2009).
22. http://all-craft.missouri.edu/.
23. Courtesy of the National Renewable Energy Laboratory.
24. E. Poirier, R. Chahine, P. Bnard, D. Cossement, L. Lafi, E. Mlancon, T.K. Bose, and S. Desilets, *Appl. Phys. A* **78**, 961 (2004).
25. P. Pfeifer, *et al.*, private communication (unpublished).
26. See also: C.L. Pint, M.W. Roth, and C. Wexler, *Phys. Rev. B* **73**, 085422 (2006); M.W. Roth, C.L. Pint, and C. Wexler, *Phys. Rev. B* **71**, 155427 (2005).
27. A. Ferre-Vilaplana, *J. Chem. Phys.* **122**, 104709 (2005).
28. M. J. Frisch, G. W. Trucks, H. B. Schlegel *et al.*, GAUSSIAN03, Revision C.02, Gaussian, Inc., Wallingford, CT, 2004.
29. C. Møller and M. S. Plesset, *Phys. Rev.* **46**, 618 (1934).
30. W. J. Stevens, H. Basch, and M. Krauss, *J. Chem. Phys.* **81**, 6026 (1984).
31. R. Krishnan, J. S. Binkley, R. Seeger, and J. A. Pople, *J. Chem. Phys.* **72**, 650 (1980).
32. M. Endo, T. Hayashi, S.-H. Hong, T. Enoki, and M. S. Dresselhaus, *J. Appl. Phys.* **90**, 5670 (2001).

Part F
Superconductivity

GENERALIZED BOSE-EINSTEIN CONDENSATION IN SUPERCONDUCTIVITY

MANUEL de LLANO*

*Physics Department, University of Connecticut,
Storrs, CT 06269, USA*
dellano@servidor.unam.mx

Received 14 October 2009

Unification of the BCS and the Bose-Einstein condensation (BEC) theories is surveyed in detail via a generalized BEC (GBEC) finite-temperature statistical formalism. Its major difference with BCS theory is that it can be diagonalized *exactly*. Under specified conditions it yields the precise BCS gap equation for all temperatures as well as the precise BCS zero-temperature condensation energy for all couplings, thereby suggesting that a BCS condensate is a BE condensate in a *ternary* mixture of kinematically independent unpaired electrons coexisting with equally proportioned weakly-bound two-electron and two-hole Cooper pairs. Without abandoning the electron-phonon mechanism in moderately weak coupling it suffices, in principle, to reproduce the unusually high values of T_c (in units of the Fermi temperature T_F) of 0.01-0.05 empirically reported in the so-called "exotic" superconductors of the Uemura plot, including cuprates, in contrast to the low values of $T_c/T_F \leq 10^{-3}$ roughly reproduced by BCS theory for conventional (mostly elemental) superconductors. Replacing the characteristic phonon-exchange Debye temperature by a characteristic magnon-exchange one more than twice in size can lead to a simple interaction model associated with spin-fluctuation-mediated pairing.

Keywords: Cooper pairing; Boson-fermion models; Bose-Einstein condensation.

1. Introduction

Boson-fermion (BF) models of superconductivity (SC) as a Bose-Einstein condensation (BEC) go back to the mid-1950's,[1-4] pre-dating even the BCS-Bogoliubov theory.[5,6] Although BCS theory only contemplates the presence of "Cooper correlations" of single-particle states, BF models[1-4,7-18] posit the existence of actual bosonic CPs. With two[17,18] exceptions, however, all BF models neglect the effect of *hole* CPs included on an equal footing with electron CPs to give the "complete" *ternary* (instead of binary) BF model that constitutes the generalized Bose-Einstein condensation (GBEC) theory to be analyzed here in greater detail.

---

*Permanent Address: Instituto de Investigaciones en Materiales, Universidad Nacional Autónoma de México, Apdo. Postal 70-360, 04510 México, DF, MEXICO.

2. Ternary Boson-fermion Gas Hamiltonian

The GBEC[17,18] is described by the Hamiltonian $H = H_0 + H_{int}$ that describes a *ternary* boson-fermion gas where

$$H_0 = \sum_{\mathbf{k_1},s_1} \epsilon_{\mathbf{k_1}} a^+_{\mathbf{k_1},s_1} a_{\mathbf{k_1},s_1} + \sum_{\mathbf{K}} E_+(K) b^+_{\mathbf{K}} b_{\mathbf{K}} - \sum_{\mathbf{K}} E_-(K) c^+_{\mathbf{K}} c_{\mathbf{K}} \qquad (1)$$

and $\mathbf{K} \equiv \mathbf{k_1} + \mathbf{k_2}$ is the CP center-of-mass momentum (CMM) wavevector. Without the last term (1) is essentially the starting Hamiltonian of Ref. 8 as well as that of Friedberg and T.D. Lee[12–15] which describe a *binary* boson-fermion gas. Here $\epsilon_{\mathbf{k_1}} \equiv \hbar^2 k_1^2/2m$ are the single-electron, and $E_\pm(K)$ the 2e-/2h-CP *phenomenological,* energies. Also, $a^+_{\mathbf{k_1},s_1}$ $(a_{\mathbf{k_1},s_1})$ are creation (annihilation) operators for electrons and similarly $b^+_{\mathbf{K}}$ $(b_{\mathbf{K}})$ and $c^+_{\mathbf{K}}$ $(c_{\mathbf{K}})$ for 2e- and 2h-CP bosons, respectively. As suggested by the original Cooper-pair problem,[19] the b and c operators proposed depend only on $\mathbf{K}$ and so are *distinct* from the BCS-pair operators depending on both $\mathbf{K}$ *and* the relative wavevector $\mathbf{k} \equiv \frac{1}{2}(\mathbf{k_1} - \mathbf{k_2})$; see Ref. 5 Eqs. (2.9) to (2.13) for the particular case of $\mathbf{K} = 0$ and shown there *not* to satisfy the ordinary Bose commutation relations. Nonetheless, CPs are objects easily be seen to obey Bose-Einstein statistics as, in the thermodynamic limit, an indefinitely large number of distinct $\mathbf{k}$ values correspond to a given $\mathbf{K}$ value characterizing the energy levels $E_+(K)$ or $E_-(K)$ given in (1). Only this is needed to ensure a BEC (or macroscopic occupation of a given state that appears below a certain fixed $T = T_c$). This was found[17][18] numerically *a posteriori* in the GBEC theory. Also, the BCS gap equation is recovered for equal numbers of both kinds of pairs, in the $\mathbf{K} = 0$ state and in all $\mathbf{K} \neq 0$ states taken collectively, and in weak coupling, regardless of CP overlaps. The precise familiar BEC T_c formula emerges[17] in strong coupling when i) 2h-CPs are ignored (whereupon the Friedberg-T.D. Lee model[12–15] equations are recovered) and ii) one switches off the BF interaction defined below in (2). The 2e- and 2h-CPs postulated in (1) are idealized objects with the same mass but opposite charges. Thus, they do not include the crucial effects on them of the ionic-crystalline band-structure such as the "dressing" and "undressing" properties connecting particles and holes extensively discussed by Hirsch[20] as a fundamental property of matter that he calls "electron-hole asymmetry."

The interaction Hamiltonian H_{int} consists of four distinct BF interaction *single* vertices each with two-fermion/one-boson creation or annihilation operators. Each vertex is reminiscent of the Fröhlich electron-phonon interaction. Here H_{int} depicts how unpaired electrons (or holes) combine to form the 2e- (and 2h-CPs) assumed in a d-dimensional system of size L, namely

$$H_{int} = L^{-d/2} \sum_{\mathbf{k},\mathbf{K}} f_+(k) \{ a^+_{\mathbf{k}+\frac{1}{2}\mathbf{K},\uparrow} a^+_{-\mathbf{k}+\frac{1}{2}\mathbf{K},\downarrow} b_{\mathbf{K}} + a_{-\mathbf{k}+\frac{1}{2}\mathbf{K},\downarrow} a_{\mathbf{k}+\frac{1}{2}\mathbf{K},\uparrow} b^+_{\mathbf{K}} \}$$

$$+L^{-d/2} \sum_{\mathbf{k,K}} f_-(k)\{a^+_{\mathbf{k}+\frac{1}{2}\mathbf{K},\uparrow} a^+_{-\mathbf{k}+\frac{1}{2}\mathbf{K},\downarrow} c^+_{\mathbf{K}} + a_{-\mathbf{k}+\frac{1}{2}\mathbf{K},\downarrow} a_{\mathbf{k}+\frac{1}{2}\mathbf{K},\uparrow} c_{\mathbf{K}}\}. \tag{2}$$

The energy form factors $f_\pm(k)$ in (2) are taken as in Refs. 17 and 18 where the associated quantities E_f and $\delta\varepsilon$ are *new* phenomenological dynamical energy parameters (in addition to the positive BF vertex coupling parameter f) that replace the previous such $E_\pm(0)$, through the relations $E_f \equiv \frac{1}{4}[E_+(0) + E_-(0)]$ and $\delta\varepsilon \equiv \frac{1}{2}[E_+(0) - E_-(0)] \geqslant 0$ where $E_\pm(0)$ are the (empirically *unknown*) zero-CMM energies of the 2e- and 2h-CPs, respectively.

We refer to E_f as the "pseudoFermi" energy. It serves as a convenient energy scale and is not to be confused with the usual Fermi energy $E_F = \frac{1}{2}mv_F^2 \equiv k_B T_F$ where T_F is the Fermi temperature. The Fermi energy E_F equals $\pi\hbar^2 n/m$ in 2D and $(\hbar^2/2m)(3\pi^2 n)^{2/3}$ in 3D, with n the total number-density of charge-carrier electrons, while E_f is similarly related to another density n_f which serves to scale the ordinary density n. The two quantities E_f and E_F, and consequently also n and n_f, coincide *only* when perfect 2e/2h-CP symmetry holds as in the BCS instance.

3. Reduced Hamiltonian and its Exact Diagonalization

The interaction Hamiltonian (2) can be further simplified by keeping only the $\mathbf{K} = 0$ terms, namely

$$H_{int} \simeq L^{-d/2} \sum_{\mathbf{k}} f_+(k)\{a^+_{\mathbf{k}\uparrow} a^+_{-\mathbf{k}\downarrow} b_0 + a_{-\mathbf{k}\downarrow} a_{\mathbf{k}\uparrow} b^+_0\}$$

$$+L^{-d/2} \sum_{\mathbf{k}} f_-(k)\{a^+_{\mathbf{k}\uparrow} a^+_{-\mathbf{k}\downarrow} c^+_0 + a_{-\mathbf{k}\downarrow} a_{\mathbf{k}\uparrow} c_0\} \tag{3}$$

which allows *exact* diagonalization as follows. One can now apply the Bogoliubov "recipe" (see e.g. Ref. 21 p. 199 ff.) of replacing in the full Hamiltonian (1)-(2) all zero-CMM 2e- and 2h-CP boson creation and annihilation operators by their respective c-numbers $\sqrt{N_0(T)}$ and $\sqrt{M_0(T)}$, where $N_0(T)$ and $M_0(T)$ are the number of zero-CMM 2e- and 2h-CPs, respectively. One eventually seeks the highest temperature, say T_c, above which e.g. $N_0(T_c)$ or $M_0(T_c)$ vanish and below which one and/or the other is nonzero. Note that T_c calculated thusly can, in principle, turn out to be zero, in which case there is no BEC, but this will not turn out to be for the BCS model interaction to be employed here. If the number operator is

$$\hat{N} \equiv \sum_{\mathbf{k}_1,s_1} a^+_{\mathbf{k}_1,s_1} a_{\mathbf{k}_1,s_1} + 2\sum_{\mathbf{K}} b^+_{\mathbf{K}} b_{\mathbf{K}} - 2\sum_{\mathbf{K}} c^+_{\mathbf{K}} c_{\mathbf{K}} \tag{4}$$

the simplified $\hat{H} - \mu\hat{N}$ is now entirely *bilinear* in the a^+ and a operators, as it already was in the boson operators b and c. It can thus be diagonalized exactly via a Bogoliubov-Valatin transformation[22,23]

$$a_{\mathbf{k},s} \equiv u_k \alpha_{\mathbf{k},s} + 2sv_k \alpha^{\dagger}_{-\mathbf{k},-s} \tag{5}$$

to new operators $\alpha^{\dagger}$ and α and where $s = \pm\frac{1}{2}$. The transformation (5) *exactly* diagonalizes (1) plus (3) to the fully bilinear form

$$
\hat{H} - \mu\hat{N} \simeq \sum_{\mathbf{k},s} [\underbrace{\xi_k \left(u_k^2 - v_k^2\right) + 2\Delta_k u_k v_k}_{\equiv E_k}]\alpha^{\dagger}_{\mathbf{k},s}\alpha_{\mathbf{k},s}
$$

$$
+ \sum_{\mathbf{k},s} 2s\overbrace{[\xi_k u_k v_k - \frac{1}{2}\Delta_k \left(u_k^2 - v_k^2\right)]}^{\equiv 0} \left(\alpha^{\dagger}_{\mathbf{k},s}\alpha^{\dagger}_{-\mathbf{k},-s} + \alpha_{\mathbf{k},s}\alpha_{-\mathbf{k},-s}\right)
$$

$$
+ \sum_{\mathbf{k},s} 2\left[\xi_k v_k^2 + \Delta_k u_k v_k\right]
$$

$$
+ [E_+(0) - 2\mu] N_0 + \sum_{\mathbf{K}\neq 0} [E_+(K) - 2\mu]\, b^{\dagger}_{\mathbf{K}} b_{\mathbf{K}}
$$

$$
+ [2\mu - E_-(0)] M_0 + \sum_{\mathbf{K}\neq 0} [2\mu - E_-(K)]\, c^{\dagger}_{\mathbf{K}} c_{\mathbf{K}}. \tag{6}
$$

with $\xi_k \equiv \epsilon_k - \mu$. The term set equal to zero in (6) is justified as this merely fixes the coefficient, say v_k, that was restricted only by $u_k^2 + v_k^2 = 1$ which in turn follows from the requirement that both the a *and* α operators obey Fermi anticommutation relations. There are no products such as $\alpha^{\dagger}_{\mathbf{k},s}\alpha^{\dagger}_{-\mathbf{k},-s}$ remaining, nor any other nonbilinear terms, as with[24] the BCS two-vertex, *four*-fermion Hamiltonian[5] that neglects other than $K = 0$ pairings

$$
H = \sum_{\mathbf{k},s} \epsilon_k a^+_{\mathbf{k},s} a_{\mathbf{k},s} - V \sum_{\mathbf{k},\mathbf{k}',s} a^+_{\mathbf{k}',s} a^+_{-\mathbf{k}',-s} a_{-\mathbf{k},-s} a_{\mathbf{k},s} \tag{7}
$$

where $-V \leqslant 0$ and the second summation is restricted by $E_F - \hbar\omega_D \leqslant \hbar^2 k^2/2m \equiv \epsilon_k, \epsilon_{k'} \leqslant E_F - \hbar\omega_D$ if the attractive pairing interaction mechanism is phonon exchange. Replacing the characteristic phonon-exchange Debye temperature $\hbar\omega_D/k_B$ of around $400K$, with k_B the Boltzmann constant, by a characteristic magnon-exchange one of around $1000K$ can lead to a simple interaction model associated with spin-fluctuation-mediated pairing.[25]

Thus, the GBEC formalism is *not* restricted to weak coupling. Eigenstates of the now fully diagonalized simplified $\hat{H} - \mu\hat{N}$ given by (6) are

$$
| ...n_{\mathbf{k},s}...N_{\mathbf{K}}...M_{\mathbf{K}}...\rangle =
$$

$$
\prod_{\mathbf{k},s} \left(\alpha^{\dagger}_{\mathbf{k},s}\right)^{n_{\mathbf{k},s}} \prod_{\mathbf{K}\neq 0} \frac{1}{\sqrt{N_K!}} \left(b^{\dagger}_{\mathbf{K}}\right)^{N_{\mathbf{K}}} \prod_{\mathbf{K}\neq 0} \frac{1}{\sqrt{M_K!}} \left(c^{\dagger}_{\mathbf{K}}\right)^{M_{\mathbf{K}}} | \mathbf{O}\rangle
$$

where the three exponents $n_{\mathbf{k},s}$, $N_{\mathbf{K}}$ and $M_{\mathbf{K}}$ are occupation numbers. Here $|\mathbf{O}\rangle$ is the vacuum state for the "bogolon" fermion quasiparticles created by $\alpha^\dagger_{\mathbf{k},s}$ with the gapped dispersion energy rewritten below in (10). It is simultaneously a vacuum state for 2e-CP and 2h-CP boson, creation and annihilation operators—which is to say that $|\mathbf{O}\rangle$ is defined by $\alpha_{\mathbf{k},s}\,|\mathbf{O}\rangle \equiv b_{\mathbf{K}}\,|\mathbf{O}\rangle \equiv c_{\mathbf{K}}\,|\mathbf{O}\rangle \equiv 0$.

4. Thermodynamic Potential

With the Hamiltonian explicitly diagonalized, one can now straightforwardly construct the thermodynamic potential $\Omega \equiv -PL^d$ for the GBEC, with L^d the system "volume" and P its pressure, which is defined as (Ref. 21, p. 228)

$$\Omega(T, L^d, \mu, N_0, M_0) = -k_B T \ln\left[\operatorname{Tr}\exp\{-\beta(H - \mu\hat{N})\}\right] \qquad (8)$$

where "Tr" stands for "trace." Inserting (1) plus (2) into (8)[17] one obtains after some algebra an explicit expression for $\Omega(T, L^d, \mu, N_0, M_0)/L^d$ (see Ref. 26 Eq. 10). In $d = 3$ one usually has

$$N(\epsilon) \equiv \frac{m^{3/2}}{2^{1/2}\pi^2\hbar^3}\sqrt{\epsilon} \quad \text{and} \quad M(\varepsilon) \equiv \frac{2m^{3/2}}{\pi^2\hbar^3}\sqrt{\varepsilon} \qquad (9)$$

for the (one-spin) fermion DOS at energies $\epsilon = \hbar^2 k^2/2m$ and the boson DOS for an *assumed quadratic*[1] boson dispersion $\varepsilon = \hbar^2 K^2/2(2m)$, respectively. The latter is an assumption to be lifted later so as to include Fermi-sea effects which change the boson dispersion relation from quadratic to *linear* as mentioned before. Finally, the relation between the resulting fermion spectrum $E(\epsilon)$ and fermion energy gap $\Delta(\epsilon)$ is of the form

$$E(\epsilon) = \sqrt{(\epsilon - \mu)^2 + \Delta^2(\epsilon)} \qquad (10)$$
$$\Delta(\epsilon) \equiv \sqrt{n_0}\,f_+(\epsilon) + \sqrt{m_0}\,f_-(\epsilon). \qquad (11)$$

This last expression for the gap $\Delta(\epsilon)$ implies a simple T-dependence rooted in the 2e-CP $n_0(T) \equiv N_0(T)/L^d$ and 2h-CP $m_0(T) \equiv M_0(T)/L^d$ number densities of BE-condensed bosons, i.e., $\Delta(T) = \sqrt{n_0(T)}\,f_+(\epsilon) + \sqrt{m_0(T)}\,f_-(\epsilon)$.

If hole pairs are ignored the resulting relation $\Delta(T) = f\sqrt{n_0(T)}$ has recently been generalized[27] to include nonzero-$\mathbf{K}$ pairs beyond the expression (3). This leads to a generalized gap $E_g(\lambda, T)$ defined as

$$E_g(\lambda, T) = \sqrt{2\hbar\omega_D V n_B(\lambda, T)} \equiv f\sqrt{n_B(\lambda, T)} \qquad (12)$$

where $n_B(\lambda, T)$ is the *net* number density of CPs, both in and above the condensate, in the BF mixture which in Ref. 27 for simplicity was taken as a binary one. The generalized gap $E_g(\lambda, T)$ accommodates pseudogap phenomena.[28] Indeed, a depairing or pseudogap temperature $T^* \geqslant T_c$ ensues as the solution to $E_g(\lambda, T^*) = 0$

whereas the critical superconducting temperature T_c is the solution to $\Delta(T_c) = 0$. Thus, a simple physical interpretation follows, namely, that T^* is the temperature below which pairs form but incoherently while the lower T_c is the temperature below which the pairs constitute a coherent phase.

5. Ternary Gas Helmholtz Free Energy

The Helmholtz free energy is by definition

$$F(T, L^d, \mu, N_0, M_0) \equiv \Omega(T, L^d, \mu, N_0, M_0) + \mu N.$$

Minimizing it with respect to N_0 and M_0, and simultaneously fixing the total number N of electrons by introducing the electron chemical potential μ in the usual way, specifies an *equilibrium state* of the system at fixed volume L^d and temperature T. The necessary conditions for an equilibrium state are thus

$$\partial F/\partial N_0 \;=\; 0, \qquad \partial F/\partial M_0 \;=\; 0, \qquad \text{and} \qquad \partial \Omega/\partial \mu \;=\; -N \quad (13)$$

where N evidently includes both paired and unpaired CP fermions.

6. Three GBEC Equations

Some algebra then leads to the three coupled transcendental Eqs. (7)-(9) of Ref. 17. These can be rewritten somewhat more transparently as: a) two *"gap-like equations"*

$$[2E_f + \delta\varepsilon - 2\mu(T)] = \frac{1}{2}f^2 \int_{E_f}^{E_f+\delta\varepsilon} d\epsilon N(\epsilon) \frac{\tanh \frac{1}{2}\beta\sqrt{[\epsilon - \mu(T)]^2 + f^2 n_0(T)}}{\sqrt{[\epsilon - \mu(T)]^2 + f^2 n_0(T)}} \quad (14)$$

and

$$[2\mu(T) - 2E_f + \delta\varepsilon] = \frac{1}{2}f^2 \int_{E_f-\delta\varepsilon}^{E_f} d\epsilon N(\epsilon) \frac{\tanh \frac{1}{2}\beta\sqrt{[\epsilon - \mu(T)]^2 + f^2 m_0(T)}}{\sqrt{[\epsilon - \mu(T)]^2 + f^2 m_0(T)}} \quad (15)$$

as well as b) a single *"number equation"*

$$2n_B(T) - 2m_B(T) + n_f(T) = n. \quad (16)$$

This last relation ensures charge conservation in the ternary mixture. In general $n \equiv N/L^d$ is the total number density of electrons, $n_f(T)$ that of the *unpaired* electrons, while $n_B(T)$ and $m_B(T)$ are respectively those of 2e- and 2h-CPs in *all* bosonic states, ground plus excited. The latter turn out to be

$$n_B(T) \equiv n_0(T) + \int_{0+}^{\infty} d\varepsilon M(\varepsilon) \left(\exp \beta[2E_f + \delta\varepsilon - 2\mu + \varepsilon] - 1\right)^{-1} \quad (17)$$

$$m_B(T) \equiv m_0(T) + \int_{0+}^{\infty} d\varepsilon M(\varepsilon) \left(\exp \beta[2\mu + \varepsilon - 2E_f + \delta\varepsilon] - 1\right)^{-1} \quad (18)$$

which are clear manifestations of the bosonic nature of both kinds of CPs. One also obtains for the number density of unpaired electrons at any T

$$n_f(T) \equiv \int_0^\infty d\epsilon N(\epsilon)[1 - \frac{\epsilon - \mu}{E(\epsilon)} \tanh \frac{1}{2}\beta E(\epsilon)] = 2 \sum_k v_k^2(T) \qquad (19)$$

where $v_k^2(T) \equiv \frac{1}{2}[1 - (\epsilon_k - \mu)/E_k] \xrightarrow[T \to 0]{} v_k^2$ with E_k being given by (10) is precisely the BCS-Bogoliubov T-dependent coefficient that is linked with $u_k(0) \equiv u_k$ through $v_k^2 + u_k^2 = 1$ of the BCS trial wavefunction

$$| \mathbf{O} \rangle \equiv \prod_k (u_\mathbf{k} + v_\mathbf{k} a^+_{\mathbf{k}\uparrow} a^+_{-\mathbf{k}\downarrow}) | O \rangle \qquad (20)$$

where $| O \rangle$ is the ordinary vacuum. The zero-T version of the two amplitude coefficients v_k and u_k originally appeared in (20) and shortly afterwards in the Bogoliubov-Valatin canonical transformation. Next, one picks $\delta\varepsilon = \hbar\omega_D$ as well as identifies nonzero $f_+(\epsilon)$ and nonzero $f_-(\epsilon)$ with $f \equiv \sqrt{2\hbar\omega_D V}$ but such that $f_+(\epsilon)f_-(\epsilon) = 0$. Assuming $n_0(T) = m_0(T)$ and adding together (14) and (15) gives the precise BCS gap *provided* one can identify E_f with μ. This in turn is guaranteed if $n_B(T) = m_B(T)$, namely, if (17) and (18) are set equal to each other so that the arguments of the two exponentials become identical.

Self-consistent (at worst, numerical) solution of the *three coupled equations* (14) to (16) yields the three thermodynamic variables of the GBEC formalism

$$n_0(T, n, \mu), \quad m_0(T, n, \mu) \quad \text{and} \quad \mu(T, n). \qquad (21)$$

Proving *a posteriori* the existence of a *nonzero* T_c associated with these expressions vindicates the GBEC theory.

All told, the three GBEC equations (14) to (16) subsume *five* different theories as special cases. The vastly more general GBEC theory has been applied and gives sizeable enhancements in T_cs over BCS theory that emerge[29] by admitting, apparently for the first time, departures from the very special case of perfect 2e/2h-pair symmetry in the mixed phase.

7. Conclusion

In conclusion, five statistical continuum theories of superconductivity, including both the BCS and BEC theories, are contained as special limiting cases within a single generalized Bose-Einstein condensation (GBEC) model. This model includes, for the first time, along with unpaired electrons, both two-electron and two-hole pair-condensates in freely variable proportions. The BCS and BEC theories are thus completely *unified* within the GBEC. The BCS condensate emerges directly from the GBEC *as a BE condensate* through the condition for phase equilibria when both total 2e- and 2h-pair number, as well as their condensate, densities are *equal* at a given T and coupling—provided the coupling is weak enough so that the electron chemical potential μ roughly equals the Fermi energy E_F. The ordinary BEC T_c-formula, on the other hand, is recovered from the GBEC when hole pairs are completely neglected, the BF coupling f is made to vanish, and the limit of zero unpaired electrons is taken, this implying very strong interelectron coupling. The practical outcome of the BCS-BEC unification via the GBEC is *enhancement* in T_c, by more than two orders-of-magnitude in 3D. These enhancements in T_c fall within empirical ranges for 2D and 3D "exotic" SCs, whereas BCS T_c values remain low and within the empirical ranges for conventional, elemental SCs using standard interaction-parameter values. Lastly, room temperature superconductivity is possible for a material with a Fermi temperature $T_F \lesssim 10^3 K$, with the *same* interaction parameters used in BCS theory for conventional SCs.

Acknowledgments

I thank S.K. Adhikari, V.C. Aguilera-Navarro, A.S. Alexandrov, P.W. Anderson, J.F. Annett, J. Batle, M. Casas, J.R. Clem, J.D. Dow, D.M. Eagles, M. Fortes, M. Grether, B.L. Györffy, K. Levin, T.A. Mamedov, P.D. Mannheim, F.J. Sevilla, M.A. Solís, W.C. Stwalley, S. Tapia, V.V. Tolmachev, O. Rojo, J.J. Valencia, A.A. Valladares, H. Vucetich and J.A. Wilson for conversations and/or for providing material prior to its publication. UNAM-DGAPA-PAPIIT (Mexico) grant IN106908 is thanked for partial support. I also thank the Physics Department of the University of Connecticut, Storrs, CT, USA, for its gracious hospitality during my sabbatical year, and CONACyT (Mexico) for support thereto.

References

1. J.M. Blatt, *Theory of Superconductivity* (Academic, New York, 1964).
2. M.R. Schafroth, Phys. Rev. **96**, 1442 (1954).
3. M.R. Schafroth, S.T. Butler, and J.M. Blatt, Helv. Phys. Acta **30**, 93 (1957).
4. M.R. Schafroth, Sol. State Phys. **10**, 293 (1960).
5. J. Bardeen, L.N. Cooper and J.R. Schrieffer, Phys. Rev. **108**, 1175 (1957).
6. N.N. Bogoliubov, JETP **34**, 41 (1958).
7. J. Ranninger and S. Robaszkiewicz, Physica B **135**, 468 (1985).
8. J. Ranninger, R. Micnas, and S. Robaszkiewicz, Ann. Phys. Fr. **13**, 455 (1988).
9. R. Micnas, J. Ranninger, and S. Robaszkiewicz, Rev. Mod. Phys. **62**, 113 (1990).
10. R. Micnas, S. Robaszkiewicz, and A. Bussmann-Holder, Phys. Rev. B **66**, 104516 (2002).
11. R. Micnas, S. Robaszkiewicz, and A. Bussmann-Holder, Struct. Bond **114,** 13 (2005).
12. R. Friedberg and T.D. Lee, Phys. Rev. B **40**, 6745 (1989).
13. R. Friedberg, T.D. Lee, and H.-C. Ren, Phys. Rev. B **42**, 4122 (1990).
14. R. Friedberg, T.D. Lee, and H.-C. Ren, Phys. Lett. A **152**, 417 and 423 (1991).
15. R. Friedberg, T.D. Lee, and H.-C. Ren, Phys. Rev. B **45**, 10732 (1992).
16. M. Casas, A. Rigo, M. de Llano, O. Rojo, and M.A. Solís, Phys. Lett. A **245**, 5 (1998).
17. V.V. Tolmachev, Phys. Lett. A **266**, 400 (2000).
18. M. de Llano and V.V. Tolmachev, Physica A **317**, 546 (2003).
19. L.N. Cooper, Phys. Rev. **104**, 1189 (1956).

20. J.E. Hirsch, arXiv:0901.4099 and extensive refs. therein.
21. A.L. Fetter and J.D. Walecka, *Quantum Theory of Many-Particle Systems* (McGraw-Hill, New York, 1971).
22. N.N. Bogoliubov, N. Cim. **7**, 794 (1958).
23. J.G. Valatin, N. Cim. **7**, 843 (1958).
24. G. Rickayzen, *Theory of Superconductivity* (Interscience, NY, 1965).
25. T. Dahm *et al.*, Nature Physics **5**, 217 (2009).
26. S.K. Adhikari, M. de Llano, F.J. Sevilla, M.A. Solís, and J.J. Valencia, Physica C **453**, 37 (2007).
27. T.A. Mamedov and M. de Llano, to be published.
28. T. Timusk and B. Statt, Rep. Prog. Phys. **62**, 61 (1999).
29. M. Grether and M. de Llano, Physica C **460**, 1141 (2007).

KOHN ANOMALY ENERGY IN CONVENTIONAL SUPERCONDUCTORS EQUALS TWICE THE ENERGY OF THE SUPERCONDUCTING GAP: HOW AND WHY?

RANJAN CHAUDHURY

S.N. Bose National Centre For Basic Sciences,
Sector-3, Block-JD, Salt Lake,
Kolkata- 700098, India
ranjan@bose.res.in

MUKUNDA P. DAS

Department of Theoretical Physics,
Research School of Physics and Engineering,
The Australian National University, Canberra,
ACT 0200, Australia
mukunda.das@anu.edu.au

Received 27 October 2009

Kohn anomaly occurs in metals as a weak but discernible kink in the phonon spectrum around $2k_F$ arising out of screened Coulombic interaction. Over the years this has been observed in a number of normal metallic systems. As a major surprise however, the recent neutron spin-echo experiments on elemental (conventional) superconductors Pb and Nb reveal a very important and striking relation that Kohn (anomaly) energy equals twice the energy of the superconducting gap. In this paper we explore the microscopic origin of this novel phenomenon and discuss its implication to the standard model BCS Theory.

Keywords: Kohn anomaly; Kohn singularity; superconducting gap.

1. Introduction

The Bardeen-Cooper-Schrieffer (BCS) pairing theory for superconductivity has been quite successful for the interpretation of most of the experimental data obtained from the conventional low-temperature superconductors. Nevertheless, the theoretical formulation of the BCS suffers from several limitations. Most of the low-temperature superconductors are based on the phonon-mediated attractive pairing interaction. In the simple BCS calculational framework the pairing interaction $V_{k,k'}$ is simplified by a square well model. This incorporates the static electronic screening effects in the long wave-length limit of the Random Phase Approximation (RPA) as appropriate to a metal; however it neglects the detailed phonon spectrum of the material. More importantly, the effects of various special characteristics of the electronic response functions either in the normal metallic phase or in the superconducting phase, particularly on the phonon energy and the linewidth are generally neglected. Although Brockhouse *et al.*[1,2] made the first experimental observation of Kohn anomaly[3] in normal metallic systems, such studies on superconductors were carried out only very recently, as late as in 2008. From the theoretical perspective too, the BCS model and the Kohn anomaly did not seem to belong to the same platform and were treated as rather two independent entities.

Occurrence of Kohn anomaly in a metallic system, normal or superconducting, depends on the nature of screening of the electric field of the ionic lattice. The latter determines the energy and lifetimes of phonons. It is worthwhile to mention that in the strong coupling version of the pairing (Eliashberg) theory for the conventional superconductors, the role of phonons are explicit and are worked out in detail (see for example, Marsiglio and Carbotte[4]). However, its implication to Kohn anomaly seems to have scanty discussion/analysis in the literature. In this context Scalapino[5] notes -"Although the effective pairing interaction involves phonons, it is difficult to see any direct connection between the Kohn anomaly energy and the energy gap."

The recent neutron spin-echo experiments by Aynajian *et al.*[6] on elemental conventional superconductors Pb and Nb in the superconducting phase, raises some interesting question on the conventional wisdom. It clearly brings out a limitation of the conventional treatment of the superconducting phase and the pairing mediated by the phononic modes, for not including all the characteristics of the dielectric response function of the system. More elaborately, this highlights the utmost importance of the restructuring and refinement of the conventional BCS/Eliashberg scheme with incorporation of the salient features of the electronic (quasi-particle) spectrum in the superconducting phase. Besides, these new experiments also motivate one to theoretically calculate the phonon lifetime or linewidth in the superconducting phase, invoking the electron-phonon interaction under the Fermi golden rule scheme.

In a simple demonstration as done here, we would like to examine if the wavevector at which the phonon linewidth exhibits the maximum, truly corresponds to the renormalized phonon energy of 2Δ (Δ being the superconducting gap), as observed experimentally by Aynajian *et al.*[6] Interestingly, the phonon dispersion function also shows an anomaly at around the same energy.[6]

2. Kohn Anomaly In A Normal Metal

The Kohn singularity[3] in the normal electron response function in a metal at low temperature, is a consequence of the sharp discontinuity of the electron distribution function in the momentum space at k_F, the Fermi wave-vector. This holds true for an ideal Fermi gas as well as for the Fermi liquid model (a model successfully describing real simple metals). Mathematically, it is manifested as the k-space gradient of the longitudinal static dielectric function exhibiting a logarithmic divergence at $2k_F$. This result can be readily derived from the RPA treatment of the weakly interacting 3-dimensional electron gas, as was first shown by Kohn[3] on the basis of the Lindhard response function[7,8] for the electron-hole polarizability. Interestingly, when the combined system of interacting electrons and phonons are considered and treated under the RPA, the phonon spectrum gets renormalized. Moreover, the dispersion function of the renormalized phonons exhibits a non-analyticity viz. the k-space gradient of the dispersion function diverges at $2k_F$. This property of the phonons in normal metals is the well-known "Kohn anomaly". The origin of this anomaly is undoubtedly the Kohn singularity of the normal electron response function. The Kohn anomaly has been experimentally observed and confirmed in the normal phase of metals such as Pb,[1] where the electron-phonon coupling is rather strong.[7]

Mathematically, the phenomena of Kohn singularity and Kohn anomaly can be briefly described in the following way:- The static electronic polarizability function $\Pi(q)$ for the normal electrons in a 3d metal is given in RPA as

$$\Pi(q) = \frac{N(0)}{4x}[2x + (1 - x^2)ln|\frac{(1 + x)}{(1 - x)}|] \tag{1}$$

where x is the reduced wave vector and is given as $x = \frac{q}{2k_F}$; k_F being the Fermi wavevector and $N(0)$ is the electronic density of states at the Fermi surface for one kind of spin. The corresponding longitudinal dielectric function is given as

$$\epsilon(q) = 1 + \frac{4\pi e^2}{q^2}\Pi(q) \tag{2}$$

The above electronic dielectric function can be shown to possess the property that its derivative with respect to q , the wave vector, diverges at $q = 2k_F$. This result is the well known Kohn singularity. Although, the above analysis is based upon a treatment corresponding to zero temperature, it is realistically valid for all temperatures $T << T_F$.

When one studies a coupled electron-phonon system in a 3d normal metal within the RPA and the resulting screening of the phonon modes is performed with the above dielectric function, the phonon dispersion function is found to display a singularity as well. More precisely, the q-space derivative of the real part of the dynamic dielectric function corresponding to the electronic component, diverges at q near $2k_F$. This causes the screened phonon dispersion function itself to exhibit non-analyticity. These may be described mathematically as,

$$\frac{\partial[Real(\epsilon(q, \omega_q))]}{\partial q} = \infty \tag{3}$$

for $q \approx 2k_F$, where ω_q is the screened phonon frequency and is expected to be much smaller than the frequency corresponding to the Fermi energy E_F .

The above electronic dielectric function when incorporated in the analysis of the coupled electron-phonon system within RPA, gives rise to the phenomenon of Kohn anomaly in the normal metal.

3. Kohn Anomaly In A Conventional Superconductor

A much more challenging task in condensed matter physics is the theoretical exploration starting from a microscopic model to answer the question whether the Kohn anomaly can also occur in the superconductors. The recent experimental observations of Aynajian *et al.*[6] with clear signature of the Kohn anomaly in the superconducting phases of *Pb* and *Nb* have provided fresh impetus to initiate theoretical investigations. Here we establish the microscopic origin of these effects by presenting our results based on some simple (at the level of RPA) preliminary calculations and arguments.

First of all, we calculate the electronic polarizability for a superconductor, as modelled by BCS, under the mean field treatment. This can be done in a straightforward way by making use of Bogoliubov transformation to diagonalise the mean field BCS Hamiltonian in terms of the non-interacting fermionic quasi-particles. One can then apply the standard linear response theory to calculate the electronic response functions.

The detailed and elaborate calculations involving the above approach lead to the result that indeed the static electronic (quasi-particle-quasi-hole) polarizability in a "BCS-superconductor" exhibits singularity at $q = 2k_F$ under certain conditions. Mathematically, it is manifested in the following expressions

$$2k_F \frac{\frac{\partial \Pi_s(q,0)}{\partial q}}{\Pi_s(2k_F,0)} = 2ln\frac{2}{k_F\xi} \tag{4}$$

where, Π_s is the electronic polarizability in the superconducting phase, ξ is the superconducting coherence length and the numerator (the derivative) is also evaluated at $q = 2k_F$.

The above expression shows that the polarizability diverges logarithmically at the wave-vector $2k_F$, when the superconducting coherence length becomes very large or extremely small with respect to the lattice spacing. This is a very genuine manifestation of Kohn-like singularity! Thus, the true Kohn-like singularity behaviour is expected to occur both in the conventional weak coupling superconductors as well as in the novel superconductors characterized by the real space pairing, independent of the mechanism for the pairing interaction. It is also worthwhile to remark that we do recover the usual Kohn singularity appropriate to the normal metal from our expression, in the limiting case of the superconducting gap becoming vanishingly small.

The next very important issue is the theoretical investigation of the consequences for the phonon spectrum in the presence of the above Kohn-like singularity. The aim obviously is to examine if there exists a Kohn anomaly in the superconducting phase. The microscopic input required for this study appears in the form of an additional term describing the coupling between the Bogoliubov quasi-particles and the phonons, in the usual BCS Hamiltonian. The objectives are two-fold. First of all, it is very important to investigate whether the Kohn-like singularity leads to a non-analyticity in the phonon-dispersion function. Secondly, it also enables us to calculate the linewidths of the phonon modes under this condition to examine whether the theoretical linewidths exhibit the same behaviour as shown by the experimental ones. It is quite remarkable that the experimental linewidths are found to acquire a maximum at the phonon energy equal to 2Δ. This probably implies a very strong coupling between the phonon modes and the quasi-particles in the superconducting phase.

The coupling between the Bogoliubov quasi-particles and the phonons can be derived from the usual electron-phonon interaction by taking into account the Bogoliubov transformation matrix elements. It may be recalled that these transformation matrix elements connect the quasi particle operators (of the superconducting phase) to the normal electron operators. Thus, the effective Hamiltonian for studying the Kohn anomaly in the superconducting phase becomes very similar to that used in the normal metallic phase, except for the presence of the appropriate BCS coherence factors[8] (see also[9]). These factors denoted by 'm' and 'n', are given by the following expressions:-

$$m(\mathbf{k}, \mathbf{k}') = u_{\mathbf{k}'} v_{\mathbf{k}} + u_{\mathbf{k}} v_{\mathbf{k}'} \tag{5}$$

and

$$n(\mathbf{k}, \mathbf{k}') = u_{\mathbf{k}'} u_{\mathbf{k}} - v_{\mathbf{k}'} v_{\mathbf{k}} \tag{6}$$

where u and v are the well known Bogoliubov coefficients; the quantity $\mathbf{q} = \mathbf{k}' - \mathbf{k}$ denotes the wavevector of the phonon mediating the pairing interaction with $\mathbf{k}$ and $\mathbf{k}'$ being the electron wave-vectors. The detailed nature of the Kohn anomaly in the superconducting phase is therefore also decided by the wave-vector dependence of these coherence factors. The screened phonon frequency spectrum $\tilde{\omega}_{\mathbf{q}}$ produced by the dynamic (frequency dependent) dielectric response of the quasi- particles (ϵ_s) in the superconducting phase, can also be easily determined under the RPA. Interestingly, the minimum quasi-particle-quasi-hole pair creation energy 2Δ is hidden here through the expressions for the dynamic electronic polarizability and is the prime source for the location (position) of the occurrence of Kohn anomaly in the superconducting phase. The renormalized phonon spectrum is expected to exhibit all the principal features of Kohn anomaly seen experimentally in the phonon dispersion function ($\omega_{\mathbf{q}}^{exp}$ vs. $\mathbf{q}$) .

The above formalism is also powerful enough to perform a calculation of the quasi-particle-phonon scattering rate in the superconducting phase, using the Fermi golden rule. The linewidth of the phonon modes ($\gamma_{\mathbf{q}}$) is directly proportional to the scattering rate ($\frac{1}{\tau}$), which in turn can be calculated by making use of the quasi-particle-phonon interaction matrix elements and the average occupation numbers of these particles. Hence the contribution due to the above process can be estimated quantitatively at various temperatures in the entire $\mathbf{q}$-space. This provides a very direct route for the comparison of the theoretical predictions with the experimental results of Brockhouse *et al.*[2] and Aynajian *et al.*[6] and further strengthens the basis of our formalism.

4. Summary

In summary, we have tried to put forward and describe in this short note the important microscopic processes responsible for the Kohn anomaly observed experimentally in the superconducting phases of some of the conventional superconductors. The confirmed appearance of this anomaly in the superconducting phases of real materials has tremendous consequences for the microscopic theories as well. It brings out the genuine need for a modification and an improvement over the conventional approach followed within the BCS pairing theory.

For the superconductors based on the phonon mediated pairing, the BCS theory assumes the phonon spectrum to remain essentially unchanged when the metal becomes superconductor. The occurrence of Kohn anomaly at the phonon energy of two times the superconducting gap Δ in the experiment of Aynajian *et al.*, challenges this assumption. These studies further imply that in the weak coupling BCS approach, the conventional static square well model for the phonon-induced effective attraction $V_{eff}(\omega = \zeta - \zeta')$ between the electrons (ζ's are the effective single electron energies measured from the chemical potential in the normal state) must be replaced by a more appropriate interaction kernel. This new realistic interaction function will have to include the contributions due to the presence of the expected non-analyticity in the phonon spectrum in the superconducting phase itself. Thus, the calculational scheme becomes much more self-consistent than in the usual treatment. This requires very sophisticated many-body approach to handle it.

Last but not the least, the incorporation of Kohn anomaly in the modelling of the BCS pairing interaction would be of immense importance towards achieving the goal of correct estimation of T_c for real superconducting materials, driven by mechanism based on phonon mediated attractive interaction.

Further work involving detailed calculations is in progress.

Acknowledgments

One of us (MPD) acknowledges the hospitality of The Institute of Physics, Bhubaneswar, India, where a part of this work was done.

References

1. B. N. Brockhouse, K. R. Rao and A. D. B. Woods, *Phys. Rev. Lett.* **7** 93 (1961).
2. B. N. Brockhouse, T. Arase, G. Caglioti, K. R. Rao and A. D. B. Woods *Phys. Rev.* **128** 93 (1962).
3. W. Kohn *Phys. Rev. Lett.* **2** 393 (1959) .
4. F. Marsiglio and J.P. Carbotte, in *Superconductivity: Conventional And Unconventional Superconductors*, ed. K.H. Bennemann and J.B. Ketterson (Springer Verlag, 2008), pp. 73-162.

5. D. J. Scalapino *Science* **319** 1492 (2008).
6. P. Aynajian, T. Keller, L. Boeri, S. M. Shapiro, K. Habicht and B. Keimer *Science* **319** 1509 (2008) .
7. D. Pines *Elementary Excitations In Solids*, (Perseus Books Publishing, Massachusetts, 1999) .
8. J. R. Schrieffer *Theory of Superconductivity* (Perseus Books Publishing, Massachusetts, 1999) .
9. L. Tewordt *Phys. Rev.* **127** 371 (1962).

COLLECTIVE EXCITATIONS IN SUPERCONDUCTORS AND SEMICONDUCTORS IN THE PRESENCE OF A CONDENSED PHASE

ZLATKO KOINOV

*Department of Physics and Astronomy, University of Texas at San Antonio,
San Antonio, TX 78249, USA*
zlatko.koinov@utsa.edu

Received 18 September 2009

We first study the spectrum of the excitonic polaritons in semiconductors in the presence of a condensed phase. With the help of the Hubbard-Stratonovich transformation we propose an exact mapping of the extended Hubbard model onto an excitonic-polariton model describing the light propagation in semiconductors. This approach allows us to derive exact formulas for spin-spin and charge-charge correlation functions for spin-triplet superconductivity.

Keywords: Excitonic polariton; Hubbard model; spin and charge correlation functions.

1. Introduction

There has been a considerable interest, during the last decade, in the study of the extended Hubbard model as a model for high-temperature superconductivity. The extended Hubbard Hamiltonian is defined by

$$H = -t \sum_{<i,j>,\sigma} \psi_{i,\sigma}^{\dagger}\psi_{j,\sigma} - \mu \sum_{i,\sigma} \widehat{n}_{i,\sigma} + U \sum_{i} \widehat{n}_{i,\uparrow}\widehat{n}_{i,\downarrow} - V \sum_{<i,j>\sigma\sigma'} \widehat{n}_{i,\sigma}\widehat{n}_{j,\sigma'}, \quad (1)$$

where μ is the chemical potential. The Fermi operator $\psi_{i,\sigma}^{\dagger}$ ($\psi_{i,\sigma}$) creates (destroys) a fermion on the lattice site i with spin projection $\sigma = \uparrow, \downarrow$ along a specified direction, and $\widehat{n}_{i,\sigma} = \psi_{i,\sigma}^{\dagger}\psi_{i,\sigma}$ is the density operator on site i. The symbol $\sum_{<ij>}$ means sum over nearest-neighbor sites. The first term in (1) is the usual kinetic energy term in a tight-binding approximation, where t is the single electron hopping integral. Depending on the sign of U, the third term describes the on-site repulsive or attractive interaction between electrons with opposite spin. We assume that $V > 0$, so the last term is expect to stabilize the pairing by bringing in a nearest-neighbor attractive interaction. The lattice spacing is assumed to be $a = 1$ and N is the total number of sites.

From theoretical point of view, the major difficulties in the model are due to (i) the absence of completely reliable approximation schemes to handle the many-body nature of the problem, and (ii) the need of sufficient computational power to perform realistic size lattice numerical simulations. Different many-body methods can be employed to study the possibility for the Hamiltonian (1) to give rise to a super-conducting instability. Among them are the mean-field analysis of pairing followed by the generalized random phase approximation (GRPA), the fluctuation-exchange (FLEX) method, the quantum Monte Carlo (QMC) simulations, the cluster per-turbation theory, the cellular dynamical mean-field theory, and the two-particle self-consistent (TPSC) approach. Except for the QMC method, all of the above many-body approaches must choose an approximation for the single-particle mass operator Σ. The accuracy of the approximations is crucial for understanding of physical properties of the Hubbard model because of the self-consistent relations between single- and two-particle quantities: the kernel $\delta\Sigma/\delta G$ of the Bethe-Salpeter (BS) for the spectrum of the collective excitations does depend on the mass oper-ator, and the mass operator itself depends on the two-particle Green function K. Strictly speaking, in the case of Hubbard model the mass operator is a sum of two terms: the Hartree term Σ_H and the Fock term Σ_F. In many papers related to the above methods, the existence of the Hartree term has been neglected. A very im-portant question naturally arises here as to what extent the self-consistent relations between single- and two-particle quantities are affected by neglecting the existence of the Hartree term.

The same question arises in the polariton theory of light propagation in semicon-ductors. Since superconductivity due to the formation of Cooper pairs and corre-sponding analogous phenomenon of Bose-Einstein condensation (BEC) of polaritons (or excitons), both phenomena are two manifestations of the same effect, namely the spontaneous breaking gauge invariance, one could expect that very similar equa-tions and conclusions should occur in both models. In the polariton BEC we have photons in a semiconductor (or semiconductor microcavity) strongly coupled to electronic excitations (excitons), thus creating polaritons with a very small effective mass (10^{-5} of the free electron mass). The spectrum of collective excitations (po-laritons) manifest itself as common poles of the two-particle Green function K and the photon Green function D. The later satisfies the well-known Dyson equation $D^{-1} = D^{(0)-1} - \widetilde{\Pi}$, where $D^{(0)}$ and $\widetilde{\Pi}$ are the bare photon propagator and the photon self-energy, respectively. We have two algorithms to calculate the polariton spectrum. The first one is a direct solution of the BS equation for the poles of the self-consistent relations between single- and two-particle quantities. Since the ker-nel of the BS equation is expressed as a functional derivative of the mass operator (with Hartree and Fock terms included) with respect to the function G , and the mass operator itself does depend on the two-particle Green function, one has to solve self-consistently the Schwinger-Dyson equation for the single-particle Green function and the BS equation. This is a very complicated task, so it is easy to use a second algorithm which is based on the fact that the photon self-energy $\widetilde{\Pi}$ (or dielectric function) and the exciton two-particle Green function $\widetilde{K}$, both have com-mon poles. By definition, exciton Green function satisfies the BS equation when the Hartree term of the mass operator is neglected. To obtain the poles of $\widetilde{K}$ we again have to solve self-consistently the Schwinger-Dyson equation for the single-particle Green function and the BS equation for $\widetilde{K}$ keeping only the Fock term in the expression for the mass operator. But, it is also possible to get information about the energy of these common poles by measuring the positions of the peaks in the exciton absorption spectrum, so we can construct an expression for the photon self-energy by using some measurable parameters. The next step is to find the spectrum of the collective excitation by solving Maxwell equations.

It is possible to establish one-to-one correspondence between the extended Hubbard model and the polariton model. The simplest method to obtain this correspondence is to use the Hubbard-Stratonovich transformation, which allows us to transform the extended Hubbard model to a model in which the narrow-band *free electrons* are coupled to a boson field due to some spin-dependent mechanism. Depside of the existence of an exact mapping of the extended Hubbard model onto the polariton model, there is no way to construct an expression for the boson self-energy by using some measurable parameters. The only method to calculate the spectrum of the collective excitations of the Hubbard model is to use the first algorithm. In other words, if the spectrum of the collective excitations is calculated by solving only the BS equation, we have to use mass operator written as a sum of Hartree and Fock terms.

In what follows, we shall first provide an overview of polariton theory of light propagation in semiconductors. We then use the same equations to analyze the spectrum of the collective excitations of the extended Hubbard model.

2. Polariton Spectrum in the Presence of a Condensed Phase

The model that will be assumed is described by the action $S = S_0^{(e)} + S_0^{(A)} + S^{(e-A)}$, where: $S_0^{(e)} = \overline{\psi}(y)G^{(0)-1}(y;x)\psi(x)$, $S_0^{(A)} = \frac{1}{2}A_\alpha(z)D_{\alpha\beta}^{(0)-1}(z,z')A_\beta(z')$ and $S^{(e-A)} = \overline{\psi}(y)\Gamma_\alpha^{(0)}(y,x \mid z)\psi(x)A_\alpha(z)$. The boson (photon) field $A_\alpha(z)$ and the fermion (electron) fields $\overline{\psi}(y)$ and $\psi(x)$ are quantized according to commutation and anticommutation rules. In a non-relativistic model α and β label the three Cartesian coordinates. The composite variables $y = \{\mathbf{r}, u\}$, $x = \{r', u'\}$ and $z = \{\mathbf{R}, v\}$ are defined as follows: $\mathbf{r}, r, \mathbf{R}$ are position vectors, and according to imaginary-time (Matsubara) formalism the variable u, u', v range from 0 to $\hbar\beta = \hbar/(k_B T)$. Here T is the temperature and k_B is the Boltzmann constant. Throughout this paper we set $\hbar = 1$ and we use the summation-integration convention: that repeated variables are summed up or integrated over.

We consider the case of a direct-gap semiconductor with non-generate and isotropic bands. For simplicity, we assume that the electron and the hole masses are the same, and therefore, the dispersion relations of the electrons and holes with respect to the corresponding chemical potential $\mu_{e,h} = (E_g \pm \mu)/2$ are: $\epsilon_e(\mathbf{k}) = \mathbf{k}^2/2m + E_g - \mu_e$ and $\epsilon_h(\mathbf{k}) = -(\mathbf{k}^2/2m + \mu_h)$, respectively. Here, $\mu = \mu_e - \mu_h$ is the exciton chemical potential, E_g is the semiconductor gap. The chemical potential μ is determined self-consistently as a nontrivial function of the density n of the excitations in the system.

The inverse propagator $G^{(0)-1}(y;x)$ for non-interacting electrons in a periodic lattice potential. The inverse photon propagator $D_{\alpha\beta}^{(0)-1}(z,z')$ is in a gauge, when the scalar potential equals zero. In non-relativistic model the electron-photon bare vertex is defined as $\Gamma_\alpha^{(0)}(y,x \mid z) = \frac{\delta(u-v)\delta(u-u')}{cV}\sum_{\mathbf{q}} e^{\imath\mathbf{q}\cdot\mathbf{R}} < \mathbf{r} \mid \widehat{j}_\alpha(\mathbf{q}) \mid \mathbf{r}' >$. Here V is the volume of the crystal, c is the speed of light, and $\widehat{\mathbf{j}}(\mathbf{q})$ denotes the single-particle current operator.

In field theory the expectation value of a general operator $\widehat{O}(u)$ is expressed as a functional integral over the boson field A and the Grassmann fermion fields

$$< \widehat{T}_u(\widehat{O}(u)) >= \frac{1}{Z[J,M]} \int D\mu[\overline{\psi},\psi,A]\widehat{O}(u)e^{[J_\alpha(z)A_\alpha(z)-\overline{\psi}(y)M(y;x)\psi(x)]}\big|_{J=M=0},$$

where the symbol $< \ ... \ >$ means that the thermodynamic average is made, and $\widehat{T}_u$ is an u-ordering operator. J, M are the sources of the boson and fermion fields, respectively. The functional $Z[J,M]$ is defined as $Z[J,M] = \int D\mu[\overline{\psi},\psi,A]e^{[J_\alpha(z)A_\alpha(z)-\overline{\psi}(y)M(y;x)\psi(x)]}$, where the functional measure $D\mu[\overline{\psi},\psi,A] = DAD\overline{\psi}D\psi\exp{(S)}$ satisfies the condition $\int D\mu[\overline{\psi},\psi,A] = 1$.

The fundamental point in our approach is that all quantities of interest could be expressed in terms of the corresponding Green functions, which themselves are expectation values of time-ordered product of field operators. To obtain a particular product of field operators, we can use functional differentiation of $Z[J,M]$ over the corresponding sources. The functional Z generates all types of Feynman graphs, both connected and disconnected. However, it is useful to introduce a new generating functional $W[J,M] = \ln Z[J,M]$ that generates just the connected Green functions. By means of the functional $W[J,M]$ we introduce the following functions (after the functional differentiation one should set $J = M = 0$): boson Green function: $D_{\alpha\beta}(z,z') = -\frac{\delta^2 W}{\delta J_\alpha(z)\delta J_\beta(z')}$; single-fermion Green function: $G(x;y) = -\frac{\delta W}{\delta M(y;x)}$; two-particle fermion Green function: $K\begin{pmatrix} x & y' \\ y & x' \end{pmatrix} = -\frac{\delta G(x;y)}{\delta M(y';x')}$, and boson-fermion vertex function: $\Gamma_\alpha(y,x \mid z) = -\frac{\delta G^{-1}(y;x)}{\delta J_\beta(z')}D_{\beta\alpha}^{-1}(z',z)$.

As a consequence of the fact that the measure $D\mu[\overline{\psi},\psi,A]$ is invariant under the translations $\overline{\psi} \to \overline{\psi}+\delta\overline{\psi}$, $A \to A+\delta A$, one can derive the so-called Schwinger-Dyson equations:

$$D_{\alpha\beta}^{(0)-1}(z,z')R_\beta(z') + G(1;2)\Gamma_\alpha^{(0)}(2;1 \mid z) + J_\alpha(z) = 0, \tag{2}$$

$$G^{-1}(1;2) - G^{(0)-1}(1;2) + \Sigma(1;2) + M(1;2) = 0, \tag{3}$$

$$\Sigma(1;2) = \Sigma_H(1;2) + \Sigma_F(1;2), $$

$$\Sigma_H(1;2) = -\Gamma_\alpha^{(0)}(1;2 \mid z)R_\alpha(z) = \Gamma_\alpha^{(0)}(1;2 \mid z)G(4;3)\Gamma_\beta^{(0)}(3;4 \mid z')D_{\alpha\beta}^{(0)}(z,z'), \tag{4}$$

$$\Sigma_F(1;2) = -\Gamma_\alpha^{(0)}(1;3 \mid z)G(3;4)\Gamma_\beta(4;2 \mid z')D_{\alpha\beta}(z,z'). \tag{5}$$

Here, $R_\alpha(z) = \delta W/\delta J_\alpha(z)$ and the mass operator Σ is a sum of Hartree Σ_H and

Fock Σ_F parts. The complex indices $1, 2, ...$ are defined as follows: $1 = \{\mathbf{r}_1, u_1\}$, $2 = \{\mathbf{r}_2, u_2\},...$

Using the method of Legendre transforms, we go over from the functional $W[J, M]$ to a new functional $V[R, G] = W[J, M] - J_\alpha(z)R_\alpha(z) + M(1;2)G(2;1)$. After the Legendre transform, J and M must be considered as functionals of R and G. By means of the Legendre transform, we can derive the following set of exact equations:[1]

$$D_{\alpha\beta}(z, z') = D^{(0)}_{\alpha\beta}(z, z') + D^{(0)}_{\alpha\gamma}(z, z'')\Gamma^{(0)}_\gamma(1; 2 \mid z'')\times$$
$$K\begin{pmatrix} 2\ 3 \\ 1\ 4 \end{pmatrix} \Gamma^{(0)}_\delta(3; 4 \mid z''')D^{(0)}_{\delta\beta}(z''', z'), \tag{6}$$

$$G(1;3)\Gamma_\beta(3; 4 \mid z')G(4;2)D_{\beta\alpha}(z', z) = K\begin{pmatrix} 1\ 3 \\ 2\ 4 \end{pmatrix} \Gamma^{(0)}_\beta(3; 4 \mid z')D^{(0)}_{\beta\alpha}(z', z), \tag{7}$$

$$G(1;3)\Gamma_\alpha(3; 4 \mid z)G(4;2) = \widetilde{K}\begin{pmatrix} 1\ 3 \\ 2\ 4 \end{pmatrix} \Gamma^{(0)}_\alpha(3; 4 \mid z), \tag{8}$$

$$K\begin{pmatrix} 1\ 3 \\ 2\ 4 \end{pmatrix} = \widetilde{K}\begin{pmatrix} 1\ 3 \\ 2\ 4 \end{pmatrix} + \widetilde{K}\begin{pmatrix} 1\ 5 \\ 2\ 6 \end{pmatrix}\Gamma^{(0)}_\alpha(5; 6 \mid z)D^{(0)}_{\alpha\beta}(z, z')\Gamma^{(0)}_\beta(7; 8 \mid z')K\begin{pmatrix} 8\ 3 \\ 7\ 4 \end{pmatrix}. \tag{9}$$

$$\Gamma_\alpha(1; 2 \mid z) = \Gamma^{(0)}_\alpha(1; 2 \mid z) + \frac{\delta\Sigma_F(1; 2)}{\delta G(3; 4)}G(3; 5)G(6; 4)\Gamma_\alpha(5; 6 \mid z) \tag{10}$$

The Green function $\widetilde{K}$, which we shall call the exciton Green function, satisfies the following BS equation:

$$\widetilde{K}^{-1}\begin{pmatrix} 1\ 3 \\ 2\ 4 \end{pmatrix} = K^{(0)-1}\begin{pmatrix} 1\ 3 \\ 2\ 4 \end{pmatrix} - \frac{\delta\Sigma_F(1; 2)}{\delta G(3; 4)}. \tag{11}$$

Here $K^{(0)-1}\begin{pmatrix} 1\ 3 \\ 2\ 4 \end{pmatrix} = G^{-1}(1; 3)G^{-1}(4; 2)$ is the two-particle free propagator constructed from a pair of fully dressed single-particle Green functions. Equations (8) and (10) allow us to write the following equation for the vertex function:

$$\Gamma_\alpha(1; 2 \mid z) = \Gamma^{(0)}_\alpha(1; 2 \mid z) + \frac{\delta\Sigma_F(1; 2)}{\delta G(3; 4)}\widetilde{K}\begin{pmatrix} 3\ 5 \\ 4\ 6 \end{pmatrix}\Gamma^{(0)}_\alpha(5; 6 \mid z). \tag{12}$$

According to Eq. (12), vertex function Γ and the Green function of electronic excitations $\widetilde{K}$, both have common poles.

It is known from the Green function theory that the single-particle excitations in the system are provided by the poles of the Fourier transform of the single-particle

Green function. Similarly, the spectrum of the two-particle excitations (or collective modes) $\omega(\mathbf{Q})$ can be obtained by locating the positions of the poles of the Fourier transform of the two-particle fermion Green function:

$$K \begin{pmatrix} 1\ 3 \\ 2\ 4 \end{pmatrix} \approx \sum_{\omega_p} \exp\left[-\imath\omega_p(u_1 - u_3)\right] \frac{\Psi^{\mathbf{Q}}(\mathbf{r}_2, \mathbf{r}_1; u_2 - u_1)\Psi^{\mathbf{Q}*}(\mathbf{r}_3, \mathbf{r}_4; u_4 - u_3)}{\imath\omega_p - \omega(\mathbf{Q})}, \quad (13)$$

where $\Psi^{\mathbf{Q}}(\mathbf{r}_{i_2}, \mathbf{r}_{i_1}; u_1 - u_2)$ is the BS amplitudes. The symbol $\sum_{\omega_p}$ is used to denote $\beta^{-1}\sum_p$. For boson fields we have $\omega_p = (2\pi/\beta)p; p = 0, \pm 1, \pm 2, \dots$. The BS amplitude, as well as the spectrum of the two-particle excitations, both can be obtained from the following BS equation (written in symbolic form) $\left[K^{(0)-1} - \frac{\delta\Sigma_H}{\delta G} - \frac{\delta\Sigma_F}{\delta G}\right]\Psi = 0$. From the exact equations (5) and (7) follows a relationship between the Fock part of the mass operator and the two-particle Green function:

$$\Sigma_F(1; 2) = -\Gamma_\alpha^{(0)}(1; 6|z)D_{\alpha\beta}^{(0)}(z, z')\Gamma_\beta^{(0)}(4; 5|z')K \begin{pmatrix} 5\ 3 \\ 4\ 6 \end{pmatrix} G^{-1}(3; 2). \quad (14)$$

The last two equations along with the Schwinger-Dyson equation (3) form a set of equations that should be solved self-consistently. This is the first algorithm mentioned in the Introduction.

Since equations (6) and (12) hold, the two-particle excitations in the model manifest themselves not only as poles of the two-particle fermion Green function (13), but as poles of the Fourier transforms of the boson Green function $D_{\alpha\beta}(z, z')$ as well. To find the poles of the boson Green function we star from the Dyson equation for the boson Green function:

$$D_{\alpha\beta}^{-1}(z, z') = D_{\alpha\beta}^{(0)-1}(z, z') - \widetilde{\Pi}_{\alpha\beta}(z, z'), \quad (15)$$

where $\widetilde{\Pi}_{\alpha\beta}(z, z')$ is the proper self-energy of the boson field. From the exact equations (6)-(10) follows that the proper self-energy is given by

$$\widetilde{\Pi}_{\alpha\beta}(z, z') = \Gamma_\alpha^{(0)}(1; 2 \mid z)\widetilde{K} \begin{pmatrix} 2\ 3 \\ 1\ 4 \end{pmatrix} \Gamma_\beta^{(0)}(3; 4 \mid z')$$
$$= \Gamma_\alpha^{(0)}(1; 2 \mid z)G(2; 3)G(4; 1)\Gamma_\beta(3; 4 \mid z'). \quad (16)$$

According to Eqs. (12) and (16), the proper self-energy $\widetilde{\Pi}$, the Green function $\widetilde{K}$ and the vertex function Γ must have common poles. These poles are noting but the excitonic excitations in the model, which manifest themselves as peaks in the absorption spectrum. Let $E_l(\mathbf{Q})$ and $\mathbf{Q}$ denote the energy and momentum of these two-particle excitations. Close to $E_l(\mathbf{Q})$ one can write:

$$\widetilde{K}\begin{pmatrix} 1 & 3 \\ 2 & 4 \end{pmatrix} \approx \sum_{\omega_p} \exp\left[-\imath\omega_p(u_1 - u_3)\right] \frac{\Phi^{l\mathbf{Q}}(\mathbf{r}_2, \mathbf{r}_1; u_2 - u_1)\Phi^{l\mathbf{Q}*}(\mathbf{r}_3, \mathbf{r}_4; u_4 - u_3)}{\imath\omega_p - \omega_{l\mathbf{Q}}},$$

$$(17)$$

where $\omega_{l\mathbf{Q}} = E_l(\mathbf{Q}) - \mu$, μ is the chemical potential. The wave function $\Phi^{l\mathbf{Q}}(\mathbf{r}_2, \mathbf{r}_1; u_1 - u_2)$ describes two-particle electronic excitations in the model, and $F(\imath\omega_p)$ is a term regular at $\imath\omega_p \to \omega_{l\mathbf{Q}}$. The wave function of the electronic excitations can be written in the following form $\Phi^{l\mathbf{Q}}(\mathbf{r}_2, \mathbf{r}_1; u_1 - u_2) = e^{[\imath\frac{1}{2}\mathbf{Q}\cdot(\mathbf{r}_1 + \mathbf{r}_2)]}\phi^{l\mathbf{Q}}(\mathbf{r}_1 - \mathbf{r}_2; u_1 - u_2)$. Due to the form of the bare vertex function $\Gamma_\alpha^{(0)}(y; x \mid z)$ in non-relativistic QED, we have to take into account only the equal "time" $u_1 = u_2$ wave functions $\phi^{l\mathbf{Q}}(\mathbf{r}_1 - \mathbf{r}_2; 0) = \sum_{\mathbf{k}} \exp\{\imath\mathbf{k}.(\mathbf{r}_1 - \mathbf{r}_2)\}\phi^l(\mathbf{k}, \mathbf{Q})$, where $\phi^l(\mathbf{k}, \mathbf{Q})$ is the equal "time" two-particle wave function in $\mathbf{k}$-representation. Using Eqs. (16) and (17), and taking into account the expression for the bare vertex in non-relativistic QED we obtain:

$$\widetilde{\Pi}_{\alpha\beta}(\mathbf{Q}, \omega) = \frac{1}{Vc^2} \sum_l \left[\frac{j_\alpha^l(\mathbf{Q})j_\beta^{l*}(\mathbf{Q})}{\omega - \omega_{l\mathbf{Q}} + \imath 0^+} - \frac{j_\alpha^{l*}(\mathbf{Q})j_\beta^l(\mathbf{Q})}{\omega + \omega_{l\mathbf{Q}} + \imath 0^+} \right]. \quad (18)$$

Here and $j_\alpha^l(\mathbf{Q})$ is the current density. With the help of the Dyson equation (15) we find that the vanishing of the following 3×3 determinant $det||D_{\alpha\beta}^{(0)-1}(\mathbf{Q}, \omega) - \widetilde{\Pi}_{\alpha\beta}(\mathbf{Q}, \omega)|| = 0$ is the equation for the spectrum of the collective excitations $\omega(\mathbf{Q})$ in the model. With the help of the dielectric tensor $\varepsilon_{\alpha\beta}(\mathbf{Q}, \omega) = \delta_{\alpha\beta} - (4\pi c^2/\omega^2)\widetilde{\Pi}_{\alpha\beta}(\mathbf{Q}, \omega)$, we can rewrite the last equation in more familiar form $det||\frac{\omega^2}{c^2}\varepsilon_{\alpha\beta}(\mathbf{Q}, \omega) - \delta_{\alpha\beta}Q^2 + Q_\alpha Q_\beta|| = 0$. In cubic crystals, close to an isolated pole $E_l(\mathbf{Q})$, we can approximate the dielectric tensor by the following expression: $\varepsilon_{\alpha\beta}(\mathbf{Q}, \omega) = \delta_{\alpha\beta}\varepsilon_b \left[1 + \frac{4\pi\beta E_l(\mathbf{Q}=0)}{E_l^2(\mathbf{Q}) - \omega^2}\right]$, where the exciton dispersion $E_l(\mathbf{Q})$, oscillator strength $4\pi\beta$ and the background dielectric constant ε_b could be obtained from experimental data. This is the second algorithm for calculating the polariton spectrum.

3. Field-Theoretical Description of the Extended Hubbard Model

In order to include the triple states we use the following Nambu fermion fields

$$\widehat{\overline{\psi}}(y) = \left(\psi_\uparrow^\dagger(y)\psi_\downarrow^\dagger(y)\psi_\uparrow(y)\psi_\downarrow(y)\right), \quad \widehat{\psi}(y) = \widehat{\overline{\psi}}^T(y) \quad (19)$$

Here the symbol "hat" over any quantity $\widehat{O}$ means that this quantity is a matrix, and T means transposed matrix.

The interaction part of the Hamiltonian (1) is quartic in the Grassmann fermion fields so the functional integrals cannot be evaluated exactly. However, we can transform the quartic Hubbard terms in (1) to a quadratic form by introducing a model system which consists of a boson field $A_\alpha(z)$ interacting with a fermion fields $\widehat{\overline{\psi}}(y)$ and $\widehat{\psi}(x)$. The spin degree of freedom of the Bose field $A_\alpha(z)$, where $\alpha = \uparrow, \downarrow$, reflects the spin-dependent nature of the Hubbard interaction. The action of our model system is assumed to be of the form $S = S_0^{(e)} + S_0^{(A)} + S^{(e-A)}$, where: $S_0^{(e)} = \widehat{\overline{\psi}}(y)\widehat{G}^{(0)-1}(y;x)\widehat{\psi}(x)$, $S_0^{(A)} = \frac{1}{2}A_\alpha(z)D_{\alpha\beta}^{(0)-1}(z,z')A_\beta(z')$ and $S^{(e-A)} = \widehat{\overline{\psi}}(y)\widehat{\Gamma}_\alpha^{(0)}(y,x \mid z)\widehat{\psi}(x)A_\alpha(z)$. Here, we have introduced composite variables $y = \{\mathbf{r}_i, u\} = \{i, u\}$, $x = \{\mathbf{r}_{i'}, u'\} = \{i', u'\}$, $z = \{\mathbf{r}_j, v\} = \{j, v\}$ and $z' = \{\mathbf{r}_{j'}, v'\} = \{j', v'\}$, where $\mathbf{r}_i, \mathbf{r}_{i'}, \mathbf{r}_j$ and $\mathbf{r}_{j'}$ are the lattice site vectors.

The action $S_0^{(e)}$ describes the fermion (electron) part of the system. The inverse Green function of free electrons $\widehat{G}^{(0)-1}(y;x)$ is 4×4 diagonal with respect to the spin indices matrix and has its usual form:

$$\widehat{G}^{(0)-1}(y;x) = \sum_{\mathbf{k}} \sum_{\omega_m} e^{[i\mathbf{k}\cdot(\mathbf{r}_i - \mathbf{r}_{i'}) - \omega_m(u-u')]}$$

$$\begin{pmatrix} G_{\uparrow,\uparrow}^{(0)-1}(\mathbf{k}, i\omega_m) & 0 & 0 & 0 \\ 0 & G_{\downarrow,\downarrow}^{(0)-1}(\mathbf{k}, i\omega_m) & 0 & 0 \\ 0 & 0 & -G_{\uparrow,\uparrow}^{(0)-1}(\mathbf{k}, i\omega_m) & 0 \\ 0 & 0 & 0 & -G_{\downarrow,\downarrow}^{(0)-1}(\mathbf{k}, i\omega_m) \end{pmatrix};$$

$$G_{\uparrow,\uparrow}^{(0)-1}(\mathbf{k}, i\omega_m) = [i\omega_m - (\epsilon(\mathbf{k}) - \mu)]^{-1},$$
$$G_{\downarrow,\downarrow}^{(0)-1}(\mathbf{k}, i\omega_m) = [i\omega_m + (\epsilon(\mathbf{k}) - \mu)]^{-1}.$$

Here μ is the electron chemical potential, and $\epsilon(\mathbf{k})$ is the Fourier transform of the hoping integral, i.e. $-t_{ij} = 1/N \sum_{\mathbf{k}} \exp[i\mathbf{k}\cdot(\mathbf{r}_i - \mathbf{r}_j)]\,\epsilon(\mathbf{k})$. The symbol $\sum_{\omega_m}$ is used to denote $\beta^{-1}\sum_m$. For fermion fields we have $\omega_m = (2\pi/\beta)(m + 1/2); m = 0, \pm 1, \pm 2, \ldots$.

The action $S_0^{(A)}$ describes the boson field. The bare boson propagator provides an instantaneous spin-dependent interaction:

$$D_{\alpha\beta}^{(0)}(z, z') = \delta(v - v')\left[U\delta_{jj'}\delta_{\alpha\overline{\beta}} - 2V_{<jj'>}(\delta_{\overline{\alpha},\beta} + \delta_{\alpha\beta})\right]$$
$$= \frac{1}{N}\sum_{\mathbf{k}}\sum_{\omega_p} e^{\{i[\mathbf{k}\cdot(\mathbf{r}_j - \mathbf{r}_{j'}) - \omega_p(v-v')]\}}D_{\alpha\beta}^{(0)}(\mathbf{k}; i\omega_p).$$

Here $D_{\alpha\beta}^{(0)}(\mathbf{k}; i\omega_p) = U\delta_{\overline{\alpha},\beta} - V(\mathbf{k})(\delta_{\overline{\alpha},\beta} + \delta_{\alpha,\beta})$, $\overline{\alpha}$ is complimentary of α, and $V(\mathbf{k}) = 4V(\cos k_x + \cos k_y)$ is the nearest-neighbor interaction in momentum space. The symbol $V_{<jj'>}$ is equal to V if j and j' sites are nearest neighbors, and zero otherwise.

The interaction between the fermion and the boson fields is described by the action $S^{(e-A)}$. The bare vertex $\widehat{\Gamma}_\alpha^{(0)}(y_1; x_2 \mid z) = \widehat{\Gamma}_\alpha^{(0)}(i_1, u_1; i_2, u_2 \mid j, v)$ is a 4×4 matrix which elements are as follows:

$$\left(\widehat{\Gamma}_\alpha^{(0)}(y_1; x_2 \mid z)\right)_{n_1 n_2} = \delta(u_1 - u_2)\delta(u_1 - v)\delta_{i_1 i_2}\delta_{i_1 j}\delta_{n_1 n_2}$$

$$[\delta_{\alpha\uparrow}(\delta_{1n_1} - \delta_{3n_1}) + \delta_{\alpha\downarrow}(\delta_{2n_1} - \delta_{4n_1})], \{n_1, n_2\} = 1, 2, 3, 4. \tag{20}$$

The relation between the extended Hubbard model and our model system can be found by applying the Hubbard-Stratonovich transformation for the electron operators:

$$\int \mu[A] \exp\left[\widehat{\overline{\psi}}(y)\widehat{\Gamma}_\alpha^{(0)}(y; x|z)\widehat{\psi}(x)A_\alpha(z)\right]$$

$$= \exp\left[-\tfrac{1}{2}\widehat{\overline{\psi}}(y)\widehat{\Gamma}_\alpha^{(0)}(y; x|z)\widehat{\psi}(x)D_{\alpha,\beta}^{(0)}(z, z')\widehat{\overline{\psi}}(y')\widehat{\Gamma}_\beta^{(0)}(y'; x'|z')\widehat{\psi}(x')\right]. \tag{21}$$

From the Hubbard-Stratonovich transformation follows that the extended Hubbard model can be mapped onto the model system described by the action (3). In other words, the spectrum of the single-particle excitations, as well as the spectrum of the collective modes of the extended Hubbard model, both are exactly the same as the corresponding two spectra of our model system.

In field theory the expectation value of a general operator $\widehat{O}(u)$ is expressed as a functional integral over the boson field A and the Grassmann fermion fields $\widehat{\overline{\psi}}$ and $\widehat{\psi}$. As in the exciton case, we introduce the sources of the boson J and fermion fields $\widehat{M}$. The fermion source $\widehat{M}(y; x)$ is a 4×4 matrix. It is convenient to introduce complex indices $1 = \{n_1, x_1\}$, and $2 = \{n_2, y_2\}$ where, $x_1 = \{\mathbf{r}_{i_1}, u_1\}$, $y_2 = \{\mathbf{r}_{i_2}, u_2\}$ and $\{n_1, n_2\} = 1, 2, 3$ and 4. Thus, the matrix element $\left(\widehat{M}(y; x)\right)_{n_1 n_2} = M_{n_1 n_2}(y_1; x_2)$ can be written as $M(1; 2)$. The functional $Z[J, M]$ is defined by $Z[J, M] = \int D\mu[\widehat{\overline{\psi}}, \widehat{\psi}, A]e^{[J_\alpha(z)A_\alpha(z) - \widehat{\overline{\psi}}(y)\widehat{M}(y;x)\widehat{\psi}(x)]}$, where the functional measure $D\mu[\widehat{\overline{\psi}}, \widehat{\psi}, A] = DAD\widehat{\overline{\psi}}D\widehat{\psi}\exp(S)$ satisfies the condition $\int D\mu[\widehat{\overline{\psi}}, \widehat{\psi}, A] = 1$. We define a functional derivative $\delta/\delta M(2; 1)$, and depending on the spin degrees of freedom there are sixteen possible derivatives.

As in the excitonic case, all Green functions related to system under consideration can be expressed in terms of the functional derivatives of the generating functional of the connected Green functions $W[J, M] = \ln Z[J, M]$. By means of the functional $W[J, M]$, we define the Green and vertex functions of the extended Hubbard model in a similar way as in the exciton case. The definition of the boson Green function is identical to the exciton case. The single-electron Green function $\widehat{G}(x_1; y_2)$ is a 4×4 matrix whose elements are $G_{n_1 n_2}(x_1; y_2) = -\delta W/\delta M_{n_2 n_1}(y_2; x_1)$. Depending on the two spin degrees of freedom, there exist eight "normal" Green functions and eight "anomalous" Green functions. We introduce Fourier transforms of the "normal" $G_{\sigma_1, \sigma_2}(\mathbf{k}, u_1 - u_2) = - < \widehat{T}_u(\psi_{\sigma_1, \mathbf{k}}(u_1)\psi_{\sigma_2, \mathbf{k}}^\dagger(u_2)) >$, and "anomalous" $F_{\sigma_1, \sigma_2}(\mathbf{k}, u_1 - u_2) = - < \widehat{T}_u(\psi_{\sigma_1, \mathbf{k}}(u_1)\psi_{\sigma_2, -\mathbf{k}}(u_2)) >$ one-particle Green functions, where $\{\sigma_1, \sigma_2\} = \uparrow, \downarrow$. Here $\psi_{\uparrow, \mathbf{k}}^+(u), \psi_{\uparrow, \mathbf{k}}(u)$ and $\psi_{\downarrow, \mathbf{k}}^+(u), \psi_{\downarrow, \mathbf{k}}(u)$ are the creation-annihilation Heisenberg operators. The final form of single-particle Green function is given by

$$\widehat{G}(1;2) = \frac{1}{N}\sum_{\mathbf{k}}\sum_{\imath\omega_m} e^{\{\imath[\mathbf{k}\cdot(\mathbf{r}_{i_1}-\mathbf{r}_{i_2})-\omega_m(u_1-u_2)]\}} \begin{pmatrix} \widehat{G}(\mathbf{k},\imath\omega_m) & \widehat{F}(\mathbf{k},\imath\omega_m) \\ \widehat{F}^{\dagger}(\mathbf{k},\imath\omega_m) & -\widehat{G}(-\mathbf{k},-\imath\omega_m) \end{pmatrix}.$$

Here, $\widehat{G}$ and $\widehat{F}$ are 2×2 matrices whose elements are G_{σ_1,σ_2} and F_{σ_1,σ_2}, respectively. The two-particle Green function is defined by:

$$K\begin{pmatrix} n_1, x_1 \; n_3, y_3 \\ n_2, y_2 \; n_4, x_4 \end{pmatrix} = K\begin{pmatrix} 1\ 3 \\ 2\ 4 \end{pmatrix} = -\frac{\delta G_{n_1 n_2}(x_1;y_2)}{\delta M_{n_3 n_4}(y_3;x_4)}; \tag{23}$$

The vertex function $\widehat{\Gamma}_\alpha(2;1\mid z)$ is a 4×4 matrix whose elements are $\left(\widehat{\Gamma}_\alpha(i_2,u_2;i_1,u_1\mid v,j)\right)_{n_2 n_1} = -\frac{\delta G^{-1}_{n_1 n_2}(i_2,u_2;i_1,u_1)}{\delta J_\beta(z')}D_{\beta\alpha}^{-1}(z',z).$

4. Spectrum of the Collective Excitations

Spectrum of the two-particle excitations (or collective modes) $\omega(\mathbf{Q})$ can be obtained by locating the positions of the common poles of the Fourier transform of the two-particle fermion Green function (23) and the Fourier transform of the boson Green function $D_{\alpha\beta}$. In other words, collective modes are defined by the solutions of the BS equation for the function K or the Dyson equation for the boson function D. The BS equation is $K^{-1}\Psi = \left[K^{(0)-1} - I\right]\Psi = 0$, where I is the kernel, and

$$K^{(0)-1}\begin{pmatrix} n_1, i_1, u_1 \; n_3, i_3, u_3 \\ n_2, i_2, u_2 \; n_4, i_4, u_4 \end{pmatrix} = G^{-1}_{n_1 n_3}(i_1,u_1;i_3,u_3)G^{-1}_{n_4 n_2}(i_4,u_4;i_2,u_2)$$

is the two-particle free propagator constructed from a pair of fully dressed single-particle Green functions. The kernel $I = \frac{\delta\Sigma_H}{\delta G} + \frac{\delta\Sigma_F}{\delta G}$ depends on the functional derivative of the Fock contribution to the mass operator. Since the Fock term itself depends on the two-particle Green function K, we have to solve self-consistently a set of two equations, namely the BS equation and the Dyson equation $G^{-1} = G^{(0)-1} - \Sigma$ for the single-particle Green function.

Similar obstruction arises if we start from the Dyson equation (15) for boson Green function. The obstruction now is that the proper self-energy depends on the vertex function Γ, or on the two-particle Green function $\widetilde{K}$. The Green function $\widetilde{K}$, which we shall call the Green function of electronic excitations, satisfies BS equation (11). According to Eq. (16), the proper self-energy $\widetilde{\Pi}$, the Green function $\widetilde{K}$ and the vertex function Γ must have common poles. Let $E_l(\mathbf{Q})$ and $\mathbf{Q}$ denote the energy and momentum of these common poles. Close to $E_l(\mathbf{Q})$ one can write:

$$\widetilde{K}\begin{pmatrix} 1\ 3 \\ 2\ 4 \end{pmatrix} \approx \sum_{\omega_p} e^{-\imath\omega_p(u_1-u_3)}\frac{\Phi^{l\mathbf{Q}}_{n_1,n_2}(\mathbf{r}_{i_2},\mathbf{r}_{i_1};u_2-u_1)\Phi^{l\mathbf{Q}*}_{n_3,n_4}(\mathbf{r}_{i_3},\mathbf{r}_{i_4};u_4-u_3)}{\imath\omega_p - \omega_{l\mathbf{Q}}}, \tag{24}$$

where $\Phi^{l\mathbf{Q}}_{n_1,n_2}(\mathbf{r}_{i_2},\mathbf{r}_{i_1};u_1-u_2)$ are the BS amplitudes:

$$\Phi_{n_1,n_2}^{l\mathbf{Q}}(\mathbf{r}_{i_2},\mathbf{r}_{i_1};u_1-u_2)=\exp\left[\imath\frac{1}{2}\mathbf{Q}.(\mathbf{r}_{i_1}+\mathbf{r}_{i_2})\right]\phi_{n_1,n_2}^{l\mathbf{Q}}(\mathbf{r}_{i_1}-\mathbf{r}_{i_2};u_1-u_2).$$

Due to the form of the bare vertex $\widehat{\Gamma}^{(0)}$, we have to take into account only the equal "time" $u_1=u_2$ amplitudes:

$$\phi_{n_1,n_2}^{l\mathbf{Q}}(\mathbf{r}_{i_1}-\mathbf{r}_{i_2};0)=\frac{1}{N}\sum_{\mathbf{k}}\exp\{\imath\mathbf{k}.(\mathbf{r}_{i_1}-\mathbf{r}_{i_2})\}\phi_{n_1,n_2}^{l}(\mathbf{k},\mathbf{Q}),$$

where $\phi_{\downarrow\uparrow}^{l}(\mathbf{k},\mathbf{Q})$, $\phi_{n_1 n_2}^{l}(\mathbf{k},\mathbf{Q})$ is the equal "time" two-particle wave functions in $\mathbf{k}$-representation. By means of (24) and (16) we obtain:

$$\widetilde{\Pi}_{\alpha\beta}(\mathbf{Q},\omega)=\sum_{l}\left[\frac{\varphi_{\alpha}^{l\mathbf{Q}}\varphi_{\beta}^{*l\mathbf{Q}}}{\omega-\omega_{l\mathbf{Q}}+\imath 0^{+}}-\frac{\varphi_{\alpha}^{*l\mathbf{Q}}\varphi_{\beta}^{l\mathbf{Q}}}{\omega+\omega_{l\mathbf{Q}}+\imath 0^{+}}\right],\qquad(25)$$

where $\omega_{l\mathbf{Q}}=E_l(\mathbf{Q})-\mu$, and $\varphi_{\alpha}^{l\mathbf{Q}}=\delta_{\alpha\uparrow}\left(\phi_{1,1}^{l\mathbf{Q}}(0;0)-\phi_{3,3}^{l\mathbf{Q}}(0;0)\right)+\delta_{\alpha\downarrow}\left(\phi_{2,2}^{l\mathbf{Q}}(0;0)-\phi_{4,4}^{l\mathbf{Q}}(0;0)\right)$. From Eq. (15) follows that the spectrum of the collective excitations $\omega(\mathbf{Q})$ could be obtained assuming the vanishing of the following 2×2 determinant:

$$det\parallel\delta_{\alpha,\beta}-(U-V(\mathbf{Q}))\,\widetilde{\Pi}_{\overline{\alpha}\beta}(\mathbf{Q},\omega)+V(\mathbf{Q})\widetilde{\Pi}_{\alpha\beta}(\mathbf{Q},\omega)\parallel=0.\qquad(26)$$

By solving Eq. (26) (with the help of $\widetilde{\Pi}_{\alpha\beta}=\widetilde{\Pi}_{\beta\alpha}$) we find two different types of collective modes. The first one is a solution of the following equation:

$$0=1+U\left[\widetilde{\Pi}_{\uparrow\uparrow}(\mathbf{Q},\omega)-\widetilde{\Pi}_{\uparrow\downarrow}(\mathbf{Q},\omega)\right].\qquad(27)$$

Note that the V interaction is included indirectly in Eq. (27) through BS amplitudes and poles of the function $\widetilde{K}$.

The second type of collective modes satisfies the following equation:

$$0=1-[U-2V(\mathbf{Q})]\left[\widetilde{\Pi}_{\uparrow\uparrow}(\mathbf{Q},\omega)+\widetilde{\Pi}_{\uparrow\downarrow}(\mathbf{Q},\omega)\right].\qquad(28)$$

The two types of collective modes can be related to the poles of the so-called general response function $\Pi_{\alpha\beta}$, defined as follows:

$$\Pi_{\alpha\beta}(z;z')=$$

$$\left(\widehat{\Gamma}_{\alpha}^{(0)}(i_1,u_1;i_2,u_2\mid z)\right)_{n_1 n_2}K\left(\begin{matrix}n_2,i_2,u_2\;n_3,i_3,u_3\\n_1,i_1,u_1\;n_4,i_4,u_4\end{matrix}\right)\left(\widehat{\Gamma}_{\beta}^{(0)}(i_3,u_3;i_4,u_4\mid z')\right)_{n_3 n_4}.$$

By means of the response function, we can rewrite the BS and Dyson equations as follows:

$$K=K^{(0)}+K^{(0)}\Gamma_{\alpha}^{(0)}\Pi_{\alpha\beta}\Gamma_{\beta}^{(0)}K^{(0)},\qquad D=D^{(0)}+D^{(0)}\Pi D^{(0)}.$$

The Fourier transforms of the general response function $\Pi_{\alpha\beta}(\mathbf{Q};\omega)$ and the proper self-energy $\widetilde{\Pi}_{\alpha\beta}(\mathbf{Q};\omega)$ are connected by the following equation:

$$\Pi_{\alpha\beta}(\mathbf{Q};\omega) = \widetilde{\Pi}_{\alpha\beta}(\mathbf{Q};\omega) + \widetilde{\Pi}_{\alpha\gamma}(\mathbf{Q};\omega)\left(U - V(\mathbf{Q})\right)\Pi_{\overline{\gamma}\beta}(\mathbf{Q};\omega)$$
$$-\Pi_{\alpha\gamma}(\mathbf{Q};\omega)V(\mathbf{Q})\widetilde{\Pi}_{\gamma\beta}(\mathbf{Q};\omega), \qquad (29)$$

The first type of collective modes manifests itself as a pole of the dynamical spin response function (or spin susceptibility) $\chi_{ss}(\mathbf{Q};\omega) = 2\left[\Pi_{\uparrow\uparrow}(\mathbf{Q};\omega) - \Pi_{\uparrow\downarrow}(\mathbf{Q};\omega)\right]$, and therefore, gives rise to the spin instabilities. By means of (29) we find that the spin correlation function is given by

$$\chi_{ss}(\mathbf{Q};\omega) = \frac{\widetilde{\chi}_{ss}(\mathbf{Q};\omega)}{1 + \widetilde{\chi}_{ss}(\mathbf{Q};\omega)U/2}, \qquad (30)$$

where $\widetilde{\chi}_{ss}(\mathbf{Q};\omega) = 2\left[\widetilde{\Pi}_{\uparrow\uparrow}(\mathbf{Q};\omega) - \widetilde{\Pi}_{\uparrow\downarrow}(\mathbf{Q};\omega)\right]$.

The second collective mode manifests itself as a pole of the dynamical charge response function (or charge susceptibility) $\chi_{cc}(\mathbf{Q};\omega) = 2\left[\Pi_{\uparrow\uparrow}(\mathbf{Q};\omega) + \Pi_{\uparrow\downarrow}(\mathbf{Q};\omega)\right]$. This response functions is determined by the following equation:

$$\chi_{cc}(\mathbf{Q};\omega) = \frac{\widetilde{\chi}_{cc}(\mathbf{Q};\omega)}{1 - \widetilde{\chi}_{cc}(\mathbf{Q};\omega)\left[U/2 - V(\mathbf{Q})\right]}, \qquad (31)$$

where $\widetilde{\chi}_{cc}(\mathbf{Q};\omega) = 2\left[\widetilde{\Pi}_{\uparrow\uparrow}(\mathbf{Q};\omega) + \widetilde{\Pi}_{\uparrow\downarrow}(\mathbf{Q};\omega)\right]$.

5. Spin Correlation Function in GRPA

Within the GRPA, one should replace $\widetilde{K}$ in (16) by $K^{(0)}$, thus obtaining the free response functions $\chi^{(0)}$ instead of the exact $\widetilde{\chi}$ expressions. In other words, the exact relations (30) and (31) in the GRPA are given by

$$\chi_{ss}(\mathbf{Q};\omega) = \frac{\chi_{ss}^{(0)}(\mathbf{Q};\omega)}{1 + \chi_{ss}^{(0)}(\mathbf{Q};\omega)U/2}, \qquad (32)$$

In GRPA $\widetilde{\Pi}_{\alpha\beta}^{(0)}(j,v;j,v') = \widetilde{\Pi}_{\alpha\beta}^{(0)}(\mathbf{r}_j - \mathbf{r}_{j'};v - v')$ and its Fourier transform $\widetilde{\Pi}_{\alpha\beta}^{(0)}(\mathbf{Q};\imath\omega_p)$, and therefore the spin correlation function assumes the form:

$$\chi_{ss}^{(0)}(\mathbf{Q};\omega) = \sum_{\omega_m}\sum_{\mathbf{k}}Tr\left\{\widehat{\alpha}_z\widehat{G}(\mathbf{k};\imath\omega_m)\widehat{\alpha}_z\widehat{G}(\mathbf{k}+\mathbf{Q};\imath\omega_p + \imath\omega_m)\right\}, \qquad (33)$$

where $\widehat{\alpha}_z$ is 4×4 matrix:

$$\widehat{\alpha}_i = \begin{pmatrix} \sigma_i & 0 \\ 0 & \sigma_y\sigma_i\sigma_y \end{pmatrix}, i = x, y, z.$$

Here, σ_i are the Pauli matrices.

The spin susceptibility defined by Eq. (33) is the $\chi_{zz}^{(0)}$ component of the more general spin susceptibility χ_{ij} which in GRPA has the following form:[2]

$$\chi_{ij}^{(0)}(\mathbf{Q};\omega) = -\frac{1}{4}\sum_{\omega_m}\sum_{\mathbf{k}} Tr\left\{\widehat{\alpha}_i\widehat{G}(\mathbf{k};\imath\omega_m)\widehat{\alpha}_j\widehat{G}(\mathbf{k}+\mathbf{Q};\imath\omega_p+\imath\omega_m)\right\} \qquad (34)$$

The missing factor $1/4$ in (33) comes from the fact that the two-particle Green function in the extended Hubbard model is one-forth part of our two-particle Green function. Similarly, the single-particle Green function in the extended Hubbard model is one-half of the single-particle Green function. It should be mentioned that there is a missing factor of $1/2$ in Eq. (10) in Ref. [3] (see our Eq. (32)).

According to Ref. [3], a resonance peak (that could be detected by inelastic neutron-scattering experiments) should appear in $Im\chi_{zz}^{(0)}$ in triplet superconductors such as Sr_2RuO_4.

6. Conclusion

We have established an one-to-one correspondence between the extended Hubbard model and the polariton model for light propagation in crystals. This mapping allows us to obtain relationships between the singe-particle and two-particle Green functions and to derive exact results for spin-spin and charge-charge correlation functions of the extended Hubbard model.

References

1. Z. Koinov, Phys. Rev. B **72**, 085203 (2005).
2. W.F. Brinkman, J.W. Serene, and P.W. Anderson, Phys. Rev. A **10**, 2386 (1974).
3. M. Yakiyama and Y. Hasegawa, Phys. Rev. B **67**, 2345 (2003).

THERMAL EXPANSION OF FERROMAGNETIC SUPERCONDUCTORS: POSSIBLE APPLICATION TO UGe$_2$

NOBUKUNI HATAYAMA* and RIKIO KONNO†

Kinki University Technical College, 2800 Arima-cho
Kumano-shi, Mie 519-4395, Japan
*hatayama@ktc.ac.jp
†r-konno@ktc.ac.jp

Received 14 November 2009

We investigate the temperature dependence of thermal expansion of the ferromagnetic triplet superconductors and their thermal expansion coefficients below the superconducting transition temperature of a majority spin conduction band. The free energy of the ferromagnetic superconductors derived by Linder *et al.* is used. The superconducting gaps in the A2 phase of ^{3}He and with a node in UGe$_2$ are considered. By applying Takahashi's method to the free energy, i.e. by taking into account the volume dependence of the free energy explicitly, the temperature dependence of the thermal expansion and the thermal expansion coefficients is studied below the superconducting transition temperature of the majority spin conduction band. We find that we have anomalies of the thermal expansion in the vicinity of the superconducting transition temperatures and that we have divergence of the thermal expansion coefficients are divergent at the superconducting transition temperatures. The Grüneisen's relation between the temperature dependence of the thermal expansion coefficients and the temperature dependence of the specific heat at low temperatures is satisfied.

Keywords: Thermal expansion; ferromagnetic superconductors; Grüneisen's relation.

1. Introduction

The ferromagnetic superconductors have attracted many researchers since they found in UGe$_2$ under pressures,[1] URhGe[2,3] and UCoGe[4] at ambient pressures. We have the superconducting transition temperature $T_{sc} = 0.8K$ and the Cürie temperature $T_C = 32K$ at around 1.2GPa in UGe$_2$. The saturation magnetization is about $1\mu_B$. We have the superconducting gap of the up-spin conduction band with a line node and no superconducting gap of the down-spin conduction band in UGe$_2$ experimentally.[5]

Recently, Linder *et al.* derive the free energy of the ferromagnetic superconductors based on the single band model.[6,7] In this model, there are a superconducting gap of the up-spin band and that of the down-spin band because the conduction band is split into an up-spin band and a down-spin band in the ferromagnetic state below T_C. Correspondingly, there are the superconducting transition temperature $T_{sc,\uparrow}$ of the up-spin conduction band and that $T_{sc,\downarrow}$ of the down-spin conduction band. Linder *et al.* compare the free energy of the ferromagnetic superconductors

316

with the free energy of the pure ferromagnetic state, that of the unitary superconducting state, and that of the paramagnetic state below $T_{sc,\uparrow}$. In other words, they
found that the free energy of the ferromagnetic superconductors is lower than that
of the pure ferromagnetic state, that of the unitary superconducting state, and that
of the paramagnetic state below $T_{sc,\uparrow}$. They found that the coexistent state of the
superconductivity and the ferromagnetism is stable.

On the other hand, theory of the thermal expansion due to spin fluctuations has
developed. The Grünneisen's relation between the temperature dependence of the
thermal expansion coefficient and that of the magnetic specific heat at low temperatures is not satisfied in the conventional theories. [8,9] Takahashi *et al.* resolved this
problem. They consider the same free energy as that used in the magnetic specific
heat.[10,11] By taking into account the volume V dependence of the free energy explicitly, they derive the thermal expansion and its coefficient. They found that the
Grüneisen's relation is satisfied.

Clark makes contrasting collective versus topological scenarios for the quantum
critical point. He investigates the temperature dependence of the thermal expansion
within the scenarios.[12]

However, by our using the free energy derived by Linder *et al.*[6,7] thermal expansion of ferromagnetic superconductors and its coefficient have not been discussed.
We investigate the temperature dependence of thermal expansion of ferromagnetic
superconductors and its coefficient by applying Takahashi's method to the free energy derived by Linder *et al.*, i.e. by taking into account the volume dependence of
the free energy explicitly in this paper.

This paper is organized as follows. In the next section, thermal expansion of the
ferromagnetic superconductors and its coefficient will be derived. In section 3, the
numerical results of the thermal expansion and its coefficient will be given with the
superconducting gaps in A2 phase of ^{3}He and that with a node of the up-spin band
where it is appropriate in UGe$_2$. In section 4, the results will be summarized.

2. The Derivation of Thermal Expansion of Ferromagnetic Superconductors and its Coefficient

We begin with the following free energy[6,7]:

$$F_{coexist}/N = F_0/N + F_T/N, \tag{1}$$

$$F_0/N = \frac{IM^2}{2} + \frac{1}{2\pi}\int_0^{2\pi} d\theta \sum_\sigma \frac{\Delta_\sigma^2(\theta)}{2g} - \frac{1}{2\pi}\sum_\sigma \int_0^{2\pi} d\theta \int_0^{E_F} d\varepsilon N(\varepsilon)\frac{E_\sigma(\varepsilon,\theta)}{2}, \tag{2}$$

$$F_T/N = -\frac{T}{2\pi}\sum_\sigma \int_0^{2\pi} d\theta \int_0^\infty d\varepsilon N(\varepsilon)\ln(1 + e^{-E_\sigma(\varepsilon,\theta)/T}) \tag{3}$$

where F_0 is the ground state energy and F_T is the thermal part of the free energy.
E_F is the Fermi energy. N is the number of magnetic atoms, $N(\epsilon)$ is the density

of state, and $\Delta_\sigma(\theta)$ is the superconducting gap of the spin σ conduction band. θ is an azimuthal angle in the k_{Fx}-k_{Fy} plane and k_F is the Fermi wave number. g is the effective attractive pairing coupling constant. I is the on-site Coulomb coupling constant. M is the magnetization. ϵ is the kinetic energy of electrons. $E_\sigma(\varepsilon, \theta)$ is given by

$$E_\sigma(\varepsilon, \theta) = \sqrt{(\varepsilon - \sigma I M - E_F)^2 + \Delta_\sigma^2(\theta)}. \tag{4}$$

Thermal expansion of the ferromagnetic superconductors is obtained as follows:

$$\omega = -K\frac{\partial F_{coexist}}{\partial V}. \tag{5}$$

where K is the compressibility. The thermal expansion is

$$\omega/(NE_F) = \omega_0/(NE_F) + \omega_T/(NE_F). \tag{6}$$

ω_0 comes from the part of the ground state energy. ω_T originates from the thermal part of the free energy. ω_0 is

$$\omega_0 = -K\{\frac{1}{2}\frac{E_F}{I}(\frac{\partial \ln I}{\partial V})\tilde{M}^2 + \frac{1}{2}E_F\frac{\partial}{\partial g}(\frac{1}{g})\sum_\sigma \frac{1}{2\pi}\int_0^{2\pi} d\theta \tilde{\Delta}_\sigma^2(\theta)$$

$$-N(0)E_F\frac{\partial \ln N(0)}{\partial V}\sum_\sigma \frac{1}{2\pi}\int_0^{2\pi} d\theta \int_0^1 dx \frac{\tilde{E}_\sigma^2(x,\theta)}{2}$$

$$-\frac{1}{2}\sum_\sigma A_\sigma \frac{1}{2\pi}\int_0^{2\pi} d\theta(\sqrt{(-\sigma\tilde{M})^2} - \sqrt{(-\sigma\tilde{M}-1)^2 + \tilde{\Delta}_\sigma^2(\theta)})\} \tag{7}$$

where $\tilde{M} = IM/E_F$, $N(0)$ is the density of states at the Fermi energy, and $\tilde{\Delta}_\sigma(\theta) = \Delta_\sigma(\theta)/E_F$. ω_T is

$$\omega_T = -K\frac{T}{T_F}N(0)E_F\sum_\sigma[\frac{1}{2\pi}\int_0^{2\pi} d\theta(-\frac{\partial \ln N(0)}{\partial V}\int_0^\infty dx \ln(1 + e^{-\frac{T_F}{T}\tilde{E}_\sigma(x,\theta)})$$

$$+A_\sigma \ln(1 + e^{-\frac{T_F}{T}\tilde{E}_\sigma(0,\theta)}))] \tag{8}$$

with

$$A_\sigma = \frac{\partial x}{\partial V} - \sigma\frac{\partial \ln I}{\partial V}\tilde{M} - \frac{\partial \ln E_F}{\partial V} \tag{9}$$

where $x = \epsilon/E_F$. $E_\sigma(x, \theta)$ is

$$\tilde{E}_\sigma(x, \theta) = \sqrt{(x - \sigma\tilde{M} - 1)^2 + \tilde{\Delta}_\sigma^2(\theta)}. \tag{10}$$

The corresponding thermal expansion coefficient is given by

$$\alpha = \frac{\partial \omega}{\partial T}. \tag{11}$$

The Grüneisen's relation between the temperature dependence of the thermal expansion coefficient and that of the specific heat at low temperatures is satisfied because the expression of the free energy used in the present work is the same as that of the free energy used in the derivation of the specific heat.

We assume that the Cürie temperature T_C is much larger than the superconducting transition temperatures and that T_C is much lower than the Fermi energy E_F throughout this paper. Correspondingly, the magnetization is constant. This assumption is valid in UGe$_2$.

In order to obtain the temperature dependence of the thermal expansion and its coefficient, we need the temperture dependence of the superconducting gaps. In the next subsections, the temperature dependence of the superconducting gaps used by Linder et al. is given.

2.1. The superconducting gaps in A2 phase of liquid ³He

The superconducting gaps in A2 phase of liquid ^{3}He is

$$\Delta_{k\sigma\sigma} = -\sigma\Delta_{\sigma,0}\sin\phi \tag{12}$$

where $\Delta_{\sigma,0}$ is the superconducting order parameter with the spin σ conduction band. We shall consider $\sin\phi = 1$. The superconducting order parameters at $T = 0[K]$ are obtained

$$\Delta_{\sigma,0}(0) = 2E_0\exp(-1/c\sqrt{1+\sigma\tilde{M}(0)}). \tag{13}$$

where the weak coupling constant $c = gN(0)/2$ and E_0 is the cutoff energy. E_0 and c are set to 0.01 and 0.2, respectively. The temperature dependence of the superconducting order parameters is obtained as follows:

$$\Delta_{\sigma,0}(T) = \Delta_{\sigma,0}(0)\tanh(1.74\sqrt{T_{sc,\sigma}/T - 1}) \tag{14}$$

with

$$T_{sc,\sigma} = 1.13E_0\exp(-1/c\sqrt{1+\sigma\tilde{M}(\tilde{T}_{sc,\sigma})}. \tag{15}$$

where $T_{sc,\sigma}$ is the superconducting transition temperature of the spin σ band.

2.2. The superconducting gap of the up-spin conduction band with a line node

Harada *et al.* show that there is the superconducting gap of the up-spin conduction band with the line node experimentally. Therefore, we consider the following superconducting gap:

$$\Delta_\sigma(\theta) = \begin{cases} \Delta_0\cos\theta(\sigma = \uparrow) \\ 0(\sigma = \downarrow) \end{cases} \tag{16}$$

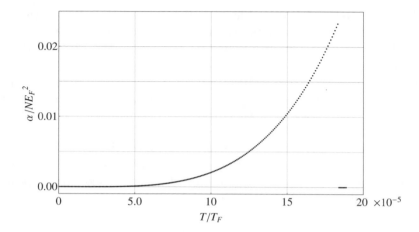

Fig. 1.   The reduced temperature dependence of thermal expansion coefficient with the supercon-ducting gaps in A2 phase of liquid ^{3}He when $I/E_F = 0.1$, $\frac{\partial \ln I}{\partial V} = 0.1$, $E_F \frac{\partial}{\partial V}\left(\frac{1}{g}\right) = 0.1$, $\frac{\partial x}{\partial V} = 0.1$, $N(0)E_F = 0.1$, and $N(0)E_F \frac{\partial \ln N(0)}{\partial V} = 0.1$.

The superconducting order parameter at $T = 0K$ is

$$\Delta_0(0) = 2.426 E_0 \exp(-1/c\sqrt{1 + \tilde{M}(0)}) \tag{17}$$

The temperature dependence of the superconducting order parameter $\Delta_0(T)$ is obtained

$$\Delta_0(T) = \Delta_0(0) \tanh(1.70\sqrt{T_{sc}/T - 1}) \tag{18}$$

with

$$T_{sc} = 1.13 E_0 \exp(-1/c\sqrt{1 + \sigma \tilde{M}(\tilde{T}_{sc})}) \tag{19}$$

where T_{sc} is the superconducting transition temperature.

In the next section the numerical results will be presented.

3. Results

From Eqs.(6-11), the thermal expansion of the ferromagnetic superconductors and its coefficient are obtained.

3.1. *The thermal expansion coefficient with the superconducting gaps in A2 phase of liquid* ^{3}He

The temperature dependence of thermal expansion of ferromagnetic superconductors in this case is given by Ref. 13. Fig. 1 shows the temperature dependence of

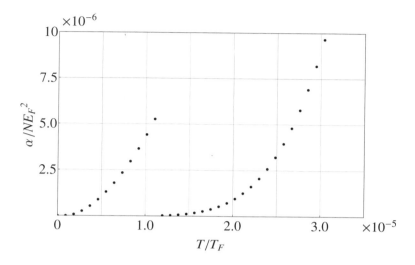

Fig. 2.   The reduced temperature dependence of the thermal expansion coefficient with the super-conducting gaps in A2 phase of liquid ^{3}He at very low temperatures when $I/E_F = 0.1$, $\frac{\partial \ln I}{\partial V} = 0.1$, $E_F \frac{\partial}{\partial V}(\frac{1}{g}) = 0.1$, $\frac{\partial x}{\partial V} = 0.1$, $N(0)E_F = 0.1$, and $N(0)E_F \frac{\partial \ln N(0)}{\partial V} = 0.1$.

the corresponding thermal expansion coefficient. Fig. 2 shows the temperature dependence of the thermal expansion coefficient at very low temperatures. From Fig. 1 and Fig. 2, the thermal expansion coefficient is divergent at both of the super-conducting transition temperature of the up-spin conduction band and that of the down-spin conduction band. At the low temperature limit, from Eqs.(7) and (11), the thermal expansion coefficient is

$$\alpha \propto \sum_\sigma 1.74 \cdot \frac{\tilde{T}_{sc,\sigma}}{\tilde{T}^2} \cdot \frac{1}{2\sqrt{\frac{\tilde{T}_{sc,\sigma}}{\tilde{T}} - 1}} \exp\left(-3.48\sqrt{\frac{\tilde{T}_{sc,\sigma}}{\tilde{T}} - 1}\right) \tag{20}$$

From Eq.(20), the thermal expansion coefficient increases with the temperature rise exponentially. On the other hand, when T goes to $T_{sc,\sigma}$, the thermal expansion coefficient is got

$$\alpha \propto \sum_\sigma 1.74 \cdot \frac{\tilde{T}_{sc,\sigma}}{\tilde{T}^2} \cdot \frac{1}{2\sqrt{\frac{\tilde{T}_{sc,\sigma}}{\tilde{T}} - 1}} \tag{21}$$

This shows that the thermal expansion coefficient is divergent at both of the super-conducting transition temperature of the down-spin conduction band and that of the up-spin conduction band.

In the next subsection, we investigate the temperature dependence of thermal expansion of ferromagnetic superconductors and its coefficient with the superconducting gap of the up-spin conduction band with the line node.

3.2. *The thermal expansion with the superconducting gap of the up-spin conduction band with a line node and its coefficient*

From Eqs.(6-11) and (16-19), thermal expansion of ferromagnetic superconductors with the superconducting gap of the up-spin band with the line node and its coefficient are obtained. Fig. 3 shows the temperature dependence of the thermal expansion. In the vicinity of the superconducting transition temperature, we have the anomaly of the thermal expansion. In order to clarify the behavior of the thermal expansion at the low temperature limit, Eq.(18) is expanded about the small T/T_{sc}

$$\Delta_0(T) \cong \Delta_0(0)(1 - \exp(-3.4\sqrt{\frac{T_c}{T} - 1})). \tag{22}$$

The thermal expansion is obtained at the low temperature limit from Eq. (7)

$$\omega \propto \Delta_0^2(0)(1 - 2\exp(-3.4\sqrt{\frac{T_c}{T} - 1})). \tag{23}$$

From Eq.(23), the thermal expansion increases with temperature rise exponentially. When the temperature approaches the superconducting temperature, the thermal expansion is

$$\omega \propto -\Delta_0^2(0)(\frac{T_c}{T} - 1). \tag{24}$$

Fig.4 shows the temperature dependnece of the corresponding thermal expansion coefficient. From Fig.4 the thermal expansion coefficient is divergent at the superconducting transition temperature. In order to clarify this behavior, the superconducting order parameter is expanded at the low temperature limit and around the superconducting transition temperature. At the low temperature limit, the thermal expansion coefficient is

$$\alpha \propto 3.4 \cdot \frac{\tilde{T}_{sc}}{\tilde{T}^2} \cdot \frac{1}{2\sqrt{\frac{\tilde{T}_{sc}}{\tilde{T}} - 1}} \exp(-3.4\sqrt{\frac{\tilde{T}_{sc}}{\tilde{T}} - 1}) \tag{25}$$

from Eq.(11). On the other hand, when the temperature goes to the superconducting transition temperature, the thermal expansion coefficient is

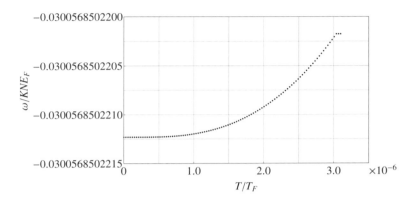

Fig. 3. The reduced temperature dependence of the thermal expansion with the superconducting gap of the up-spin conduction band with a line node when $I/E_F = 0.1$, $\frac{\partial \ln I}{\partial V} = 0.1$, $E_F \frac{\partial}{\partial V}(\frac{1}{g}) = 0.1$, $\frac{\partial x}{\partial V} = 0.1$, $N(0)E_F = 0.1$, and $N(0)E_F \frac{\partial \ln N(0)}{\partial V} = 0.1$.

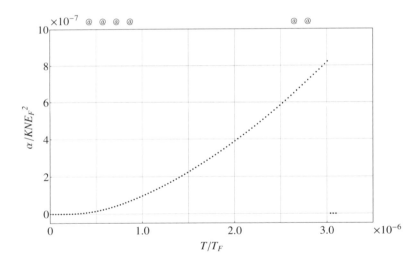

Fig. 4. The reduced temperature dependence of the thermal expansion coefficient with the superconducting gap of the up-spin conduction band with a line node when $I/E_F = 0.1$, $\frac{\partial \ln I}{\partial V} = 0.1$, $E_F \frac{\partial}{\partial V}(\frac{1}{g}) = 0.1$, $\frac{\partial x}{\partial V} = 0.1$, $N(0)E_F = 0.1$, and $N(0)E_F \frac{\partial \ln N(0)}{\partial V} = 0.1$.

$$\alpha \propto 1.7 \cdot \frac{\tilde{T}_{sc}}{\tilde{T}^2} \cdot \frac{1}{2\sqrt{\frac{\tilde{T}_{sc}}{\tilde{T}} - 1}}. \tag{26}$$

From Eq.(26), the thermal expansion coefficient is divergent at the superconducting transition temperature.

4. Summary

Firstly we have investigated the temperature dependence of the thermal expansion coefficient of the ferromagnetic superconductors with the superconducting gaps in A2 phase of ^{3}He. Secondly, we have studied the temperature dependence of the ferromagnetic superconductors with the superconducting gap of the up-spin conduction band with the line node and with no superconducting gap of the down-spin conduction band where this assumption is appropriate in UGe$_2$. In both cases, the thermal expansion coefficients are divergent at the superconducting transition temperatures. The thermal expansion of the superconducting gap with the line node increases with the temperature rise exponentially. The Grüneisen's relation between the temperature dependence of the thermal expansion coefficient and that of the specific heat at low temperatures is satisfied because the expression of the free energy used in the present work is the same as that of the free energy used in the derivation of the specific heat.

Acknowledgments

The authors would like to thank Y. Takahashi, M. Kanno, and O. Stockert for stimulating conversations. One of the author (R. K.) is grateful to K. Grube, H. v. Löhneysen, S. Abe, A de Visser, D. Aoki, V. P. Mineev, E. J. Patino, and M. de Llano for stimulating conversations. This work is supported by the Kinki University Technical College Research Fund.

References

1. S. S. Saxena, P. Agarwal, K. Ahilan, F. M. Grosch, R. K. Haselwimmer, M. J. Steiner, E. Pugh, I. R. Walker, S. R. Julian, P. Monthoux, G. G. Lonzarich, A. Huxley, I. Sheikin, D. Braithwaite, J. Flouque, *Nature* **406**, 587 (2000).
2. D. Aoki, A. Huxley, E. Ressouche, D. Braithwaite, J. Flouque, J. P. Brison, E. Lhotel, C. Paulsen, *Nature* **413**, 613 (2001).
3. F. Levy, I. Sheikin, B. Grenier, C. Marcenat, A. Huxley, *J. Phys.:Condensed Matter* **21**, 164211 (2009), and references therein.
4. N. T. Huy, A. Gasparini, D. E. de Nijs, Y. K. Huang, J. C. P. Klasse, T. Gortenmulder, A. de Visser, A. Hamann, T. Görlach, H. v. Löhneysen, *Phys. Rev. Lett.* B **99**, 067006 (2007), and references therein.
5. A. Harada, S. Kawasaki, H. Mukuda, Y. Kitaoka, Y. Haga, E. Yamamoto, K. M. Itoh, E. E. Haller, H. Harima, *Phys. Rev.* B **75**, 140502(R) (2007).
6. J. Linder, A. Sudbo, *Phys. Rev.* B **76**, 054511((2007), and references therein.
7. J. Linder, I. B. Sperstad, A. H. Nevidomskyy, Mario Cuoco, A. Sudbo, *Phys. Rev.* B **77**, 184511 (2008).
8. T. Moriya, K. Usami, *Solid State Commnun.* **34**, 95 (1980).
9. E. P. Wohlfarth, *Physica* B**91**, 305 (1977).
10. Y. Takahashi, H. Nakano, *J. Phys.:Condensed Matter* **18**, 521 (2006).
11. R. Konno, N. Hatayama, Y. Takahashi, H. Nakano, *J. Phys. Conf. Ser.* **150**, 042100 (2009).

12. J. W. Clark, to be submitted in *the Proc. the 33rd Int. Workshop on Condensed Matter Theories.*
13. N. Hatayama, R. Konno, to be published in *J. Phys. Conf. Ser.*

GENERALIZED SUPERCONDUCTING GAP IN A BOSON-FERMION MODEL

T. A. MAMEDOV

Faculty of Engineering, Baskent University, 06530 Ankara, TURKEY and
Institute of Physics, Academy of Sciences of Azerbaijan, 370143 Baku, Azerbaijan*
tmamedox@baskent.edu.tr

M. DE LLANO

Physics Department, University of Connecticut, Storrs, CT 06269 USA and
Instituto de Investigaciones en Materiales, Universidad Nacional Autónoma de México, Apdo.
Postal 70-360, 04510 México, DF, Mexico*
dellano@servidor.unam.mx

Received 8 October 2009
Revised 10 January 2010

A quantum-statistical binary gas mixture model consisting of positive-energy *resonant* bosonic Cooper electron pairs in chemical and thermal equilibrium with single unpaired electrons yields, via two-time retarded Green functions, an analytic expression for a dimensionless coupling λ- and temperature T-dependent generalized energy gap $E_g(\lambda, T)$ in the single-electron spectrum. The new gap gives a reasonable description of overdoped $Bi_2Sr_2CuO_{6+\delta}$ (Bi2201).

Keywords: Cuprate superconductivity; boson-fermion models; Bose-Einstein condensation; pseudogap.

1. Introduction

Unusual normal state properties of high-temperature superconductors (HTSCs) suggest the opening well above the critical temperature T_c of a so-called *pseudogap*[1] in the electronic spectrum. This has given rise to an intense debate on the origin of this phenomenon and led to a revival of some traditional superconductivity (SC) scenarios. In particular, boson-fermion (BF) models which posit the existence of actual bosonic Cooper pairs (CPs) as possible "preformed" CPs have been reexamined. Here one assumes that a subsystem of single charge carriers attract each other as in the BCS theory[2] via a constant interaction strength $-V \leq 0$ when lying (in 3D) within the spherical shell $E_F - \hbar\omega_D \leq \epsilon \leq E_F + \hbar\omega_D$ about the Fermi energy E_F of the ideal Fermi gas, with $\hbar\omega_D$ the Debye-frequency energy. The system of electrons then evolves into *two* dynamically interacting subsystems: pairable

---

*Permanent address.

but *unpaired* charged fermions, and individual bosonic CP entities made up of two mutually-confined electrons. The simplest grand canonical Hamiltonian

$$\mathcal{H} = \sum_{\mathbf{k},\sigma} \xi_{\mathbf{k}} a^{+}_{\mathbf{k}\sigma} a_{\mathbf{k}\sigma} + \sum_{\mathbf{K}} \mathcal{E}_{\mathbf{K}} b^{+}_{\mathbf{K}} b_{\mathbf{K}} + H_{int} \qquad (1)$$

describing a binary mixture of fermions interacting with bosons as suggested in Refs. 3, 4 has been applied in an effort to understand HTSCs. The first two terms on the rhs of (1) are respectively the Hamiltonians of free (pairable but unpaired) fermions and of composite-boson CPs. Here $a^{+}_{\mathbf{k}\sigma}$ and $a_{\mathbf{k}\sigma}$ are fermion creation and annihilation operators for individual electrons of momenta $\mathbf{k}$ and spin $\sigma = \uparrow$ or $\downarrow$ while $b^{+}_{\mathbf{K}}$ and $b_{\mathbf{K}}$ are postulated[5,6] to be bosonic operators associated with CPs of definite total, or center-of-mass momentum (CMM), wavevector $\mathbf{K} \equiv \mathbf{k}_1 + \mathbf{k}_2$. Fermion $\xi_{\mathbf{k}} \equiv \epsilon_{\mathbf{k}} - \mu$ and boson $\mathcal{E}_{\mathbf{K}}$ energies are measured from μ and 2μ, respectively, where μ is the fermionic chemical potential. If L is the system size in d dimensions, boson formation/disintegration processes are then driven by an interaction Hamiltonian

$$H_{int} \equiv \frac{f}{L^{d/2}} \sum_{\mathbf{q},\mathbf{K}} \left(b^{+}_{\mathbf{K}} a_{\mathbf{q}+\mathbf{K}/2\uparrow} a_{-\mathbf{q}+\mathbf{K}/2\downarrow} + h.c. \right) \qquad (2)$$

where $f \geq 0$ is a phenomenological BF (*two*-fermion/one-boson) one-vertex form-factor coupling parameter which when related with the attractive interelectron (*four*-fermion) two-vertex interaction constant V of the s-wave BCS model as $f = \sqrt{2\hbar\omega_D V}$ gives[5,6] the precise BCS gap equation as well as the BCS condensation energy at $T = 0$. However, exact diagonalization in Refs. 5, 6 via a Bogoliubov-Valatin transformation relied on dropping $K \neq 0$ terms in (2) although not in (1). Here we drop this assumption and go beyond it.

Also associated with nonzero CMM is the so-called Larkin-Ovchinnikov-Fulde-Ferrell[7,8] superconducting phase which we do not investigate here.

2. Gapped Resonant Cooper Pairs

In Ref. 9, it was suggested that if CPs are considered composite bosons shifted in energy at zero $\mathbf{K}$ from E_F by a *positive* gap, it then becomes possible to exhibit *two* characteristic temperatures, T^* and T_c. First, a depairing temperature T^* below which the electronic chemical potential $\mu(\lambda, T)$ first dips below E_F and below which the first CPs begin to appear in the system. Here $\lambda \equiv N(0)V$ is the usual BCS dimensionless coupling parameter with $N(0)$ the electronic density of states (DOS) for each spin at the Fermi surface. The equation $E_F - \mu(\lambda, T^*) = 0$ then yields the T^* below which a transition occurs from the normal state with *no* composite bosons to one *with* such bosons. Second, the BEC temperature T_c at which a singularity signalling macroscopic occupation in a particular state occurs in the total number density of bosons $n_B(\lambda, T_c)$. We stress that identifying these two distinct temperatures T^* and T_c was possible in the present BF model owing *only* to assuming in (1) a gapped boson spectrum such as

$$\mathcal{E}_{\mathbf{K}} \equiv 2E_F + 2\Delta(\lambda) + \varepsilon_{\mathbf{K}} \tag{3}$$

where $\varepsilon_{\mathbf{K}}$ is a nonnegative CP excitation energy that vanishes when $K = 0$. Thus, in (3) $\mathcal{E}_{\mathbf{K}}$ is *higher* than the total energy $2E_F$ of two individual electrons before their mutual confinement into bosonic CPs occurs.

How is a boson energy $\mathcal{E}_{\mathbf{K}}$ given by (3) feasible? To motivate an answer we refer to the original Cooper problem of electrons interacting via an attractive potential $-V \leq 0$ in the presence of $N - 2$ spectator fermions filling the Fermi sea. His *negative*-energy solution (Eq. 5 of Ref. 10)

$$\varepsilon_{\mathbf{0}}^- = -2\hbar\omega_D/(e^{2/\lambda} - 1) \leq 0 \tag{4}$$

is well-known. This solution $\varepsilon_{\mathbf{0}}^-$ implies a binding energy (i.e., *below* $2E_F$) of two electrons associated with the appearance of *bound* pairs. It played a key role in understanding conventional superconductors and lead to the BCS theory.[2] However, along with the $\varepsilon_{\mathbf{0}}^- < 0$ (4) there is another, a *scarcely-known* improper solution with total energy $\mathcal{E}_0 \equiv 2E_F + \varepsilon_{\mathbf{0}}^+$ *above* $2E_F$ and rooted in the elementary fact that $\int x^{-1}dx = \ln|x|$ and not $\ln x$ as commonly assumed. It is

$$\varepsilon_{\mathbf{0}}^+ = +2\hbar\omega_D/(e^{2/\lambda} + 1) \geq 0 \tag{5}$$

which is an apparently new *positive*-energy solution noted in Ref. 11 of the Cooper equation. The two-particle state $\Psi(\mathbf{r}_1, \mathbf{r}_2)$ associated with this solution can be interpreted as a *resonant* state of two mutually-confined electrons in the continuum of single-particle fermions. Probably because of its seemingly unphysical character, the $\varepsilon_{\mathbf{0}}^+$ went unnoticed. In fact, $\Psi(\mathbf{r}_1, \mathbf{r}_2)$ which might correspond to (5) has an energy *higher* than the sum of energies of two free electrons. Such a state must be unstable and cannot be considered as describing actual bound states of two electrons.

However, this statement, obvious *in vacuo*, should be reexamined if the two-fermion attraction acts not in a vacuum but in the presence of the rest of the other $N - 2$ fermions *competing* to occupy the lowest-lying energy states. The $\Psi(\mathbf{r}_1, \mathbf{r}_2)$ then describes the resonant-like two-particle-correlations which can be the *short-lived* excitations already suggested as early as 1954 by Schafroth[12] when referring to them as "resonant states of electron pairs" in contradistinction with the "bound electron pairs" mentioned by Cooper.[10] Resonant modes were found in the more general treatment of Cooper pairing that does *not* neglect hole pairs.[13,14] For a recent review see Ref. 15 where the Bethe-Salpeter treatment for *two-particle* and *two-hole* coupled wavefunctions yielded a two-particle excitation spectrum $\varepsilon_{\mathbf{K}}$ separated, as in (3), from the Fermi sea by a gap $2\Delta(\lambda)$ where[2]

$$\Delta(\lambda) = \hbar\omega_D/\sinh(1/\lambda). \tag{6}$$

Going beyond standard RPA Traven also found[16] pair excitations with energy $\geq 2\Delta$ in the ground state of a 2D attractive Fermi gas, where 2Δ is the threshold for the decay into two fermionic quasiparticles. Similar results were also reported[17] for an attractive δ-function interfermion interaction in the 1D fermion gas. It might be suspected that excitations with a positive gap would *increase* the energy contribution from free bosons in (1) and therefore that their presence would seem unfavorable. This suspicion, however, is not confirmed on comparing the energies of the BF mixture described by (1)-(2) with that of interactionless fermions. Processes driven by (2) of continual formation of pairs and their subsequent disintegration into two unpaired electrons, and vice versa, were crucial in Ref. 11 to discover a new type of lower-energy BF gas mixture state with bosonic excitations *above* the Fermi sea of unpaired electrons. The role of processes of creation and destruction of pairs in reducing the self- energy of electrons was emphasized also in[22]. Here we sketch the derivation of an analytical expression for both the pseudogap and superconducting gap energies originating from the BF binary mixture model (1)-(2) based on bosonic CPs having the gapped spectrum (3).

3. Two-time Green-function Approach

The nature of a pseudogap as to whether or not it is of the same origin as superconductivity, may be investigated by starting from the T-dependent occupation number of unpaired electrons $n_{\mathbf{k}} \equiv \sum_{\sigma} \left\langle a_{\mathbf{k},\sigma}^{+} a_{\mathbf{k},\sigma} \right\rangle$ in a state with momentum $\mathbf{k}$. In (7) below we define *single*-angular brackets $\langle X \rangle_{\mathcal{H}}$ of an operator X as T-dependent thermal averages over $\mathcal{H}$ as given by (1); square brackets $[A, B]_{\eta}$ denote the commutator ($\eta = -1$) or anticommutator ($\eta = +1$) of operators A and B. The numbers $n_{\mathbf{k}}$ can then be found, e.g., from the infinite chain of equations

$$i\frac{d}{dt}\left\langle\left\langle A(t) \mid B(t') \right\rangle\right\rangle = i\delta(t - t')\left\langle [A(t), B(t')]_{\eta} \right\rangle$$
$$+ \left\langle\left\langle [A(t), H] \mid B(t') \right\rangle\right\rangle \tag{7}$$

for two-time retarded Green functions, designated with *double*-angular brackets, as defined in Ref. 18 Eq. (2.1b) for dynamical operators $a_{\mathbf{k}\uparrow}(t)$ and $a_{\mathbf{k}'\uparrow}^{+}(t')$ at times t and t'. In this formalism any operator $X(t)$ is of the form $X(t) = \exp(i\mathcal{H}t)X \exp(-i\mathcal{H}t)$. The Fourier transform $\langle\langle A \mid B \rangle\rangle_{\omega}$ in ω of $\langle\langle A(t) \mid B(t) \rangle\rangle$ satisfies the chain of equations (Ref. 9 Eq. A2)

$$\hbar\omega \left\langle\langle A \mid B \rangle\right\rangle_{\omega} = \left\langle [A, B]_{\eta} \right\rangle_{\mathcal{H}} + \left\langle\langle [A, \mathcal{H}]_{-} \mid B \rangle\right\rangle_{\omega}. \tag{8}$$

Knowing $\left\langle\left\langle a_{\mathbf{k}\alpha} \mid a_{\mathbf{k}'\alpha}^{+} \right\rangle\right\rangle_{\omega}$ one can find thermal-average values $\langle a_{\mathbf{k}\alpha}^{+} a_{\mathbf{k}\alpha} \rangle_{\mathcal{H}}$ from the so-called spectral density $J(\omega)$ (Ref. 18 p. 78). Choosing in (8) *first* $A \equiv a_{\mathbf{k}\uparrow}$ and *then* $A \equiv a_{\mathbf{k}\downarrow}^{+}$ and setting B equal to $a_{\mathbf{k}'\uparrow}^{+}$ one obtains after some algebra (Ref. 9

Eq. A6)

$$(\hbar\omega - \epsilon_{\mathbf{k}} + \mu)\left\langle\left\langle a_{\mathbf{k}\uparrow} \mid a_{\mathbf{k}'\uparrow}^+ \right\rangle\right\rangle_\omega = \delta_{\mathbf{k}\mathbf{k}'}$$

$$+ \frac{f^2}{L^d}\sum_{\mathbf{K},\mathbf{Q}} \frac{\langle b_{\mathbf{K}}\rangle\left\langle b_{\mathbf{Q}}^+\right\rangle}{\hbar\omega + \epsilon_{-\mathbf{k}+\mathbf{K}} - \mu}\left\langle\left\langle a_{\mathbf{k}-\mathbf{K}+\mathbf{Q}\uparrow} \mid a_{\mathbf{k}'\uparrow}^+ \right\rangle\right\rangle_\omega. \tag{9}$$

The Green functions $\left\langle\left\langle BC \mid a_{\mathbf{k}'\uparrow}^+ \right\rangle\right\rangle$ arising from the last term on the rhs of (8) are put into the form

$$\left\langle\left\langle BC \mid a_{\mathbf{k}'\uparrow}^+ \right\rangle\right\rangle_\omega = \langle B\rangle\left\langle\left\langle C \mid a_{\mathbf{k}'\uparrow}^+ \right\rangle\right\rangle_\omega$$

$$+ \left\langle\left\langle (B - \langle B\rangle)C \mid a_{\mathbf{k}'\uparrow}^+ \right\rangle\right\rangle_\omega. \tag{10}$$

Within all so-called first-order theories (Ref. 19 p. 100) higher-order Green functions of the type $\left\langle\left\langle BC \mid a_{\mathbf{k}'\uparrow}^+ \right\rangle\right\rangle_\omega$ are cast as linear combinations of the first-order Green functions $\left\langle\left\langle C \mid a_{\mathbf{k}'\uparrow}^+ \right\rangle\right\rangle_\omega$ and $\left\langle\left\langle B \mid a_{\mathbf{k}'\uparrow}^+ \right\rangle\right\rangle_\omega$. There are no terms containing functions such as $\left\langle\left\langle B \mid a_{\mathbf{k}'\uparrow}^+ \right\rangle\right\rangle_\omega$ in (10) where $B = b_{\mathbf{K}}$ or $b_{\mathbf{K}}^+$, i.e., a pure boson operator. If the chain (8) is applied to obtain an equation for $\left\langle\left\langle B \mid a_{\mathbf{k}'\uparrow}^+ \right\rangle\right\rangle_\omega$ one finds that this contains an extra f^2 with respect to terms like $\left\langle\left\langle C \mid a_{\mathbf{k}'\uparrow}^+ \right\rangle\right\rangle_\omega$ with $C = a_{\mathbf{k}\alpha}$ or $a_{\mathbf{k}\alpha}^+$ being a pure fermion operator. Besides, terms $\left\langle\left\langle B \mid a_{\mathbf{k}'\uparrow}^+ \right\rangle\right\rangle_\omega$ which could appear in (10) contain coefficients like $\langle a_{\mathbf{k}\alpha}\rangle$ and $\langle a_{\mathbf{k}\alpha}^+\rangle$. However, *nonzero* averages $\langle a_{\mathbf{k}\alpha}\rangle$ and $\langle a_{\mathbf{k}\alpha}^+\rangle$ emerging due to (2) turn out to be $O(f^3)$. Therefore, at least for a mixture with *weakly* interacting BF subsystems, i.e., small f, we expect that the contribution from $\langle C\rangle\left\langle\left\langle B \mid a_{\mathbf{k}'\uparrow}^+ \right\rangle\right\rangle_\omega$ will be $O(f^5)$. Formally, the smallness of f justifies absence of terms like $\left\langle\left\langle B \mid a_{\mathbf{k}'\uparrow}^+ \right\rangle\right\rangle_\omega$ in (10). Finally, terms of the type $\left\langle\left\langle (B - \langle B\rangle)C \mid a_{\mathbf{k}'\uparrow}^+ \right\rangle\right\rangle_\omega$ on the rhs of (10) are considered small contributions and hence neglected in obtaining (9). The last approximation is similar to Tyablikov's (Ref. 20 p. 258, Eq. 32.7) RPA which underlies *all* first-order theories (Ref. 19 pp. 108 ff.). That is, if in considering *higher than first-order* Green functions one assumes that the phases of pure boson B and pure fermion C operators vary independently, then by averaging over $\mathcal{H}$ in (10) to obtain $\left\langle\left\langle (B - \langle B\rangle)C \mid a_{\mathbf{k}'\uparrow}^+ \right\rangle\right\rangle_\omega$ this vanishes precisely because of the factor $B - \langle B\rangle$.

4. Generalized Gap

The magnitude of boson wavenumbers are defined by the energy scale of 2e-bosonic excitations which in turn can be expected to be much less than E_F. For estimates one may therefore assume that $\hbar^2 K_{\max}^2/2m_B \ll \hbar^2 k_F^2/2m \equiv E_F$. Here, the maximum value $K_{\max}$ of 2e-CP wavenumbers K and with the 2e-CP mass m_B being approximately twice the electron mass m. Hence, K and Q in (9) can be taken as much less than k_F. As to fermion wavevectors $\mathbf{k}$ and $\mathbf{k}'$ their magnitude is of order k_F. Therefore, one may assume that $k \gg K$ and $k \gg |\mathbf{K} - \mathbf{Q}|$, thanks to which $\epsilon_{-\mathbf{k}+\mathbf{K}} \simeq \epsilon_{-\mathbf{k}}$ and $\left\langle\left\langle a_{\mathbf{k}-\mathbf{K}+\mathbf{Q}\uparrow} \mid a_{\mathbf{k}'\uparrow}^{+}\right\rangle\right\rangle_{\omega} \simeq \left\langle\left\langle a_{\mathbf{k}\uparrow} \mid a_{\mathbf{k}'\uparrow}^{+}\right\rangle\right\rangle_{\omega}$, allowing one to factor all terms *except* $\langle b_{\mathbf{K}}\rangle \left\langle b_{\mathbf{Q}}^{+}\right\rangle$ outside the summation in (9). Furthermore, we assume that $\langle b_{\mathbf{K}}\rangle \left\langle b_{\mathbf{Q}}^{+}\right\rangle \simeq \langle b_{\mathbf{K}}^{+} b_{\mathbf{Q}}\rangle$ for any $\mathbf{K}$ and $\mathbf{Q}$. This approximation was taken from Ref. 19 p. 57 where it is justified in detail for $\mathbf{K} = \mathbf{Q} = 0$ when $L^d \to \infty$. Separating the terms with $\mathbf{K} = \mathbf{Q}$ in the sum over $\mathbf{K}$ and $\mathbf{Q}$ in (9) we write $\langle b_{\mathbf{K}}^{+} b_{\mathbf{K}}\rangle = \langle n_{B\mathbf{K}}\rangle$ with $\langle n_{B\mathbf{K}}\rangle$ the number of bosons with CMM $\mathbf{K}$ and ignore averages $\langle b_{\mathbf{K}}^{+} b_{\mathbf{Q}}\rangle$ with $\mathbf{K} \neq \mathbf{Q}$ which vanish *exactly* provided translational symmetry holds. Relating the phenomenological BF vertex (*"two-electron/one-boson"*) coupling parameter f in (2) with the attractive interelectron (*"four-electron"*) BCS interaction model[2] strength V via $f = \sqrt{2\hbar\omega_D V}$[5,6,21], recalling that $\lambda \equiv N(0)V$ and since $\epsilon_{\mathbf{k}} = \epsilon_{-\mathbf{k}}$, after some algebra one gets

$$\left\langle\left\langle a_{\mathbf{k}\uparrow} \mid a_{\mathbf{k}'\uparrow}^{+}\right\rangle\right\rangle_{\omega} \simeq \delta_{\mathbf{k}\mathbf{k}'} \frac{\hbar\omega + \xi_{\mathbf{k}}}{2E_{\mathbf{k}}} \left[\frac{1}{\hbar\omega - E_{\mathbf{k}}} - \frac{1}{\hbar\omega + E_{\mathbf{k}}}\right] \qquad (11)$$

where

$$E_{\mathbf{k}}(\lambda, T) \equiv \sqrt{\xi_{\mathbf{k}}^2 + E_g^2(\lambda, T)} \qquad (12)$$

with the "generalized gap" $E_g(\lambda, T)$ defined as

$$E_g(\lambda, T) = \sqrt{2\hbar\omega_D V n_B(\lambda, T)} \equiv f\sqrt{n_B(\lambda, T)} \qquad (13)$$

where $n_B(\lambda, T) \equiv L^{-d} \sum_{\mathbf{K}} \langle n_{B\mathbf{K}}\rangle$ is the *net* number density of CPs in the BF mixture at any T. This generalizes the square-root relation first reported in Ref. 3 between the ordinary gap and the BE-condensate T-dependent density $n_0(\lambda, T)$ that refers *only* to $K = 0$ bosons. Namely, $\Delta(\lambda, T) = f\sqrt{n_0(\lambda, T)}$.

Using entirely different methods, an excitation gap similar to (12) was anticipated in Ref. 22 where the properties of ultracold Fermi gases are addressed. The derivation given here provides an explicit T-dependent expression for the $E_{\mathbf{k}}(\lambda, T)$ (12) that includes contributions from condensed pairs (i.e., with CMM $K = 0$) and as well from noncondensed pairs (i.e., with CMM $K \neq 0$) that are formed by the BCS model interaction characterized by SC parameters λ and $\hbar\omega_D$. For all $T \geq T^*$ (13) is exactly zero and begins differing from zero as one lowers the temperature T below T^*. The value of $n_B(\lambda, T)$ was related in Ref. 9 Eq. (20) with the magnitude of $E_F - \mu(\lambda, T)$ driven by bosonization, so that

$$n_B(\lambda, T) \simeq N(0)[E_F - \mu(\lambda, T)]. \qquad (14)$$

Finally, applying (11) to calculate $\left\langle a_{\mathbf{k},\sigma}^{+} a_{\mathbf{k},\sigma}\right\rangle$ yields

$$n_{\mathbf{k}} = \frac{1}{2}\left\{1 - \frac{\xi_{\mathbf{k}}}{E_{\mathbf{k}}}\tanh(E_{\mathbf{k}}/2k_B T)\right\}. \qquad (15)$$

Clearly, (12) formally resembles the gapped single-fermion energies $\sqrt{\xi_{\mathbf{k}}^2 + \Delta^2(\lambda, T)}$ of the BCS theory.[2] However, the BCS gap $\Delta(\lambda, T)$ and the generalized gap $E_g(\lambda, T)$ are of entirely distinct origins. Once formed at and below T_c CPs in the BCS model are allowed to occupy only paired-two-particle states. The breakup of a CP into two unpaired fermions requires an energy $2\Delta(\lambda, T)$ which must be supplied to the system to remove a CP from the subsystem of CPs. Namely, owing to the mutual transitions between CPs the BCS state is energetically more stable compared with that of the assembly of attractively-interacting single fermions, Ref. 2 Eq. (2.42). In sharp contrast however, resonant CPs appear via processes of continual *pair-to-unpaired-2e* transformations responsible for the BF state. Because of this continuous disintegration of resonant pairs into two unpaired electrons, and vice versa, one cannot say definitely which couple of electrons in bosonic or fermionic states were involved in a given time interval. The appearance of a mixture of fermions or bosons continually transforming into each other is driven by the term (2) in (1). Occupation by an unpaired electron of a single-electron state k lying immediately above μ blocks pair-to-unpaired-2e, and vice versa, transitions that would otherwise occur. Such blocking of transitions *raises* the total system energy as shown in Ref. 11 Eq. (31); this drives the renormalization of the single-particle spectrum and gives rise to the generalized gap $E_g(\lambda, T)$.

5. Results and Discussion

In Fig. 1 the behavior of the $E_g(\lambda, T)$ (in units of $E_F \equiv k_B T_F$) is shown for $\lambda = 0.6$ (full curve) and for $\lambda = 0.8$ (dashed curve) as functions of reduced temperature T/T_F and obtained from (13)-(14) where $\Delta(\lambda)$ is given by (6).

Dots on curves refer to $T_c/T_F = 0.036$ for $\lambda = 0.6$ (full curve) and $T_c/T_F = 0.046$ for $\lambda = 0.8$ (dashed curve) as obtained from the result $T_c/T_F = \lambda\sqrt{3}\pi^{-2}[1 - \mu(\lambda, T_c)/E_F]^{1/2}$ reported in Ref. 9 Eq. 35 for 2D. From Ref. 23 p. 596 T_F in 2D superconductors ranges from $510\ K$ to $3150\ K$, and using a median value for cuprates of $\hbar\omega_D/k_B = 400\ K^{24}$ gives a Debye-to-Fermi-temperature ratio $\hbar\omega_D/E_F$ in the range 0.13 to 0.78. The qualitative behavior in T of E_g does not change radically over this range of $\hbar\omega_D/E_F$. The numerical results shown correspond to a typical value of $\hbar\omega_D/E_F = 0.35$ within the stated range. The generalized gap $E_g(\lambda, T)$ is seen not to close at and above T_c as does the coherent-state gap predicted by BCS theory for conventional superconductors. Instead, upon further heating $E_g(\lambda, T)$ evolves into a well-defined "pseudogap" which over an appreciably broad range of temperatures is roughly of the same order of magnitude as the superconducting gap itself. The pseudogap gradually fills up and finally vanishes at and above $T^*/T_F = 0.255$ for $\lambda = 0.6$ and 0.209 for $\lambda = 0.8$. As to the value of the coherent-state gap at temperatures below T_c, it remains virtually unchanged as T dips below T_c. This behavior contrasts with the gap associated with BCS superconductors which is precisely zero for all $T \geq T_c$. The behavior shown in the figure agrees qualitatively with tunneling conductance measurements (Fig. 3 in Ref. 25) of the overdoped

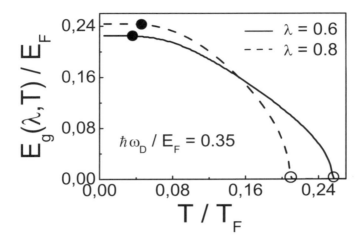

Fig. 1. Dimensionless generalized gap $E_g(\lambda, T)/E_F$ as function of dimensionless temperature T/T_F for two values of λ and for a typical cuprate value of $\hbar\omega_D/E_F = 0.35$. Dots on curves are T_c/T_F from 2D T_c-formula of Ref.[9] Eq. 35. Open circles mark T^* values deduced here and defined as $E_g(\lambda, T^*) \equiv f\sqrt{n_B(\lambda, T^*)} = 0$. Curves are otherwise qualitatively the same for any $\hbar\omega_D/E_F$ in the range 0.13 to 0.78.

$Bi_2Sr_2CuO_{6+\delta}$ (Bi2201) compound.

6. Conclusion

It has been shown that an upward shift of $2\Delta(\lambda)$ between the fermion and boson spectra leads to a BF binary gas mixture state with a generalized gap $E_g(\lambda, T)$ in the spectrum of single-fermions which develops gradually by bosonization, i.e., as the difference $E_F - \mu(\lambda, T)$ starts differing from zero at and below a temperature $T = T^* > T_c$. Both condensate (i.e., with zero center-of-mass momenta) as well as nonzero center-of-mass or noncondensate boson pairs are accounted for. A minimum energy $E_g(\lambda, T)$ is thus required to excite single fermions from the subsystem of unpaired fermions in the BF mixture. The new gap $E_g(\lambda, T)$ is consistent with overall empirical pseudogap trends in overdoped $Bi_2Sr_2CuO_{6+\delta}$ (Bi2201)[25] in particular, and in general with the results of a host of other experiments as surveyed, e.g., in Ref. 1.

Acknowledgments

UNAM-DGAPA-PAPIIT (Mexico) is thanked for partial support through grant IN106908.

References

1. T. Timusk, Sol. St. Comm. **127**, 337 (2003); T. Timusk and B. Statt, Rep. Prog. Phys. **62**, 61-122 (1999).
2. J. Bardeen, L.N. Cooper, and J.R. Schrieffer, Phys. Rev. **108**, 1175 (1957).
3. J. Ranninger, R. Micnas, and S. Robaszkiewicz, Ann. Phys. Fr. **13**, 455 (1988).
4. R. Friedberg, T.D. Lee, and H.-C. Ren, Phys. Rev. B **45**, 10732 (1992) and refs. therein.
5. V.V. Tolmachev, Phys. Lett. A **266**, 400 (2000).
6. M. de Llano and V.V. Tolmachev, Physica A **317**, 546 (2003).
7. A.J. Larkin and Y.N. Ovchinnikov, Zh. Exsp. Teor. Fiz. **47,** 1136 (1964).
8. P. Fulde and R.A. Ferrell, Phys. Rev. **135,** A550 (1964).
9. T.A. Mamedov and M. de Llano, Phys. Rev. B **75**, 104506 (1-12) (2007).
10. L.N. Cooper, Phys. Rev. **104**, 1189 (1956).
11. T.A. Mamedov and M. de Llano, Int. J. Mod. Phys. B **21** (2007) 2335.
12. M.R. Schafroth, Phys. Rev. **96**, 1442 (1954).
13. M. Fortes, M.A. Solís, M. de Llano, and V.V. Tolmachev, Physica C **364**, 95 (2001).
14. V.C. Aguilera-Navarro, M. Fortes, and M. de Llano, Sol. St. Comm. **129**, 577 (2004).
15. M. de Llano and J.F. Annett, Int. J. Mod. Phys. B **21**, 3657 (2007).
16. S.V. Traven, Phys. Rev. Lett. **73**, 3451-3454 (1994); Phys. Rev. B **51**, 3242 (1995).
17. T. Alm and P. Schuck, Phys. Rev. B **54**, 2471 (1996).
18. D.N. Zubarev, Sov. Phys. Uspekhi **3**, 320 (1960).
19. *Statistical physics and Quantum field theory* (edited by N.N. Bogoluibov, in Russian) (Nauka, Moscow 1972).
20. S.V. Tyablikov, *Methods in the Quantum Theory of Magnetism* (Plenum, NY, 1967).
21. S.K. Adhikari, M. de Llano, F.J. Sevilla, M.A. Solís, and J.J. Valencia, Physica C **453**, 37 (2007).
22. Q.J. Chen, C.C. Chien, Y. He, K. Levin, J. Supercond. Nov. Magn. **20**, 515 (2007).
23. C.P. Poole *et al.*, *Superconductivity* (Academic Press, New York, 1995), p. 596.
24. H. Ledbetter, Physica C **235-240**, 1325 (1994).
25. M. Kugler, Ø. Fischer, Ch. Renner, S. Ono, and Y. Ando, Phys. Rev. Lett. **86**, 4911 (2001).

INFLUENCE OF DOMAIN WALLS IN THE SUPERCONDUCTOR/FERROMAGNET PROXIMITY EFFECT

EDGAR J. PATIÑO

Departamento de Física, Grupo de Física de la Materia Condensada,
Universidad de los Andes, Bogotá, Colombia*
Centro de Investigación en Física de la Materia Condensada,
Corporación de Física Fundamental y Aplicada, Quito, Ecuador
epatino@uniandes.edu.co

Received 5 December 2009

This paper constitutes a short review of the main experimental investigations into the subject as they were known on November 2009. The paper is also intended to be useful as an introduction to the physics of superconductivity under the presence of domain walls. First, the author briefly discusses general aspects of Superconductor/Ferromagnet (S/F) proximity effects as these can be important depending of the ferromagnet thickness. Then the so called domain wall superconductivity (DWS) and stray field effects (SFE) are clarified. The experimental evidence on each of these effects is qualitatively discussed. The difficulties on DWS experiments are outlined. The article concludes with a discussion of some outstanding issues and desirable future work.

Keywords: Superconductivity; Ferromagnet proximity effect; Domain wall superconductivity.

1. Introduction

This paper is meant to review the most relevant experimental investigations on how domain walls of a ferromagnet influence the proximity effect of a superconductor/ferromagnet (S/F) heterostructure. Throughout this manuscript two distinct but equally important effects due to the presence of domain walls shall be outlined; stray field effects (SFE) and domain wall superconductivity (DWS). The difficulties to clearly distinguish one from the other will be explained. The importance of this review work lies on the fact that most experimental investigations on (S/F) heterostructures reported in the literature have dealt with experimental situations where the ferromagnet is in single domain state i.e. where no domain walls are present in the ferromagnet. Even though obtaining domain walls is a rather easy experimental exercise which only requires applying field in the opposite direction of saturation fields, most investigations on single domain state do not report the multi-domain state situation. This is probably due to the difficulties that involve

---

*Departamento de Física, Grupo de Física de la Materia Condensada, Universidad de los Andes, Carrera 1a. No. 18A-10/70 A.A. 4976, Bogotá, Colombia.

data interpretation for this situation. For the case of DWS it should be mentioned that given the experimental complexities this field of research is far from being understood and as it should become clear after reading this document further research is required in order to clarify this subject. In order to comprehend the effects of domain walls on superconductivity one should first clearly understand the usual proximity effect where the ferromagnet is in single domain state. Thus in the following section an outline of this effect will be presented.

2. The Basic Proximity Effect in S/F Heterostructures: Single Domain State

When Cooper pairs enter the ferromagnet the exchange field will basically try to align the spins of the electrons in the Cooper pairs (paramagnetic effect), leading to pair breaking. The magnitude of the exchange field h_{ex} multiplied the Bohr magneton [a] μ_B must be equal to the energy needed to rotate the electron spin i.e. $\mu_B h_{ex} \approx \Delta$, the pair formation energy. Thus in contrast with an applied field H which acts on the orbital states of the paired electrons, the exchange field h_{ex} acts on electron spins and thus the destruction of superconductivity due to this field is called the paramagnetic effect. This is of course true inside a ferromagnet since the exchange field may be as high as 10^3 tesla! Therefore the Cooper pairs are easily broken inside a ferromagnet. In a classic work Hauser[1] showed ferromagnetic thin films in contact with thin superconducting samples suppressed the transition temperature to a much greater extend than normal metals, as expected in the limit when $b \to 0$. At a critical thickness of the superconductor the proximity effect is strong enough to lower T_C to zero.

When coupling a superconductor and a ferromagnet together the transparency of the interface plays an important role in the proximity effect. In the case of perfect transparency a few monolayers of ferromagnetic material is enough to suppress T_C of the superconducting layer, this is because electrons of both layers intermix in the structure.[2] This can be easily seen by considering the case where the thickness of the S and F layers is smaller or of the order of the coherence lengths. In this case we can average the exchange field along the thickness of the S/F bilayer, which assumes a perfect transparent interface:

$$h_{ex}^{eff} = h_{ex} \frac{d_F}{d_S + d_F} \tag{1}$$

we can compare this equation with the energy gap $\Delta = 1.75 k_B T_C$ thus obtaining;

$$1.75 k_B T_C = \mu_B h_{ex} \frac{d_F}{d_S + d_F} \tag{2}$$

Using a $T_c \approx 10$ K and $\mu_B h_{ex} \sim 0.1$ eV, d_F can be calculated $d_F \sim 25 \times 10^{-3} d_s$. This means that only one monolayer of Fe is enough to suppress superconductivity !

---

[a]The Bohr magneton is closely equal to the spin magnetic moment of a free electron, in S.I. units it has a value of $\approx 9.27 \times 10^{-24}$ joules tesla^{-1}.

In reality this is not the case due to a finite value of transparency and acquisition of momentum of the Cooper pairs inside the ferromagnet which leads to wave function oscillations. These are energetically more favorable than monotonic decay.[3]

A non-monotonic behavior of T_C was first observed by Wong *et al.*,[4] as a function of F layer thickness. After this, a spectacular prediction of oscillations of T_C as a function of the F layer thickness in a S/F multilayered structure was proposed in the theoretical work by Radovic *et al.*[5] in 1991. In this work the proximity effect in S/F lattice structures is investigated by the method of Usadel equations. He assumes position and temperature independent exchange field in the F layers, and both metals in the dirty limit. The condensate of pairs in the ferromagnet exhibit an oscillatory behavior, damped by an exponential decay usual in S/N proximity systems. The characteristic penetration length of Cooper pairs in F layers is temperature independent,

$$\xi_F = \sqrt{\frac{4\hbar D_F}{|I|}} \tag{3}$$

where I is the exchange energy taken as a constant value. Its magnitude is much smaller than the corresponding coherence length for the normal metal (for the case of strong ferromagnets $|I| \gg k_B T_{CS}$, where T_{CS} is the superconductor's bulk transition temperature);

$$\xi_N = \sqrt{\frac{\hbar D_N}{2\pi k_B T}} \tag{4}$$

Radovic explains the non monotonic behavior of T_C observed in Ref. 4 by Wong and coworkers in terms of the so called π coupling effect, which takes place when the order parameter has a phase difference equal to π between two superconductors separated by a ferromagnet as shown in figure 3. In his calculations he showed that T_C must be higher for a system with π coupling than without it.

2.1. *The Fulde-Ferrel state*

When a superconductor is placed next to a ferromagnet the exchange field provokes oscillations of the superconducting order parameter.[3,6] Here the Larkin-Ovchinnikov-Fulde-Ferrel (LOFF)-like state has been realized[7,8] (this state is also refereed to in the literature as the FFLO state).

The exchange field of the ferromagnet provokes a first order phase transition to a new finite momentum superconducting state. The pairing is still singlet but now it is between two electrons of different energies (or fermi surfaces), and exists over a finite range of the strength of the exchange field.

As the exchange field increases, the superconducting energy gap of the depaired state, decreases and passes continuously to the normal state. In contrast to the completely paired BCS state, the LOFF state exhibits spin magnetization.

A spherically symmetric distribution of electrons is less favorable than a distribution extended along one direction perpendicular to the exchange field so Cooper pairs with shifted center of mass momenta appear and an inhomogeneous distribution function develops giving birth to the oscillations.

2.2. *Oscillations of the superconducting order parameter*

When Cooper pairs enter the ferromagnet there are two factors which contribute

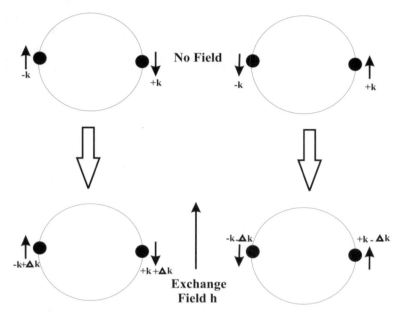

Fig. 1.   Cooper pair before and after entering a ferromagnet, the antisymmetry property requires both configurations (left and right) possible.

to their decay. First since in the ferromagnet's region the pair is not an eigenstate it becomes an evanescent state that decays exponentially over the length scale of ξ_N, the normal metal coherence length. Second in the exchange field of the ferromagnet the electrons with spin parallel to the field lower their potential energy, while electrons with spin antiparallel increase their energy by the same amount.

In order for the total electron energy to remain constant, the momenta changes by an amount + or - E_{ex}/v_f, where E_{ex} is the exchange coupling energy in the ferromagnet, and v_f is the fermi velocity. The antisymmetry property of fermions requires that the spin up and spin down electrons are interchanged in momentum space as depicted in figure 1 left and right side. The superposition of these two states into a singlet produces a wavefunction of the form $Cos(Q(x_1+x_2))\Phi(x_1-x_2)$ where Φ is the original wavefunction of the Cooper pair in the superconductor (with electrons at x_1 and x_2) and Q is the center of mass momentum $2E_{ex}/v_f$. Q changes the phase of $\varphi = Qx_F$ of the superconducting wave function that increases linearly

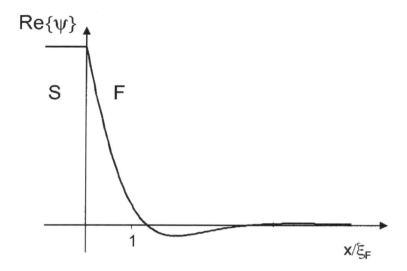

Fig. 2. Exponentially damped Oscillations of the real part of the superconducting order parameter into the ferromagnet induced by proximity effect.[10]

with distance x_F into the ferromagnet from the interface. In the case of a S/F/S trilayer, depending of the ferromagnet thickness, the oscillations may interfere destructively (0 phase coupling) or constructively (π phase coupling Fig. 3). This leads to oscillations in the transition temperature in S/F hybrids.[b] Exponentially damped oscillations of the real part of the order parameter have been confirmed experimentally by Kontos *et al.*[10] in S/F bilayer structures, induced into the ferromagnet by the proximity effect (see figure 2).

2.3. *Oscillations of T_C*

The oscillations of the order parameter in S/F multilayers can bring about a π phase difference between the superconducting layers separated by a ferromagnet as shown in Fig. 3.

Since then, various experiments have been made in the subject, including multilayer structures prepared using molecular beam epitaxy (MBE),[11] and sputtering techniques.[12] This last work by Jiang *et al.* was claimed to be the first experimental evidence of π coupling). These results are plotted in Fig. 4 and show the superconducting transition temperature T_C versus d_{Gd} in Nb/Gd multilayers, for thickness of Nb of 600 Å and 500 Å. The results show oscillations of the transition temperature

---

[b]In the case of induced triplet states in the ferromagnet layer, predicted by Volkov *et al.*,[9] no oscillations of the order parameter neither oscillations of T_C as a function of the ferromagnet's thickness are expected.

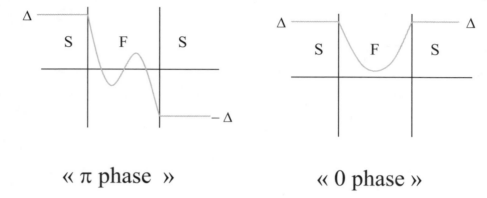

Fig. 3.   Phase coupling representation by A. Buzdin.

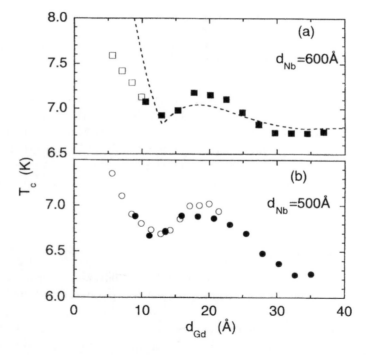

Fig. 4.   Superconducting transition temperature T_C versus d_{Gd} in Nb/Gd multilayers with (a) $d_{Nb} \approx 60$ nm and (b) 50 nm. Dashed line in (a) is a fit with theory of Radovic et al.[5]

depending on the ferromagnet's thickness. The dashed line is a fit with the theory of Radovic et al.[5] Later on Muhge et al.[13,14] carried out experiments in three layered structures of Fe/Nb/Fe, and bilayers of Nb/Fe[15] where step like behavior of T_C was found. In this structure since there was only one superconducting layer π coupling was ruled out. Explanations like the existence of a dead magnetic intermediate layer at the interface which may induce repulsion of electrons and pair breaking were given. The thickness of this layer was suggested as being in correlation with the preparation method used. Furthermore Patiño et al.[3] found that another important factor for oscillations of the order parameter to occur, is the roughness of the interface. This would have direct effect on T_C measurements. Later work by Aarts et al.[16] studied multilayer systems without interdiffusion at the interface. They found the dependence of T_C with F-layer thickness and magnetic moment of the F-layer atoms. This work points out the importance of low transparency in the oscillatory effect of the temperature and found that the penetration depth of Cooper pairs is inversely proportional to the effective magnetic moment of the F-layer atoms.

3. Domain Wall Superconductivity (DWS) vs Stray Fields Effects (SFE)

The previously described investigations of S/F heterostructures have been reported for the case when the ferromagnet is saturated i.e. it is uniformly magnetized in a single domain state. What happens when the ferromagnet is in a multiple domain state? This state is achieved in a ferromagnet for example by applying external magnetic field in the opposite direction to magnetization. The ferromagnet gradually starts to break into domains and domain walls reaching its maximum number at coercive fields.

Firstly lets clearly define what we shall understand by domain wall superconductivity (DWS). This idea was originally proposed in studies of ferromagnetic superconductors. One of the first theoretical investigations in DWS was carried out by Buzdin[17] and refers to the situation where cooper pairs are localized at domain walls. Here superconductivity is favored due to a reduction of the average exchange field at a domain wall. According to this definition and based on the historical origins of this concept it is important to point out that this concept has sometimes been misused in the literature[18-21] when referring to situations where the observed effects on superconductivity are due to the influence of stray magnetic fields from domain walls[19-21] or magnetic dots[18] that can be described classically and not as the result of a reduction of a exchange field which is a quantum effect.

One can argue that there can be experimental configurations in which both exchange field effects and stray fields are simultaneously present. Throughout this document the later situation shall be called "stray field effects" (SFE) in order to clearly distinguish from DWS. As it will be discussed in the next section in certain experimental configurations this distinction is not straightforward, since both effects can be present simultaneously, making it difficult to determine which effect is dominant over the other.

With this in mind lets first discuss DWS in more detail. Generally the Cooper pair size is much smaller than the domain size D ($\xi_S << D$) therefore the Cooper pairs within the domain wall experience a uniform exchange field h. On the other hand at the domain walls the magnetic moment and exchange field rotates as shown in Fig. 5. This rotation of the exchange field leads to a reduction in the average value of exchange field sampled by the Cooper pair. Its average value has been estimated by Buzdin[22] in the following way. Lets consider the case of a domain wall of width w. When a Cooper pair, of size ξ_S, enters the domain wall (Fig. 5) each electron in the pair experiences different exchange field directions as shown by the different directions of the arrows. The angle α of rotation can be estimated of the order $\alpha = \frac{\xi_S}{w}\pi$ for $w > \xi_S$. The average value of exchange field h_{av} is estimated as half the sum of the two vectors of exchange field experienced by the pair

$$h_{av} = h\sin(\pi/2 - \alpha/2) \tag{5}$$

Using Taylor expansion h_{av} is found to be slightly smaller than the exchange field h far away from the domain wall

$$\frac{h - h_{av}}{h} \approx \pi/8 \left(\frac{\xi_S}{w}\right)^2 \tag{6}$$

The oscillatory behavior of Tc can be described using Radovics expression:[22]

$$Re\Psi\left(\frac{1}{2} + \frac{\rho}{t_C}\right) - \Psi\left(\frac{1}{2}\right) + \ln(t_C) = 0 \tag{7}$$

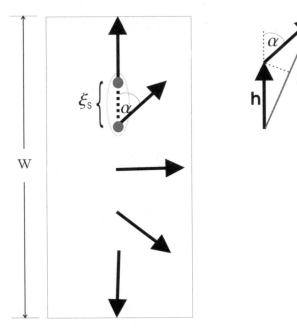

Fig. 5.   Average exchange field experienced by the Cooper pair.

where Ψ is the Digamma function, $t_C = T_C/T_{CS}$ and T_{CS} is the transition temperature in the absence of proximity effect. The previous equation summarizes the description of S/F systems using Radovic's theory. This leads to oscillations of T_C with ferromagnet thickness and a decrease of T_C with an increase of exchange field.

The solution of this equation proposed by Buzdin, in the case of finite transparency, for the critical temperature of a S/F multilayer is given by:

$$Re\Psi\left(\frac{1}{2} + \frac{1}{2\pi T_C \tau_S}\right) - \Psi\left(\frac{1}{2}\right) + \ln(t_C) = 0 \qquad (8)$$

where τ^{-1} is the new pair breaking parameter which now includes finite boundary resistance and for the "0" phase case reads:

$$\tau_{S,0}^{-1}(w > 0) = \frac{D_S}{2d_S} \frac{\sigma_M}{\sigma_S} \frac{\imath+1}{\xi_M} \frac{\tanh(\frac{\imath+1}{\xi_M}d_M)}{1 + \frac{\imath+1}{\xi_M}\xi_N \gamma_B \tanh(\frac{\imath+1}{\xi_M}d_M)} \qquad (9)$$

here the influence of the exchange field is expressed through ξ_M.

The relative decrease of the pair breaking parameter $(\Delta\tau_2^{-1}/\tau_2^{-1})$ can be estimated from Eq. 9. Assuming infinite interface transparency ($\gamma_B = 0$) and the condition $\xi_F >> d_F$ for a thin ferromagnet then $\Delta\tau_2^{-1}/\tau_2^{-1} \approx (\xi_S/w)^2$. Finally from Eq. 8 the local increase in transition temperature is given by:

$$\frac{T_C - T_{CS}}{T_{CS}} \sim \left(\frac{\xi_S}{w}\right)^2 \qquad (10)$$

where it its shown that the reduction of T_C only depends on the relative dimensions of Cooper pair size ξ_S and domain wall width w.

3.1. *Previous experimental investigations that claimed DWS*

One of the first reported experimental studies of the effect of domain walls on superconductivity was carried out by Kinsey *et al.*[23]

These were done on Nb/Co bilayers, with 15-65 nm of Nb on 54±9 nm of Co, by critical current measurements of patterned films using photolithography. The geometry of the mask used for this process included the voltage contacts perpendicular to the current track, that leaded to stray fields on the current track. In these earlier studies an increase in the critical current measurements of the patterned wire was found by applying a magnetic field of the order of the coercive field of the ferromagnet (See figure 6). This effect was attributed to a reduction of the average exchange field sampled by the Cooper pairs within the domain wall giving rise to an enhancement of superconductivity so called domain wall superconductivity (DWS).[17] However, in these experiments the effect of stray fields, from the mask geometry used, in the measurements could not be completely eliminated. This result was explained by saying that the Co layer forms antiparallel domains (at coercive fields) such that the local exchange field of the ferromagnet (acting on the superconductor) is reduced, hence the peaks observed. In another sample the effect is observed very close to T_c (i.e. 5.2 K) and by switching the field a re-entrant behavior between normal and superconducting state, is observed. However this experiment lacks of a detailed analysis of the transition temperature with and without magnetic field.

In addition he attempts to quantify the stray fields in the structure and the contribution to the effect observed, however the methods used for this aim are not clearly explained. One of the main referee's comments for these observations was that this

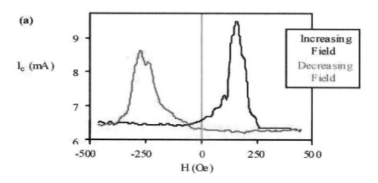

Fig. 6.   Critical current measurements versus Applied Field in Nb/Co bilayers with Nb ≈ 63 nm and Co ≈ 54 nm. by Kinsey *et al.*[23]

effect could be caused by stray fields associated to the geometry of the mask used in these experiments. Magnetic poles are induced in the field direction and "stray fields" from these poles depending on the mask geometry used to patterned the tracks, would affect the superconductor to a greater or lesser extend. If the field generated by these poles is high enough to reach H_c, it could break superconductivity thus observing peaks which would be just an artifact of the experiment and not a real physical effect of the proximity to the ferromagnet. However these fields can be eliminated using the right geometry in the design of the mask, this will be discussed in the coming section.

Even though the argument that such Nb/Co systems under varying exchange fields, could lead to effects in the critical current and transition temperature is probably correct[24] the relevance of stray fields in Kinsey's observations remains unclear and controversial. To the knowledge of the author qualitatively the same results, in Nb/Co bilayers, of this experiment have not been reproduced by other researchers.

A similar effect has been found in Nb/CuNi trilayers by Rusanov *et al.*[25] but again in this case the stray fields from mask geometry were present. Recently experiments on Nb/Py (Permalloy $Ni_{80}Fe_{20}$) by the same author assured as evidence of DWS enhancement or inplane spin switch.[26] In these experiments Rusanov found enhancement of superconductivity at coercive fields of the ferromagnet, by measuring magnetoresistance at the transition region between normal and superconducting state. Dips in magneto resistance measurements at coercive fields (Fig. 7) were attributed to be evidence of this effect. However this effect was only found for large samples (0.5 mm x 4 mm) and absent in small samples (1.5 μm x 20 μm). In addition to these dips in magneto resistance, also peaks were found as seen in Fig. 7 that were attributed to the presence of stray fields but no clear explanation of their effect on the results was given.

On perovskite structures the enhancement of the Cooper pair density has been reported in YBCO/LCMO bilayers at the coercive field of the LCMO layer relative to the saturation value of its magnetization.[27,c] Similar switching effects have been theoretically predicted by Tagirov,[24] by controlling independently the magnetization of F/S/F trilayer structures, and thereby controlling the transition temperature of the S layer. This has also been studied by Baladie and Buzdin.[28]

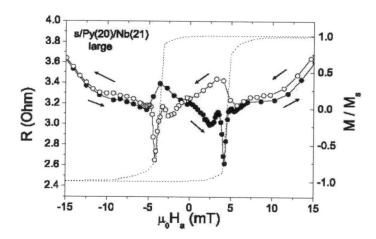

Fig. 7. Magnetoresistance (R) versus applied field H$_a$ of the large sample Py(20)/Nb(21) shows dips at coercive fields and two peaks between attributed to stray fields by Rusanov *et al.*[26] The arrows indicate field sweep direction. Right-hand scale: Magnetization M (dotted lines) normalized on the saturation magnetization M$_S$ measured at 8 K. In both cases the field is parallel to the easy axis of the magnetization.

3.2. *Experimental studies on stray field effect (SFE)*

Besides the previously described investigations were DWS is claimed to be responsible for this effects there is a second class of experiments where stray fields from the ferromagnet magnetic poles or from domain walls are clearly determined to be responsible for this effects. In this section the most relevant experimental investigations on the subject will be discussed.

---

^{c}Note that in this case the Cooper pair size is much smaller.

Ryazanov et al.[29] have carried out differential resistance measurements across an S/F bilayer. The detailed experiments consists of a S-SF-S bridge where the ferromagnet is located just at the middle of the superconducting bridge. Under this geometry I-V characteristics where measured of 10 nm Nb with small square of a 18 nm CuNi placed on top. As the magnetic field was changed magnetoresistive peaks in differential resistance measurements at coercive field values were observed as shown in Fig. 8. This effect was attributed to spontaneous vortex state formation and flow,in the superconductor, due to stray magnetic fields from the domain walls of the ferromagnet.

A theoretical work has been published by Burmistrov et al.[30] investigating

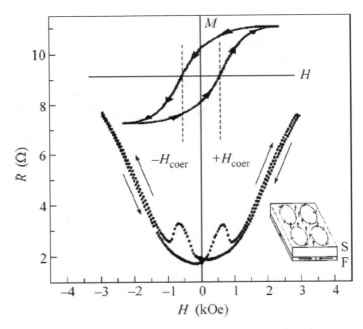

Fig. 8.   Differential resistance measurements of a CuNi(18)/Nb(9) bilayer in a S-SF-S bridge vs longitudinal magnetic field at T = 2.66 K and current of 5 μA. Arrows indicate the field-scan direction. The magnetization curve for CuNi is shown at the top. These experiment shows peaks at coercive fields of the ferromagnet by Ryazanov et al.[29]

the effect of Bloch domain walls on the current distribution in the superconductor. They also determine a lower critical value of domain wall magnetization above which vortex formation is favorable. Based on this idea experiments, performed by Steiner et al.[20] on Fe/Nb and Co/Nb bilayers, found a reduction of $T_C \approx 0.5\%$ as a function of magnetic state of the ferromagnet. Additionally Bell et al.[31] have found a double peak structure in magnetoresistance measurements of MoGe/GdNi bilayered structures as a function of the magnetic state of the ferromagnet attributed to vortex formation and flux flow.

Recently Patiño *et al.* have published a work on Nb/Co and Nb/Py (Permalloy $Ni_{80}Fe_{20}$) bilayer structures.[32] The original aim of this work was to try to reproduce Kinseys work,[23] however[32] also extends to investigate the influence of Néel and Bloch domain walls in the superconducting pairing and vortex generation in S/F hybrids. This investigation differs from previous ones in that it considers a wide range of ferromagnet's thickness and eliminates stray fields from the mask geometry. Co and Py were used as ferromagnets since the domain wall structures are fairly well understood at room temperature. In this investigation the authors found that the superconductor critical current can be controlled by the domain state of the neighboring ferromagnet. The superconductor was a thin wire of thickness $d_s \approx 2\xi_S$. Nb/Co and Nb/Py (Permalloy $Ni_{80}Fe_{20}$) bilayer structures were grown with a significant magnetic anisotropy. As seen in Fig. 9, critical current measurements of Nb/Co structures with ferromagnet thickness $d_F > 30$ nm show sudden drops in two very defined steps when the measurements are made along the hard axes direction (i.e. current track parallel to hard anisotropy axes direction). These drops disappear when they are made along the easy axis direction or when the ferromagnet thickness is below 30 nm. The drops are accompanied by vortex flux flow. In addition magnetorestistance measurements close to T_C show a sharp increase near saturation fields of the ferromagnet. Similar results are reproduced in Nb/Py bilayer structure with the ferromagnet thickness $d_F \sim 50$ nm along the easy anisotropy axes. These results are explained as being due to spontaneous vortex formation and flow induced by Bloch domain walls of the ferromagnet underneath. Patiño argues that these Bloch domain walls produce a 2D vortex-antivortex lattice structure. Due to the absence of any anomalous observations between Cobalt thicknesses 2 nm$< d_{Co} < 30$ nm[32] the authors conclude that the previous observations[23] were probably due to either stray fields from the device geometry or dipole stray fields. Previous magnetization and critical current measurements on rf-sputtered Co/Nb/Co trilayered structures have been performed by Kobayashi *et al.*[33,34] The layer thickness of the Co (d_{Co}) layer ranged between 20 and 50 nm while the Nb thickness (d_{Nb}) was kept within 50 and 200 nm. Magnetization measurements were performed on samples with lateral dimensions equal to 2×2 mm. The central peak position around the zero field of the superconducting magnetization is found to be shifted towards positive fields, in contrast to a negative peak position of a single superconducting film.[33]

This shift in the peak position is explained as being due to the anti parallel in-plane stray fields from the poles of the ferromagnet. The effective stray field strength calculated from this experiment is of the order of 200 Oe just below the transition T_C. This value coincides with the one obtained from critical current measurements by the same group on patterned stripes of 20×140 μm between voltage paths,[34] in which a pronounced dip around the zero field and a double peak structure at larger fields are observed due to inplane fields from magnetic dipoles

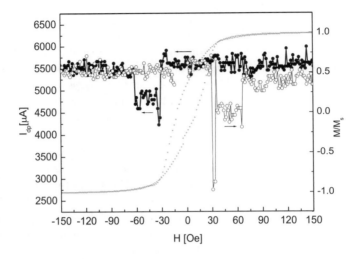

Fig. 9.   Critical current vs. in plane magnetic field of a Nb(25)/Co(42) sample taken at $T =4.2$ K with I‖H . Arrows show the field-scan direction. Superimposed is the corresponding hysteresis loop by Patiño *et al.*[32]

(See Fig. 10). The reported dipole field is much greater than a few Oe estimated from calculations and the authors argued that the large interfacial roughness might modify the distribution of the stray fields locally and increase the effective stray field. In these experiments no anomaly was found around coercive fields where the effective stray field is reversed.

Finally a different class of properties the so called field-induced superconductivity (FIS) have been realized in a superconducting Pb film covered with a magnetic dots array[18] and in 50 nm Niobium films deposited on the single crystal ferrimagnet $BaFe_{12}O_{19}$.[19],d In the first experiments the stray fields from the magnetic dots suppressed superconductivity when they are magnetized ($m \neq 0$) driving the superconductor to normal state and a reentrant behavior to superconducting state was observed when the applied field cancelled the effective field from the magnetic dots (Fig. 11).

A similar effect due to stray fields takes place in the case of the experiment of Nb film deposited on a ferrimagnet crystal (separated with an insulator)[19] with a magnetization along the perpendicular direction. Here resistance measurements near the transition temperature show an small variation due to changes of the magnetic

---

dThe author refers to his observations as evidence of DWS however these are really the result of stray fields, further analysis is given in Ref. 35.

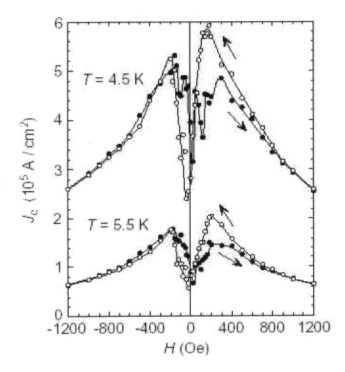

Fig. 10.   Critical Current Measurements vs magnetic field on Co(50)/Nb(50-150)/Co(50) trilayers taken at temperatures of 4.5 and 5.5 K. The arrows indicate the field direction. This measurements show two sharp peaks at large fields beyond ferromagnet's saturation by Kobayashi *et al.*[34]

state of the ferrimagnet with the sweeping field direction. These are the result of two effects the superconductivity nucleate on top of the domain walls, where the stray fields are the lowest, or nucleation of superconductivity on domain sites when the applied field equalled the stray fields from these out of plane domains. In 2008 a similar work was published on [Co/Pt]/Nb/[Co/Pt] hybrids[21] with perpendicular magnetic anisotropy were superconductivity persists at domain wall region.

4. Conclusion and Future Work

Since the work of Kinsey[23] on critical current measurements in Nb/Co bilayer structures there has been a number of reports on the influence of domain walls in the superconductor/ferromagnet proximity effect. The absence of any anomalous observations between Cobalt thicknesses 2 nm$< d_{Co} < 30$ nm found in experiments performed by Patiño *et al.*[32] permitted to conclude that previous observations[23] were probably due to either stray fields from the device geometry or dipole stray

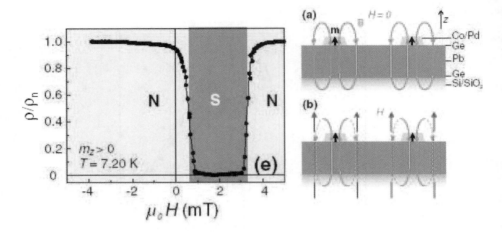

Fig. 11.   Field-induced superconductivity (FIS) in a Pb film with an array of magnetic dots.[18] Blue and yellow areas correspond to the superconducting (S) and the normal (N) states, respectively. $\rho(H)$ is shown for the magnetic states $m_z > 0$, measured at constant temperature as indicated.

fields. DWS was not observed for the case of Nb/Co bilayers. Regarding the observations of dips in magnetoresistance measurements made on Nb/Py bilayers,[26] it should be remarked these were accompanied by peaks. These imply the presence of stray fields and thus should be considered in the interpretation of the results. Recently the same author reported additional results on Nb/Py bilayers,[37] where two different thicknesses of Py were used; 20 and 60 nm. Here small dips and large peaks in magnetoresistance were respectively found. The experiment description indicates that the device geometry includes magnetic voltage contacts; thus is likely that these dips are due to stray fields. However the large peaks for a Py thickness of 60 nm are explained by the authors in Ref. 37 as the result of Bloch domain wall stray fields. This result is qualitatively similar to the one obtained in magnetoresistance measurements in Ref. 32 and previously described on Ref. 36 by the present author. The main difference in this results are the two step structure described in Refs. 32 and 36. To the knowledge of the author qualitatively the same results obtained in Ref. 26 have not been reproduced elsewhere.

Although the previous suggested mechanism in the research investigations[23,26,37] the enhancement of superconductivity at the domain walls i.e. DWS is probably correct, an unambiguous experimental evidence of DWS has not yet been successfully obtained.

In order to find it the device geometry is very important. In particular the experiment should be able to eliminate stray fields from mask geometry and whenever possible from magnetic poles of the ferromagnet as suggested by the work of Kobayashi.[33,34]

In relation to experiments where SFE are identified as responsible for the experimental observations, among these it is easy to identify three kinds; stray fields produced by magnetic dipoles, stray fields produced by Bloch Domain Walls and stray fields produced by magnetic dots so called field-induced superconductivity (FIS). Given the experimental evidence up to this date these experiments are relatively easy to reproduce.

References

1. J.J. Hauser, H.C. Theuerer and N. R. Werthamer, Phys. Rev. 142, 118 (1966)
2. I.A. Garifullin, J. Mag. Mag. Mat. 240, 571,(2002)
3. E. J. Patiño,A. Ayuela, and R.M. Nieminen, [Non-Published]
4. H.K. Wong, *et al.*, J. Low Temperature. Phys. 63, 307, (1986).
5. Z. Radov'ic, A.I. Buzdin, J. R. Clem, Phys. Rev. B ,44,759 (1991).
6. J.X. Zhu and C.S. Ting, Phys. Rev. B (Cond-mat.) **61**, 304 (2000).
7. P. Fulde, A. Ferrel, Phys. Rev. 135, A550 (1964).
8. A. Larkin, Y. Ovchinnikov, Sov. Phys. JETP 20, 762 (1965).
9. A. F. Volkov, F.S. Bergeret and K.B. Efetov, Phys. Rev. Lett., **90**, 117006 (2003)
10. T. Kontos, M. Aprili, J. Lesueur, and X. Grison, Phys. Rev. Lett. 86, 304, (2001).
11. C. Strunk, *et al.*, Phys. Rev. B 49, 4053, (1994).
12. J. S. Jiang *et al.*, Phys. Rev. Lett. 74, 314, (1995).
13. Th. Muhge, *et al.*, Phys. Rev. Lett. 77, 1857, (1996).
14. Th. Muhge, *et al.*, Phys. Rev. B 55, 8945, (1997).
15. Th. Muhge, *et al.*, Phys. Rev. B 57, 5071, (1998).
16. J. Arts, *et al.*, Phys. Rev. B 56, 2779, (1997).
17. A.I. Buzdin, L. N. Bulaevskii, and S. V. Panyukov, Sov. Phys. JETP 60, 174(1984)
18. Martin Lange, Margriet J. Van Bael, Yvan Bruynseraede, and Victor V. Moshchalkov, Phys. Rev. Lett. 90, 197006 (2003)
19. Zhaorong Yang, Martin Lange, Alexander Volodin, Ritta Szymczak and Victor V. Moshchalkov, Nature Mat. 3, 793, (2004).
20. R. Steiner and P. Ziemann, Phys. Rev. B, **74**, 94504 (2006).
21. L.Y. Zhu, T.Y. Chen, and C. L. Chien, PRL 101, 017004 (2008).
22. A. I. Buzdin, Cond-Mat 0505583, v1, (2005).
23. R.J. Kinsey, G. Burnell, and M. G. Blamire,IEEE Trans. Appl. Supercond., 11, 904, (2001).
24. L. R. Tagirov, Phys. Rev. Lett. 83, 2058, (1999).
25. A. Rusanov, M. Hesselberth, S. Habraken and J. Aarts, Physica C, 404, Issue 1-4, 322 (2004)
26. A. Yu. Ruzanov, M. Heesselberth, and J. Arts, Phys. Rev. Lett. 93, 57002 (2004).
27. M. D. Allsworth, *et al.* Appl. Phys. Lett. b 80,4196, (2002).
28. I. Baladie, and A. Buzdin, Phys. Rev. B 63, 54518, (2001).
29. V.V. Ryazanov, V. A. Oboznov, A.S. Prokof'ev, and S. V. Dubonos, JETP Lett. 77, 39 (2003).
30. I. S. Burmistrov and N.M. Chtchelkatchev, Phys. Rev. B, **72**, 144520 (2005).
31. C. Bell, S. Tursucu and J. Aarts, Phys. Rev. B, **74**, 214520 (2006).
32. E. J. Patiño, C. Bell and M. G. Blamire, Eur. Phys. J. B, **68**, 73 (2009).
33. S. Kobayashi, H. Oike, M. Takeda, and F. Itoh, Phys. Rev. B 66, 214520 (2002).
34. S. Kobayashi and F. Itoh, Journal Phys. Soc. Japan 11, 900,(2002)
35. Buzdin A., Nature Mat. 3, 751, (2004).

36. E. J. Patiño, "Study of The Influence of Domain Walls in The Superconductor/Ferromagnet Proximity Effect". Ph.D Thesis [Non-Published], Cambridge University, (2005).
37. A.Yu. Rusanov, T.E. Golikova, and S.V. Egorov. JETP Lett. **87**, 175 (2008)

SPIN SINGLET AND TRIPLET SUPERCONDUCTIVITY INDUCED BY CORRELATED HOPPING INTERACTIONS

LUIS A. PEREZ

Instituto de Física, Universidad Nacional Autónoma de México,
Apartado Postal 20-364, 01000, D.F., México
lperez@fisica.unam.mx

J. SAMUEL MILLAN

Facultad de Ingeniería, Universidad Autónoma del Carmen,
Cd. del Carmen, 24180, Campeche, México
smillan@pampano.unacar.mx

CHUMIN WANG

Instituto de Investigaciones en Materiales, Universidad Nacional Autónoma de México,
Apartado Postal 70-360, 04510, D.F., México
chumin@servidor.unam.mx

Received 15 November 2009

In this article, we show that the Bardeen-Cooper-Schrieffer (BCS) formalism applied to a Hubbard model is capable to predict the s-, p- and d-wave superconductivity within a single theoretical scheme. This study is performed on a square lattice described by a generalized Hubbard Hamiltonian, in which correlated-hopping interactions are included in addition to the repulsive Coulombic one. Within the BCS formalism using a variable chemical potential, the superconducting ground states are determined by two coupled integral equations, whose integrand functions have main contributions around the Fermi surface. We observe the existence of a maximum critical temperature for s-, p- and d-wave superconducting channels that occur at the medium, low and high electron densities, respectively. Furthermore, the p- and d-wave superconducting specific heats show a power-law temperature dependence, instead of the exponential one for the s channel. Finally, the smallness of the anisotropic single-energy-excitation-gap minima is essential for the specific heat behavior, and this fact allows to understand the experimental data obtained from Sr_2RuO_4 and $La_{2-x}Sr_xCuO_4$ superconductors.

Keywords: Superconducting gap symmetry; Hubbard model; BCS theory.

1. Introduction

The theory developed by J. Bardeen, L. N. Cooper and J. R. Schrieffer (BCS) was very successful in explaining the main features of metallic superconductors.[1] In the last two decades, the observation of d-symmetry pairing in ceramic superconductors has motivated the research of models beyond the standard BCS theory to include anisotropic superconducting gap symmetries.[2] The recent discovery of the p-wave spin-triplet superconducting state in Sr_2RuO_4 has highly enhanced this research.[3]

The two-dimensional nature present in both p and d-wave superconducting systems could be essential for understanding their peculiarities. In general, the energy spectrum of elementary excitations in solids determines the temperature dependence of their specific heat, and for a superconductor it gives information regarding to the symmetry of its superconducting state. An s-wave superconductor has an exponentially temperature-dependent electronic specific heat, while an anisotropic nodal superconducting gap leads to a power-law dependence, as occur in the cuprate superconductors and in Sr_2RuO_4.[4] For these materials, three-band Hubbard models have been proposed to describe the dynamics of the carriers on the CuO_2 and RuO_2 planes,[5,6] and the electronic states close to the Fermi energy can be reasonably well described by a single-band tight-binding model on a square lattice with second neighbour hoppings.[5,7] In this article, we study s-, p- and d-wave superconducting states within a single-band generalized Hubbard Hamiltonian, in which nearest (Δt) and next-nearest (Δt_3) neighbour correlated-hopping interactions are considered in addition of the on-site (U) Coulombic interaction.[8] Certainly, Δt and Δt_3 are always present in real materials and in spite of having small strengths, they are essential in the determination of the superconducting symmetry.

2. The Model

We start from a single-band Hubbard model with on-site Coulombic interaction (U), first- (Δt) and second-neighbour (Δt_3) correlated-hopping interactions. The corresponding Hamiltonian can be written as

$$\hat{H} = t \sum_{\langle i,j \rangle, \sigma} c_{i,\sigma}^\dagger c_{j,\sigma} + t' \sum_{\langle\langle i,j \rangle\rangle, \sigma} c_{i,\sigma}^\dagger c_{j,\sigma} + U \sum_i n_{i,\uparrow} n_{i,\downarrow}$$
$$+\Delta t \sum_{\langle i,j \rangle, \sigma} c_{i,\sigma}^\dagger c_{j,\sigma} (n_{i,-\sigma} + n_{j,-\sigma}) + \Delta t_3 \sum_{\langle i,l \rangle, \langle j,l \rangle, \sigma, \langle\langle i,j \rangle\rangle} c_{i,\sigma}^\dagger c_{j,\sigma} n_l \ , \quad (1)$$

where $c_{i,\sigma}^\dagger$ ($c_{i,\sigma}$) is the creation (annihilation) operator with spin $\sigma =\downarrow$ or $\uparrow$ at site i, $n_{i,\sigma} = c_{i,\sigma}^\dagger c_{i,\sigma}$, $n_i = n_{i,\uparrow} + n_{i,\downarrow}$, $\langle i,j \rangle$ and $\langle\langle i,j \rangle\rangle$ denote respectively the nearest-neighbor and the next-nearest-neighbor sites. This model can lead to s- and d-wave superconducting ground states without attractive density-density interactions.[9] Let us start from a square lattice with lattice parameter a, where we further consider a small distortion of its right angles in order to include the possible existence of a bulk structural distortion in Sr_2RuO_4.[10] This distortion produces changes in the second-neighbour interactions, such as t' and Δt_3 terms in Eq. (1), and their new values are $t' \pm \delta$ and $\Delta t_3 \pm \delta_3$, where $\pm$ refers to the $\hat{x} \pm \hat{y}$ direction. Performing a Fourier transform, this Hamiltonian in the momentum space becomes

$$\hat{H} = \sum_{\mathbf{k},\sigma} \varepsilon_0(\mathbf{k}) c^\dagger_{\mathbf{k},\sigma} c_{\mathbf{k},\sigma}$$

$$+ \frac{1}{N_s} \sum_{\mathbf{k},\mathbf{k}',\mathbf{q}} V_{\mathbf{k},\mathbf{k}',\mathbf{q}} c^\dagger_{\mathbf{k}+\mathbf{q},\uparrow} c^\dagger_{-\mathbf{k}'+\mathbf{q},\downarrow} c_{-\mathbf{k}'+\mathbf{q},\downarrow} c_{\mathbf{k}+\mathbf{q},\uparrow}$$

$$+ \frac{1}{N_s} \sum_{\mathbf{k},\mathbf{k}',\mathbf{q},\sigma} W_{\mathbf{k},\mathbf{k}',\mathbf{q}} c^\dagger_{\mathbf{k}+\mathbf{q},\sigma} c^\dagger_{-\mathbf{k}'+\mathbf{q},\sigma} c_{-\mathbf{k}'+\mathbf{q},\sigma} c_{\mathbf{k}+\mathbf{q},\sigma} \quad , \tag{2}$$

where N_s is the total number of sites,

$$\varepsilon_0(\mathbf{k}) = 2t \left[\cos(k_x a) + \cos(k_y a)\right] + 2t'_+ \cos(k_x a + k_y a) + 2t'_- \cos(k_x a - k_y a) \quad , \tag{3}$$

$$V_{\mathbf{k},\mathbf{k}'\mathbf{q}} = U + 2\Delta t \left[\beta(\mathbf{k}+\mathbf{q}) + \beta(-\mathbf{k}+\mathbf{q}) + \beta(\mathbf{k}'+\mathbf{q}) + \beta(-\mathbf{k}'+\mathbf{q})\right]$$
$$+ \Delta t_3^+ \left[\gamma(\mathbf{k}+\mathbf{q}, \mathbf{k}'+\mathbf{q}) + \gamma(-\mathbf{k}+\mathbf{q}, -\mathbf{k}'+\mathbf{q})\right]$$
$$+ \Delta t_3^- \left[\zeta(\mathbf{k}+\mathbf{q}, \mathbf{k}'+\mathbf{q}) + \zeta(-\mathbf{k}+\mathbf{q}, -\mathbf{k}'+\mathbf{q})\right] \quad , \tag{4}$$

and

$$W_{\mathbf{k},\mathbf{k}'\mathbf{q}} = \Delta t_3^+ \gamma(\mathbf{k}+\mathbf{q}, \mathbf{k}'+\mathbf{q}) + \Delta t_3^- \zeta(\mathbf{k}+\mathbf{q}, \mathbf{k}'+\mathbf{q}) \quad , \tag{5}$$

being

$$\beta(\mathbf{k}) = 2 \left[\cos(k_x a) + \cos(k_y a)\right] \quad , \tag{6}$$

$$\gamma(\mathbf{k}, \mathbf{k}') = 2\cos(k_x a + k'_y a) + 2\cos(k'_x a + k_y a) \quad , \tag{7}$$

$$\zeta(\mathbf{k}, \mathbf{k}') = 2\cos(k_x a - k'_y a) + 2\cos(k'_x a - k_y a) \quad . \tag{8}$$

and $2\mathbf{q}$ is the wave vector of the pair center of mass. After a standard Hartree-Fock decoupling of the interaction terms with $\mathbf{q} \neq 0$ applied to Eq. (2),[11] the reduced Hamiltonian for $\mathbf{q} = 0$ is

$$\hat{H} = \sum_{\mathbf{k},\sigma} \varepsilon(\mathbf{k}) c^\dagger_{\mathbf{k},\sigma} c_{\mathbf{k},\sigma} + \frac{1}{N_s} \sum_{\mathbf{k},\mathbf{k}'} V_{\mathbf{k},\mathbf{k}',0} c^\dagger_{\mathbf{k},\uparrow} c^\dagger_{-\mathbf{k}',\downarrow} c_{-\mathbf{k}',\downarrow} c_{\mathbf{k},\uparrow}$$

$$+ \frac{1}{N_s} \sum_{\mathbf{k},\mathbf{k}',\sigma} W_{\mathbf{k},\mathbf{k}',0} c^\dagger_{\mathbf{k},\sigma} c^\dagger_{-\mathbf{k}',\sigma} c_{-\mathbf{k}',\sigma} c_{\mathbf{k},\sigma} \quad , \tag{9}$$

where the mean-field dispersion relation is given by

$$\varepsilon(\mathbf{k}) = n\frac{U}{2} + 2(t + n\Delta t) \left[\cos(k_x a) + \cos(k_y a)\right] +$$
$$+ 2(t'_+ + 2n\Delta t_3^+) \cos(k_x a + k_y a) + 2(t'_- + 2n\Delta t_3^-) \cos(k_x a - k_y a) \quad . \tag{10}$$

Notice that the single electron dispersion relation $\varepsilon(\mathbf{k})$ is now modified by adding terms $n\Delta t$, $2n\Delta t_3^\pm$ and $nU/2$ to the hoppings t, t' and the self-energy, respectively.

3. Coupled Integral Equations

Applying the BCS formalism to Eq. (9), we obtain the following two coupled integral equations,[8,9,12] which determine the anisotropic superconducting gap $[\Delta(\mathbf{k})]$ and the chemical potential (μ) for a given temperature (T) and electron density (n),

$$\Delta(\mathbf{k}) = -\frac{1}{2N_s} \sum_{\mathbf{k}'} \frac{Z_{\mathbf{k},\mathbf{k}'}\Delta(\mathbf{k}')}{E(\mathbf{k}')} \tanh\left(\frac{E(\mathbf{k}')}{2k_BT}\right) \quad , \qquad (11)$$

and

$$n - 1 = -\frac{1}{N_s} \sum_{\mathbf{k}} \frac{\varepsilon(\mathbf{k}) - \mu}{E(\mathbf{k})} \tanh\left(\frac{E(\mathbf{k})}{2k_BT}\right) \quad , \qquad (12)$$

where the single excitation energy is given by

$$E(\mathbf{k}) = \sqrt{\left(\varepsilon(\mathbf{k}) - \mu\right)^2 + \Delta^2(\mathbf{k})} \quad , \qquad (13)$$

$Z_{\mathbf{k},\mathbf{k}'}$ and $\Delta(\mathbf{k})$ depend on the symmetry of superconducting ground states as given in Table 1.

Table 1.   Interaction potentials $(Z_{\mathbf{k},\mathbf{k}'})$ and superconducting gaps $[\Delta(\mathbf{k})]$ as functions of pairing symmetry.

Symmetry	$Z_{\mathbf{k},\mathbf{k}'}$	$\Delta(\mathbf{k})$
s-wave	$V_{\mathbf{k},\mathbf{k}',0}$	$\Delta_s + \Delta_s^* \left[\cos(k_x a) + \cos(k_y a)\right]$
p-wave	$W_{\mathbf{k},\mathbf{k}',0}$	$\Delta_p \left[\sin(k_x a) \pm \sin(k_y a)\right]$
d-wave	$V_{\mathbf{k},\mathbf{k}',0}$	$\Delta_d \left[\cos(k_x a) - \cos(k_y a)\right]$

For the s-wave case, Eq. (11) becomes *per se* two coupled equations[12]

$$\Delta_{s^*} = -4\Delta t_3 \left(I_2\Delta_{s^*} + I_1\Delta_s\right) - 4\Delta t \left(I_1\Delta_{s^*} + I_0\Delta_s\right) \quad , \qquad (14)$$

and

$$\Delta_s = -U \left(I_1\Delta_{s^*} + I_0\Delta_s\right) - 4\Delta t \left(I_2\Delta_{s^*} + I_1\Delta_s\right) \quad , \qquad (15)$$

where

$$\begin{aligned}
I_l &= \frac{1}{N_s} \sum_{\mathbf{k}} \frac{\left[\cos(k_x a) + \cos(k_y a)\right]^l}{2E(\mathbf{k})} \tanh\left(\frac{E(\mathbf{k})}{2k_BT}\right) \\
&= \frac{a^2}{8\pi^2} \int_{-\pi/a}^{\pi/a} \int_{-\pi/a}^{\pi/a} dk_x dk_y \frac{\left[\cos(k_x a) + \cos(k_y a)\right]^l}{E(\mathbf{k})} \tanh\left(\frac{E(\mathbf{k})}{2k_BT}\right) \quad , \qquad (16)
\end{aligned}$$

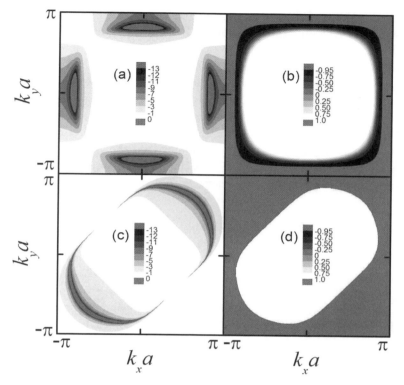

Fig. 1. (Color on line) Contour plots of integrands of (a) Eq. (17) and (b) Eq. (12) for d-wave superconducting states, while (c) Eq. (18) and (d) Eq. (12) for p-wave ones.

Notice that $\Delta t_3^{\pm}$ has no effects on the s-wave superconductivity except for the single-electron dispersion relation. For the d-channel, Eq. (11) can be written as[9]

$$
1 = \frac{4\Delta t_3 a^2}{8\pi^2} \int_{-\pi/a}^{\pi/a} \int_{-\pi/a}^{\pi/a} dk_x dk_y \frac{[\cos(k_x a) - \cos(k_y a)]^2}{E(\mathbf{k})} \tanh\left(\frac{E(\mathbf{k})}{2k_B T}\right) \ . \quad (17)
$$

Likewise, for the p-channel, Eq. (11) leads to[8]

$$
1 = \pm\frac{4\delta_3 a^2}{8\pi^2} \int_{-\pi/a}^{\pi/a} \int_{-\pi/a}^{\pi/a} dk_x dk_y \frac{[\sin(k_x a) \pm \sin(k_y a)]^2}{E(\mathbf{k})} \tanh\left(\frac{E(\mathbf{k})}{2k_B T}\right) \ , \quad (18)
$$

In general, the critical temperature (T_c) is determined by $\Delta_\alpha(T_c) = 0$, being $\alpha=s$, p or d.

Figures 1(a) and 1(b) respectively show the contour plots of the integrand functions of Eqs. (17) and (12) for $t = -1$, $t' = -0.45|t|$, $\Delta t = 0.5|t|$, $\Delta t_3 = 0.1|t|$, $\delta_3 = 0$, $U = 6|t|$, $n = 1.94$, $k_B T_c = 0.06319|t|$, and $\mu = 6.13585|t|$. Notice that the main contribution to the integral of Eq. (17) comes from the sharp walls located at the Fermi surface defined by $\varepsilon(\mathbf{k}) = \mu$, and separated by d-wave nodes. In contrast, for Eq. (12) the integrand has a step behaviour around the Fermi surface. Likewise, Figures 1(c) and 1(d) illustrate the contour plots of the integrand functions of Eqs. (18) and (12), respectively, for $t = -1$, $t' = -0.45|t|$, $\Delta t = 0.5|t|$, $\Delta t_3 = 0.1|t|$, $\delta_3 = 0.1|t|$, $U = 6|t|$, $n = 1$, $k_B T_c = 0.00093|t|$, $\mu = 3.307|t|$, and taking the upper sign in Eq. (18). Observe that similar to the previous case, the sharp walls and step behaviour are located at the Fermi surface. However, the Fermi surface is oriented along the $\hat{x} + \hat{y}$ direction, due to the magnitude and sign of δ_3. Furthermore, the step in Fig. 1(d) is sharper than in Fig. 1(b), since the p-channel T_c is smaller than the d-channel one. It would be worth mentioning that the general behaviour shown in figures 1 is not sensitive to the particular Hamiltonian parameter values and they were chosen in order to make comparisons with experimental data, as discussed in the following sections.

4. Results

Fig. 2(a) shows the critical temperature (T_c) as a function of n for s- (open circles), p- (open triangles) and d-symmetry (open squares) superconducting states with $t = -1$, $t' = -0.45|t|$, $\Delta t = 0.5|t|$, and $U = 6|t|$. For s and d symmetries we have taken $\Delta t_3 = 0.1|t|$ and $\delta_3 = 0$; whereas for p symmetry, $\Delta t_3 = 0.15|t|$ and $\delta_3 = 0.1|t|$. Notice that the maximum T_c for the d-channel is located at the optimal $n = 1.73$, similar to that observed in cuprate superconductors since the hole doping concentration x is related to $2 - n$. This relationship is taken for the lower Hubbard sub-band, because the Coulomb repulsion induces a charge-transfer gap in a half filled band.[5] In contrast, for p-channel, the maximum T_c is found around half-filling close to the expected electronic density of $n = 1.2$ for $Sr_2 RuO_4$.[13,14] Furthermore, for s-wave the maximum T_c is found for $n = 1.47$.

The mean potential energy $(\langle Z \rangle)$ can be written as[15,16]

$$\langle Z \rangle = \frac{1}{N_s^2} \sum_{\mathbf{k},\mathbf{k}'} Z_{\mathbf{k},\mathbf{k}'} u_{\mathbf{k}} v_{\mathbf{k}}^* u_{\mathbf{k}'}^* v_{\mathbf{k}'} = -\frac{1}{N_s} \sum_{\mathbf{k}} \frac{\Delta_{\mathbf{k}}^2}{2E_{\mathbf{k}}} \ , \tag{19}$$

where

$$\Delta_{\mathbf{k}} = \frac{1}{N_s} \sum_{\mathbf{k}'} Z_{\mathbf{k},\mathbf{k}'} u_{\mathbf{k}'}^* v_{\mathbf{k}'} \ , \tag{20}$$

and

$$\frac{\Delta_{\mathbf{k}}}{2E_{\mathbf{k}}} = u_{\mathbf{k}} v_{\mathbf{k}}^* \ . \tag{21}$$

In Fig. 2(b), we show $\langle Z \rangle$ as a function of n for s- (solid circles), p- (solid triangles) and d-channel (solid squares) with the same parameters of Fig. 2(a). Observe that $\langle Z \rangle$ is negative and its absolute value has almost the same electron-density dependence as T_c, mainly due to the smooth variation of the mean kinetic energy.

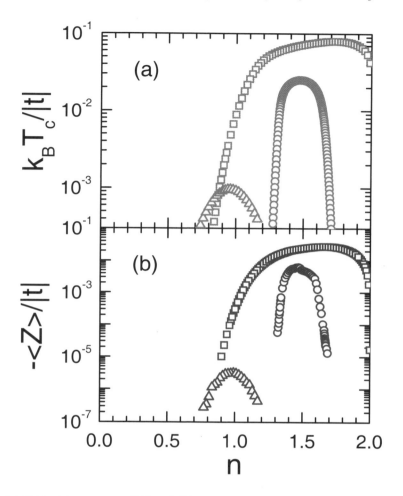

Fig. 2. (a) Critical temperature (T_c) and (b) mean potential energy ($\langle Z \rangle$) as functions of the electron density (n) for s- (solid circles), p- (solid triangles) and d-channel (solid squares).

One of the physical quantities that yields information about the symmetry of superconducting states is the electronic specific heat (C), which is highly sensitive to the low-energy excitations. The C of superconducting states is given by[15]

$$C = \frac{2k_B \beta^2}{4\pi^2} \int \int_{1BZ} f(E_{\mathbf{k}}) \left[1 - f(E_{\mathbf{k}})\right] \left[E_{\mathbf{k}}^2 + \beta E_{\mathbf{k}} \frac{dE_{\mathbf{k}}}{d\beta}\right] dk_x dk_y \quad , \qquad (22)$$

where $\beta = 1/(k_B T)$ and $f(E)$ is the Fermi-Dirac distribution. To obtain the specific heat of the normal state we take $\Delta_{\mathbf{k}}$ equal to zero.[15] In Fig. 3, two electronic densities (a) $n = 1.2$ and (b) $n = 1.94$ are chosen from Fig. 2 to calculate their d-channel electronic specific heat and compared with experimental data obtained from $La_{2-x}Sr_xCuO_4$ for $x = 0.22$ and $x = 0.1$, respectively.[17] Insets of Fig. 3 show the corresponding theoretical angular dependences of the single-excitation energy gap (Δ) defined as the minimum value of $E_\mathbf{k}$ in $\mathbf{k}$ direction.[8] The polar angle is given by $\theta = \tan^{-1}(k_y/k_x)$. Notice that for the hole overdoped regime, $n < 1.73$, Δ has a $d_{x^2-y^2}$ symmetry and in consequence C is proportional to T^2 as obtained in Ref. 18. However, for the hole underdoped regime ($n > 1.73$) Δ has a d_{xy}-like symmetry without real nodes and then, C has an exponential behaviour as occurs in an s-wave superconductor. The residual C/T value at $T = 0$ in experimental data could be due to the chemical or electronic inhomogeneity of the sample,[17,19] and this fact is not considered in the theory.

Finally, Fig. 4 shows the electronic specific heat (C) for the same p-wave system of Fig. 2 with $n = 1.0$ in comparison with the experimental data obtained from the spin-triplet superconductor Sr_2RuO_4.[20] Inset of Fig. 4 illustrates the angular dependence of the single-excitation energy gap. Notice the remarkable agreement in both, the discontinuity at T_c and the temperature dependence below T_c, as a consequence of the nature of the p-wave superconducting state. In fact, a power-law temperature dependence $C(T) \sim T^\alpha$ with $\alpha < 2$ is observed, which could be related to the wider depleted-gap region compared with the d-wave case.

5. Conclusion

We have presented a unified theory of the s-, p- and d-wave superconductivity based on the BCS formalism and a generalized Hubbard model. This approach has the advantage of being simple and general for predicting the trends of anisotropic super-conducting materials properties. Moreover, it does not require attractive density-density interactions and the superconductivity is originated by the correlated hoppings interactions. In spite of their small strength in comparison with other terms of the Hamiltonian, they determine the symmetry of the superconducting ground states. The superconducting properties are calculated by solving two coupled integral equations when the variation of the chemical potential is considered. The main contribution to the involved integrals comes from a sharp wall located around the Fermi surface, in consistence with the BCS theory.[1,15] There is a maximum T_c located at low, medium and high electron densities for p, s and d symmetry super-conductivity, respectively; in qualitative agreement with the experimental data.[2,4] In particular, positive Hall coefficients are observed[21,22] in d-wave high-T_c super-conductors indicating its superconductivity could be originated from hole carriers. On the other hand, the low-temperature behavior of C is very sensitive to the existence of nodes in the superconducting gap as well as their deepness. In particular, the results show that the d-channel C in the overdoped regime has a second-order power-law behavior, whereas in the underdoped regime C has an almost exponential temperature dependence, similar to the s-channel case, due to the absence of real nodes in the superconducting gap. In fact, the d_{xy}-like gaps without real nodes have been observed in cuprate superconductors by scanning tunneling experiments.[23] In addition, p-channel C shows a sub-second-order power-law behaviour as obtained in

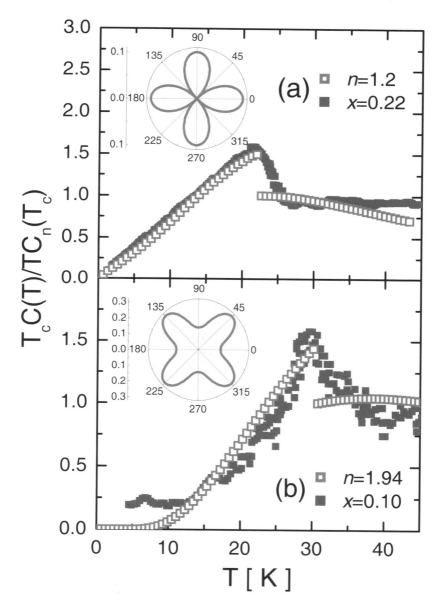

Fig. 3. Theoretical (open squares) d-wave electronic specific heat (C) versus temperature (T) for the same d-wave systems as in Fig. 2 with (a) $n = 1.2$ and (b) $n = 1.94$, in comparison with experimental data (solid squares) obtained from $La_{2-x}Sr_xCuO_4$ for (a) $x = 0.22$ and (b) $x = 0.1$.[17] Insets: Corresponding single-excitation energy gaps ($\Delta/|t|$) as a function of the polar angle (θ).

Sr_2RuO_4. We expect that this analysis could contribute the understanding of the different $C(T)$ behaviors observed in anisotropic superconductors. The present study

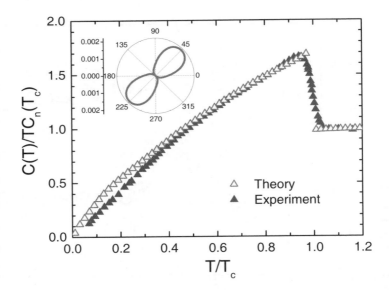

Fig. 4. Theoretical (open triangles) p-wave specific heat (C) calculated by using the same parameters of Fig. 2 with $n = 1.0$, versus temperature (T) in comparison with the experimental data (solid triangles) obtained from Sr_2RuO_4.[20] Inset: Single-excitation energy gap $(\Delta/|t|)$ as a function of the polar angle (θ).

can be extended to include the effects of external perturbations, such as magnetic fields, on the physical properties of anisotropic superconductors. This extension is currently being developed.

Acknowledgments

This work has been partially supported by CONACyT-58938, UNAM-IN113008, UNAM-IN114008 and the UNAM-UNACAR exchange project. Computations have been performed at Bakliz and KanBalam of DGSCA, UNAM.

References

1. J. Bardeen, L. N. Cooper and J. R. Schrieffer, *Phys. Rev.* **108**, 1175 (1957).
2. C. C. Tsuei and J. R. Kirtley, *Rev. Mod. Phys.* **72**, 969 (2000).
3. K. D. Nelson, Z. Q. Mao, Y. Maeno and Y. Liu, *Science* **306**, 1151 (2004).
4. A. P. Mackenzie and Y. Maeno, *Rev. Mod. Phys.* **75**, 657 (2003).
5. H. -B. Schüttler and A.J. Fedro, *Phys. Rev. B* **45**, 7588 (1992).
6. J. F. Annett, G. Litak, B. L. Gyorffy and K. I. Wysokinski, *Phys. Rev. B* **66**, 134514 (2002).
7. I. I. Mazin and D. J. Singh, *Phys. Rev. Lett.* **79**, 733 (1997).
8. J. S. Millán, L. A. Pérez and C. Wang, *Phys. Lett. A* **335**, 505 (2005).
9. L. A. Pérez and C. Wang, *Solid State Commun.* **121**, 669 (2002).
10. R. Matzdorf, Z. Fang, Ismail, J. Zhang, T. Kimura, Y. Tokura, K. Terakura, and E. W. Plummer, *Science* **289**, 746 (2000).

11. E. Dagotto, J. Riera, Y. C. Chen, A. Moreo, A. Nazarenko, F. Alcaraz, and F. Ortolani, *Phys. Rev. B* **49**, 3548 (1994).
12. J. E. Hirsch and F. Marsiglio, *Phys. Rev. B* **39**, 11515 (1989).
13. T. Oguchi, *Phys. Rev. B* **51**, 1385 (1995).
14. D. J. Singh, *Phys. Rev. B* **52**, 1358 (1995).
15. M. Tinkham, *Introduction to Superconductivity* (2nd Edition, McGraw Hill, New York, 1996).
16. G. Rickayzen, *Theory of Superconductivity* (Wiley, New York, 1965).
17. T. Matsuzaki, N. Momono, M. Oda, and M. Ido, *J. Phys. Soc. Jpn.* **73**, 2232 (2004).
18. H. Won and K. Maki, *Phys. Rev. B* **49**, 1397 (1994).
19. H.-H. Wen, Z.-Y. Liu, F. Zhou, J. Xiong, W. Ti, T. Xiang, S. Komiya, X. Sun, and Y. Ando, *Phys. Rev. B* **70**, 214505 (2004).
20. S. Nishizaki, Y. Maeno, and Z. Mao, *J. Phys. Soc. Jpn.* **69**, 572 (2000).
21. H. Y. Hwang, B. Batlogg, H. Takagi, H. L. Kao, J. Kwo, R. J. Cava, J. J. Krajewski, and W. F. Peck, Jr., *Phys. Rev. Lett.* **72**, 2636 (1994).
22. F. F. Balakirev, J. B. Betts, A. Migliori, S. Ono, Y. Ando, and G. S. Boebinger, *Nature* **424**, 912 (2003).
23. J. Kane and K. W. Ng, *Phys. Rev. B* **53**, 2819 (1996).

Part G
Statistical Mechanics, Relativistic Quantum Mechanics

BOLTZMANN'S ERGODIC HYPOTHESIS: A MEETING PLACE FOR TWO CULTURES

M. HOWARD LEE

Department of Physics and Astronomy University of Georgia
*Athens, GA 30602 USA**
and
Korea Institute for Advanced Study
Seoul 130-012, Korea
MHLee@uga.edu

Received 21 October 2009

The physical theory for the ergodic hypothesis is premised on the idea that the hypothesis is measurable by scattering experiment. Therewith it proves the hypothesis by measurable properties of the response function. Birkhoff's theorem proves the ergodic hypothesis by abstract properties of phase space, the measure and transitivity. In this work we apply the two independent approaches to a particular many-body model (spin-1/2 XY model) to see whether they arrive at the same conclusions on the ergodicity question. The two approaches are compared to gain insight into their goals.

Keywords: Ergodic hypothesis; Birkhoff's theorem.

1. Introduction: A Perspective on the Issue

Essentially all physical theories are formulated in the language of mathematics whose complexity ranges from elementary to sophisticated. Because of this practice we tend to forget that the mathematical culture is still very different from the physical culture. The mathematicians have helped us to make some of our physical theories more rigorously based. But this is not what the mathematicians normally would do in their work.

The mathematicians are preoccupied with finding theorems. Perhaps the most famous is one due to Pythagoras. These theorems are about conditions under which certain relationships hold. The theorems tend to be universal, ordinarily not tied to some specific systems. They may still provide a general description to a specific system.

The physicists go about their work differently. They do not look for theorems. They look to find solutions to some specific problems. Thus what they find is not likely to be universal, nor general. They will draw conclusions, even universal conclusions, from specific answers.

---

*Permanent Address.

These two different approaches or cultures rarely intersect. As a result, most of the time we are not even conscious of the existence of the other. There is one remarkable exception. It is in a problem proposed by Boltzmann long ago known as the ergodic hypothesis. In his attempt to formulate statistical thermodynamics Boltzmann encountered the difficult task: Calculating time averages for observables like the internal energy of a macroscopic system in thermal equilibrium. Time averages are what are being measured or observed. To do the time averages, one must first solve the relevant equations of motion, not a simple matter, even for Boltzmann evidently.

Boltzmann's idea was to circumvent this obstacle through a hypothesis. The time averages would be replaced by the ensemble averages. Why? Calculating the ensemble averages is considerably easier and, in fact, doable in many instances. But is this hypothesis universally valid?

For many years thereafter, many mathematicians became very interested in proving the validity of this hypothesis.[1] That is, to find a condition or conditions under which the hypothesis is universally valid, which is in accord with the mathematical culture referred to above.

The technical tools that are brought to bear on it seem formidable, at least in the eyes of physicists. They are not the ordinary variety that most physicists are familiar with. Perhaps that is not important. But important is to recognize that what the mathematicians would do or like to deliver is not necessarily what we could immediately use.

Suppose there are systems 1, 2, 3 and so on. We would like to know whether one or more of them are ergodic. What the mathematicians might provide for us are conditions for anyone of the systems for being ergodic. But we might be hard pressed to see how these conditions could discriminate one system from another.

As stated above they would be simply universal statements, applicable to any and all systems. If these conditions were not recognizable in physical terms, these theorems would leave us in the cold. But as far as the mathematicians are concerned, they would have proved the ergodic hypothesis to their satisfaction. For them, that would be the end.

To physicists proving the validity of the ergodic hypothesis seems so very daunting that over the years they have paid little attention to it in spite of its enormous significance. What they could have done is to follow their own traditional time-honored approach: To invent an instrument or meter, real or gedanken, with which to measure ergodicity system-by-system in systems 1, 2, 3 and so on.

Like the thermometer, such a meter could measure these different systems and determine whether ergodic one-by-one. In this way some of them could be pronounced ergodic, from which one could even draw a general conclusion. But on this problem this standard practice of physics seems to have been forgotten.

Suppose there were such a meter invented with which to determine ergodicity system-by-system. Suppose systems 1 and 2 but not 3 were measured ergodic by this meter. They must have met the conditions of the ergodic theory.

If true, in these systems we are seeing the two different cultures in action at the same time. If this coincidence should take place, the ergodic hypothesis would truly be like a meeting place of two cultures. Such an occurrence in statistical mechanics is to our knowledge a rare event.

Perhaps the foremost among all theorems on the ergodic hypothesis is one due to Birkhoff, a celebrated American mathematician, proved more than 70 years ago.[2] As with any theorem, Birkhoff's is couched in conditions. They pertain to the existence of a time average, the transitivity of phase space etc. They seem remote from all descriptions of any physical system for which the ergodic hypothesis is intended. To our knowledge, no reasonable many-body models have been ruled ergodic or otherwise by this theorem during the past 70 years of its existence. As a result, this profound work has languished in the realm of mathematics, rather than in the realm of physics, the natural and no doubt intended home.

Why there has not been comparable progress made by physicists for over a hundred years? Almost every textbook on statistical mechanics states the fundamental importance of the ergodic hypothesis. It is indeed one of the foundations of statistical mechanics. To base its foundation on something not proved seems out of character for physicists who are ordinarily highly rational. What may be the reason or reasons one might ask.

To prove the hypothesis physically, one must first solve a relevant equation of motion as earlier noted. Since this first step poses a challenge, could anyone realistically think of undertaking the second step, that of the time averaging? Over the years, most have found that the ergodic hypothesis seems to work. This fact is a powerful argument for accepting the hypothesis. It mitigates against a need for the justification. By accepting the hypothesis at its face value, equilibrium statistical mechanics has become largely synonymous with statistical mechanics. The art of obtaining time evolution, even time itself, has disappeared from the main streams of statistical mechanics. All these seem to have conspired against developing a physical theory for the ergodic hypothesis.

In the early 2000s we have given a physical theory for the ergodic hypothesis. It was developed from many-body theory, not statistical mechanics.[3] Many-body theory deals with large systems just as statistical mechanics, but it is concerned primarily with ground state properties. Modern tools of scattering experiments have spurred this field during the 1950s, giving impetus to develop a linear response theory[4] and therewith techniques of solving the Heisenberg equation of motion. Our work the method of recurrence relations is a part of this effort. It is a method, which has been demonstrated to solve the Heisenberg equation in a number of many-body models.[5]

Equipped with this technique, we have turned to calculating time averages with which to compare with ensemble averages. What has resulted is a meter. This meter, a gedanken one, has been shown to provide a physical insight into why the hypothesis can work in a specific system and why it can also fail in it. On such a system the meter, dubbed an ergometer, can unlock the mysterious conditions on

Birkhoff's theorem. This is why we call it a meeting of two cultures.

2. Statement of the Problem

A modern definition of the ergodic hypothesis (EH) may be stated as follows: Let
A be a dynamical variable of a macroscopic system denoted by a Hamiltonian **H**
and let the system be in *thermal equilibrium*. We shall further assume that **H** is
Hermitian.

Then, EH asserts that

$$\lim_{T \to \infty} \frac{1}{T} \int_0^T < \mathbf{A}(t) >=< \mathbf{A} > \tag{1}$$

where the brackets denote an ensemble average. The averages $< \mathbf{A}(t) >$, on the lhs
of (1), mean that all *initial* values of $< \mathbf{A}(t) >$ are considered. See App. A.

Thus both sides are being compared at the same temperature. At this stage
it ought be stressed that the original intent of EH was for a macroscopic system.
In statistical mechanics it means that the thermodynamic limit must be taken.
Applying EH to a finite system seems to us to be at best a misuse and at worst a
corruption of the original concept.

Is EH valid? How does one go about proving its validity? Are there some limitations to the validity of EH? Are these limitations relatively unimportant to physical
consideration? If not, how seriously should we reconsider this part of the foundations
of statistical mechanics?

2.1. Mathematical approach

Many mathematicians have studied EH,[1] almost all for classical systems. Perhaps
the most relevant for us may be the work of Birkhoff,[2] which we will briefly state
in simple terms.

It is assumed that $\mathbf{f}(t) \equiv < \mathbf{A}(t) >$, a classical function is a finite integrable
function in phase space Γ, a measure preserving space. Then it is proved that the
time average of $\mathbf{f}(t)$ exists "almost" everywhere. It means that the time average
does not exist over a set of phase points of measure zero.

The time average that exists almost everywhere is equal to the phase space
average of **f** if and only if the phase space is metrically transitive. This is said to
prove EH. We hasten to add that this proof holds for **A**, which is a classical variable
in Birkhoffs work.

3. Physical Approach

We regard (1) as a physical statement, thus open to test by measurement whether
A is a classical or quantum mechanical variable. Imagine that a macroscopic system in thermal equilibrium is scattered by a weak external probe. A scattering

process is described by two quantities: energy and momentum transfers, $\hbar\omega$ and $\hbar\mathbf{k}$, respectively.

Let a probe field couple to variable $\mathbf{A}$ of the system. Then, the lhs of (1) can represent an inelastic scattering process in which $\omega \to 0$ at a fixed $\mathbf{k}$ while the rhs an elastic scattering process at the same value of $\mathbf{k}$. Under what condition could they be the same? Since the measurement is system dependent, the conditions for the validity of EH must reflect system dependence.

Let $\mathbf{h}_A(t)$ be a time dependent probe field which couples to $\mathbf{A}$. The total energy at time t may be written as

$$\mathbf{H}'_A(t) = \mathbf{H}(\mathbf{A}, \mathbf{B}, \mathbf{C}\cdots) + \mathbf{h}_A(t)\mathbf{A}, \tag{2}$$

where $\mathbf{A}, \mathbf{B}, \mathbf{C}$ are possible dynamical variables of $\mathbf{H}$, but only $\mathbf{A}$ is excited by the probe. Observe that EH refers to one specific variable, $\mathbf{A}$ in this instance, not to the others. That is, if $\mathbf{A}$ is ergodic, the ergodicity of $\mathbf{A}$ does not necessarily imply the ergodicity of $\mathbf{B}$ or others. In other words EH is not a general statement about a system but a specific statement about a specific variable of a system.

In this physical picture a mode of a system or a particle at a local site denoted by a dynamical variable is perturbed by an external dynamical field. This perturbed energy will delocalize to other modes or other particles of the systems causing a time evolution in $\mathbf{A}$.

We can do the same for different variables $\mathbf{B}$, $\mathbf{C}$, etc., of the same system. We can repeat with other variables of another system. This is how one could test EH. It is an approach different from the math approach.

4. Physical Theory

A simple way to describe the above processes is to use linear response theory[4] by taking weak external fields. Then, EH (1) may be transformed to:

$$\lim_{T\to\infty} \frac{1}{T} \int_0^T \int_0^t \chi_A(t - t')dt'dt = \chi_A \tag{3}$$

where $\chi_A(t)$ and χ_A are, resp., the dynamic and static response functions of $\mathbf{A}$ defined by linear response theory.

We have shown[3] that if (3) is to hold, the following condition must be satisfied: $0 < \mathbf{W}_A < \infty$, where

$$\mathbf{W}_A = \int_0^\infty \mathbf{r}_A(t)dt \tag{4}$$

where $\mathbf{r}_A(t)$ is the autocorrelation function of $\mathbf{A}$.[5-7]

It is evident that if $\mathbf{W}_A$ is to be finite $\mathbf{r}_A$ must vanish sufficiently fast as $t \to \infty$. This behavior is termed irreversibility.[8,12,22-24] The ergodicity depends on two properties: (i) irreversibility. (ii) delocalizability. The second property pertains to how a perturbation energy delocalizes within a system. The two properties together constitute a sufficient condition for ergodicity according to this physical theory.

Also note that $\mathbf{W}_A$ is self-diffusivity if $\mathbf{A}$ is a velocity operator. If $\mathbf{W}_A$ is finite, it denotes normal diffusion.[9-12] In physical theory, $\mathbf{W}_A$ has been dubbed an ergometer since it can tell or detect ergodicity of $\mathbf{A}$ in a system.[13] Being a device albeit gedanken it measures the existence of ergodicity system-by-system. Its approach to establishing ergodicity is evidently very different from the math approach, as we shall further discuss below.

Recently there has been some interest in establishing ergodicity in anomalous diffusion.[18] Whether our condition (4) should or could apply in its present form to non-Hermitian systems is not yet known. When dissipation or other mechanisms driven by external sources becomes important, our condition probably is not applicable.

5. Meeting of Two Cultures

The math approach is by design universal. The physical approach by its methodology is specific. Clearly the two approaches are orthogonal. But couldn't they meet at some place? Suppose the ergometer were to pronounce a variable of a system not ergodic. Should we not find some conditions of ergodic theory not fulfilled in it? If so, it would be like a meeting of the two cultures, arrived at the same end but from orthogonal directions.[13]

Let us be a little more specific. Suppose $\mathbf{W}_A$ is calculated for a system and is found to be finite. It says that the autocorrelation function of $\mathbf{A}$ is irreversible and the delocalization complete. The ergometer's reading must mean that phase space is metrically transitive (if it is a classical problem.) Conversely if $\mathbf{W}_B$ of the same system were found not finite, it would mean that the delocalization is only partly complete. It must mean that phase space is decomposable, hence not transitive.

6. Ergometry in a Many-body Model

6.1. *Quantum spin chain*

Let us consider a well-known model: the spin-1/2 XY model in 1d,

$$\mathbf{H} = -2\lambda \sum_i \left(\mathbf{s}_i^x \mathbf{s}_{i+1}^x + \mathbf{s}_i^y \mathbf{s}_{i+1}^y \right) \tag{5}$$

where $\mathbf{s}_i^x$ and $\mathbf{s}_i^y$ are the x and y components of the spin-1/2 operator at site $\mathbf{i}$ in a one-dimensional periodic lattice, satisfying $[\mathbf{s}_i^x, \mathbf{s}_i^y] = i\hbar \mathbf{s}_i^z$, where $\mathbf{s}_i^z$ is the z component, and λ the coupling constant between the nearest neighbors. We impose periodic boundary conditions e.g. $\mathbf{s}_{N+1}^x = \mathbf{s}_1^x$.

For this spin chain model, there are two possible independent dynamical variables $\mathbf{A}$ and $\mathbf{B}$: $\mathbf{A} = \mathbf{s}_i^x$ or equivalently $\mathbf{s}_i^y$ and $\mathbf{B} = \mathbf{s}_i^z$, where site $\mathbf{i}$ may be anyone of N sites. Observe that the variable $\mathbf{B}$ is not explicitly contained in $\mathbf{H}$. Thus classically it would be a null operator but quantum mechanically it is implicitly present in the Hilbert space of the spin-1/2 operator.

At infinite temperature the autocorrelation functions of $\mathbf{A}$ and $\mathbf{B}$ are both known. For $\mathbf{A}$, $\mathbf{r}_A(t) = e^{-\lambda^2 t^2}$, in units where $\hbar = 1$.[14,15] Thus, $\mathbf{W}_A = \sqrt{\pi}/\lambda$, hence $\mathbf{A}$ for this model is ergodic. For $\mathbf{B}$, $\mathbf{r}_B(t) = [J_0(\lambda t)]^2$, J_0 is the Bessel function, thus $\mathbf{W}_B = \infty$, so that $\mathbf{B}$ of the same model is not ergodic.

Why this contrasting results in the ergodicity in the same model? Since both $\mathbf{A}$ and $\mathbf{B}$ are irreversible, the difference must arise from delocalizability: complete for $\mathbf{A}$ and not complete for $\mathbf{B}$. In terms of Birkhoff's condition, $\mathbf{A}$ must be transitive but $\mathbf{B}$ not. Can we point to their different physical mechanisms?

If external field $\mathbf{h}_A$ couples to a spin $\mathbf{s}_1^x$, the perturbation energy is delocalized from site 1 to 2, site 2 to 3 etc., because $\mathbf{s}_1^x$ and the others are not in a stationary state of $\mathbf{H}$. As time evolves, the delocalization spreads with $N \to \infty$, attaining ergodicity. One may say after Birkhoff that there can be just one invariant in it, i.e., transitive.

If external field $\mathbf{h}_B$ is now turned on, the physics is very different since the field does not directly couple to any one of the spins of the model. The effect of the field is to create a vortex-like motion along the z direction in spin space as may be seen by the following equation of motion:

$$\frac{d\mathbf{s}_1^z}{dt} = \frac{i\mathbf{J}}{\hbar}\left(\mathbf{s}_i^y \mathbf{s}_j^x - \mathbf{s}_i^x \mathbf{s}_j^y\right) \tag{6}$$

where $j = i \pm 1$. The vortex-like motion is perpendicular to the transverse direction of spin interaction. Thus the effect of the perturbation is to create motions, which are in the orthogonal direction to the transverse plane of the interaction energy. One may say after Birkhoff that this is a realization of two invariants.

A fundamental difference between $\mathbf{A}$ and $\mathbf{B}$ may also be seen by the following: If $\mathbf{B}' = \sum \mathbf{s}_i^z$, $\mathbf{B}(t) = \mathbf{B}'$, a constant of motion so that there is no time evolution in this variable. But if $\mathbf{A}' = \sum \mathbf{s}_i^x$, $\mathbf{A}'$ is not a constant of motion. Thus there is a time evolution in it.

6.2. *Classical hard sphere chain*

We now consider a many-body system of hard spheres arranged in a linear chain in periodic boundary conditions. It resembles beads on a string, hence sometimes called a bead model. Since it is a one-dimensional analog of a hard sphere, it ought be called a hard rod. The model is defined by

$$\mathbf{H} = \sum_i \left(\frac{1}{2}mv_i^2 + \phi(|x_i - x_{i+1}|)\right) \tag{7}$$

where x_i and v_i are the coordinate and velocity of ith particle. All the particles have the same mass m. If x is the magnitude of the separation distance between two particles,$\phi(x) = \infty$ if $x < a$ and $\phi(x) = 0$ if $x > a$, where a is the length of a hard rod.

Because of the hard sphere potential, particles cannot penetrate each other. When two particles collide, they merely exchange their velocities: $\mathbf{v}_1 + \mathbf{v}_2 = \mathbf{v}'_1 + \mathbf{v}'_2$, where $\mathbf{v}_2 = \mathbf{v}'_1$ and $\mathbf{v}_1 = \mathbf{v}'_2$. . Their total energy and momentum are thus conserved in this perfectly elastic collision. This model is perhaps one of the simplest to study nonequilibrium behavior. Some years ago in a remarkable paper Jepson[19] studied the dynamics of this model and obtained the velocity autocorrelation function of a single particle when $a = 0$: $< \mathbf{v}(t)\mathbf{v}(0) > / < \mathbf{v}(0)\mathbf{v}(0) >$, which is the classical analog of the autocorrelation function $\mathbf{r}$.

Jepson's solution for the velocity autocorrelation function is in the integrals of the Bessel functions. See eq (68) of Ref 19. But for $t \to \infty$, he has obtained an asymptotic solution,

$$\mathbf{r}(t \to \infty) = \left(\frac{m}{2\pi kT} \right)^{1/2} \left(-1 + \frac{5}{2\pi} \right) \frac{1}{(\rho t)^3} \tag{8}$$

where $\rho = N/L$ the number density of hard rods. Thus the velocity autocorrelation function is irreversible. In early days it was thus considered that $\mathbf{v}$ is ergodic. We have shown recently that irreversibility is not a sufficient condition for ergodicity. See ref [13]. The sufficient condition is given by (4). Since we do not have a complete closed form solution for the velocity autocorrelation function, $\mathbf{W}$ cannot be evaluated. Still it is possible to apply (8) to determine ergodicity. Since $\mathbf{r}(t \to \infty)$ goes as t^{-3}, it is clear that $\mathbf{W}$ is finite, it cannot be infinite nor zero. Hence we would conclude that the velocity of a hard rod in a many hod rod model is ergodic.

The interpretation of the ergometric theory is as follows: When two particles initially collide, the initial collision sets off a chain of collisions involving ultimately all the particles in the system. There is thus complete delocalization of the initial energy which begins the collision process. Complete delocalization means ergodicity.

Now let us see why it must be ergodic from the perspectives of Birkhoff's theorem. Consider the phase space of the coordinates of length $\mathbf{L}$. It has measure $\mathbf{1}$ since all points are included in the collisions there are no exclusions. Hence the time average exists. Since there is only one invariant phase space, the two averages must be the same.

7. Concluding Remarks

We showed that the math and physics approaches to proving ergodicity are orthogonal. Nonetheless they do arrive at the same end, giving rise to what we have termed a meeting of two cultures.

We have demonstrated this meeting explicitly through a many body model. Although the math approach due to Birkhoff[2] is meant for classical many-body systems, we find that his concepts seem more general as we find them applicable even to a quantum model.

We would argue that a theorem whose conditions cannot be measured is an impotent knowledge. If we had no means to measure angles, the Pythagorean theorem would be a statement in the math realm. In the same vein, it is the ergometer, which takes Birkhoff theorem out into the physical arena.

Birkhoff's theorem says nothing specific about what kind of system it addresses to. At a first glance it is disappointing, for this information is critical to physics. A moment's reflection would convince us however that no math theorems would or are expected to do that. A physical problem has to be solved by the physics mind molded and tempered by the heat of physical reality.

Acknowledgments

A portion of this work was completed at the Institute for Advanced Study, Seoul. I thank Prof. H. Park, vice director of the Institute for his warm hospitality. I also thank Mr. John Nagao for his assistance with the preparation of this manuscript.

Appendix A. Ensemble Averages in Ergodic Hypothesis

A modern version of the ergodic hypothesis (EH) is given as follows: If $\mathbf{A}$ is a dynamical variable of a system in thermal equilibrium, EH asserts that,

$$\lim_{T \to \infty} \frac{1}{T} \int_0^T < \mathbf{A}(t) >=< \mathbf{A} > \tag{A.1}$$

where the brackets denote ensemble averages. They may be canonical or grand ensemble averages, not just limited to microcanonical averages as originally given by Boltzmann. Since most variables are most often evaluated in canonical or grand ensembles, this more general definition is certainly preferred. In fact, many years ago Kubo had already recognized it and suggested this possibility.[16] Also see.[20,21]

The brackets on the lhs of A.1 really mean that all possible initial conditions are appropriately taken into account. Since EH is deemed to apply to a macroscopic system in thermal equilibrium, the brackets on the lhs are also the same as the ensemble averages on the rhs.

Consider, for ex., the momentum of a particle in an ideal classical gas of N identical particles in volume V, where N/V is finite in the thermodynamic limit. By one of Hamilton's equations of motion,

$$\frac{d\mathbf{p}_i(t)}{dt} = 0, \qquad i = 1, 2, ...N \tag{A.2}$$

Hence,

$$\mathbf{p}_i(t) = \mathbf{p}_i(0) \tag{A.3}$$

Evidently the initial value of ith particle or, equivalently, of any one particle can range from 0 to ∞. Hence the lhs of (A.1) becomes:

$$LHS = \frac{1}{T} \int_0^T <\mathbf{p}(t)> dt = <\mathbf{p}(0)> \frac{1}{T} \int_0^T dt$$
$$= <\mathbf{p}> = RHS \tag{A.4}$$

This elementary example shows that the lhs of (A.1) is at some temperature say T_1 and the rhs of (A.1) is also at T_1 . Thus both sides are being compared at the same temperature as they should only be.

If one were to consider a quantum system, the brackets on the lhs are all the more necessary since $\mathbf{A}$ may be an operator. The bracketed quantities in (A.1) mean:

$$<\mathbf{A}(t)> = tr\rho(t)\mathbf{A} \tag{A.5}$$

$$<\mathbf{A}> = tr\rho\mathbf{A} \tag{A.6}$$

where $\rho(t)$ and ρ are, resp., time dependent and time independent density matrices in appropriate ensembles, so that they are functions, not operators on both sides. Our work on ergometry, which illustrates this point, is aimed largely at studying the ergodicity in quantum many-body systems. The classical structure would follow by taking the classical limit on it as ordinarily done ($\hbar/kT \to 0$).

In the original version of Boltzmann's EH, the lhs does not have brackets because the trajectories of a dynamical variable such as $\mathbf{p}$ are on the surface of constant energy. That is, the system is studied in microcanonical ensembles. The rhs of (A.1) is thus a microcanonical ensemble average.

If E is fixed at all times, there can be but one equivalent initial condition. For ex., consider again the momentum of a particle in an ideal classical gas. Since $E = \sum_j p_j^2/2m$, the values of $\mathbf{p}_j$ are constrained to lie on the surface of a hypersphere of radius $\sqrt{2mE}$. Hence the initial conditions do not matter. Essentially the same may be said of an assembly of harmonic oscillators, another common problem.

Technically one has to be a little more careful here. If ergodic, the following must hold according to Khinchin [17]: The time average starting from any point $\mathbf{P}$ on an ergodic surface (i.e. the surface of constant energy) will approach a certain static value (i.e., microensem. av.) independent of almost all values of initial point $\mathbf{P}$ save sets of measure zero. It is not difficult to show that initial point $\mathbf{P}$ belonging to a set of measure zero will produce a trajectory of periodic motion.

As noted at the outset, the brackets on both sides of (A.1) may be averages of any appropriate ensembles of choice. But for general purposes especially for studying quantum many-body systems, we contend that our definition is an appropriate one.

References

1. I. E. Farquhar, Ergodic Theory in Statistical Mechanics (Wiley, NY, 1964)

2. G. D. Birkhoff, Proc. Nat. Acad.. Sci. (USA) 17, 656 (1931)
3. M.H. Lee, Phys. Rev. Lett. 87, 250601 (2001).
4. R. Kubo, Rep. Prog. Phys. 29, 255 (1966), H. Nakano, Int. J. Mod. Phys.B 7, 239 (1993). For a recent review, see A.L. Kuzemsky, Int. J. Mod. Phys. B 19, 1029 (2005).
5. U. Balucani, M.H.Lee, and V. Tognetti, Phys. Rep. 373, 409 (2003).
6. R. M. Yulmatyev, A V Mokshin and P. Hnggi, Phys. Rev. E 68, 051201 (2003). A. V. Mokshin, R. M. Yulmatyev and P. Hnggi, Phys. Rev. Lett. 95, 200601 (2005).
7. A. L. Kuzemsky, Int. J. Mod. Phys. B 21. 2821 (2007).
8. M.H. Lee, Phys. Rev. Lett. 98, 190601 (2007).
9. P. Grigolini, Quantum mechanical irreversibility and measurement (World Scientific, Singapore 1993)
10. R. Morgado, , F. A. Oliveira, G.G. Bartolouni, and A. Hansen, Phys. Rev. Lett. 89, 100601 (2002). M. H. Vainstein, I. V. L. Costa, R. Morgado, and F. A. Oliveira, Europhys. Lett 73, 726 (2006). J. D. Bao,Y Z Zhuo, F.A. Oliveira and P. Hnggi, Phys. Rev. E 74, 061111 (2006).
11. T. Prosen, J Phys A 40, 7881 (2007) and New J. Phys. 10, 1 (2008).
12. U. M. B. Marconi, A. Puglisi, L. Rondoni, and A. Vulpiani, Phys. Rep. 461, 111 (2008).
13. M.H. Lee, Phys. Rev. Lett. 98, 110403 (2007).
14. J. Florencio and M.H. Lee, Phys. Rev. B 35, 1835 (1987).
15. Th. Niemeijer, Physica (Amsterdam) 36, 377 (1967).
16. R. Kubo, J. Phys. Soc. (Jpn) 12, 570 (1957). See esp. p. 577.
17. A. I. Khinchin, Mathematical foundation of statistical mechanics (Dover, New York 1949).
18. L.C. Lapaz, R. Morgado, M.H. Vainstein, J.M. Rubi, and F.A. Oliveira, Phys. Rev. Lett. 101, 230602 (2008).
19. D.W. Jepsen, J. Math. Phys. 6, 405 (1967).
20. P. Mazur, Physica 43, 533 (1969). See esp. eqs. 8 and 58 therein.
21. E. Barouch, B. McCoy and M. Dresden, Phys. Rev. A 2, 1075 (1970). See esp. eq.8.1 therein.
22. P. Mazur and E. Montroll, J. Math. Phys. 1, 70 (1960).
23. T Prosen, Open problems in strongly correlated electron systems, eds. J. Bonca et al., Kluwer Acdemic, Amsterdam 2001, p. 317.
24. M.H. Vainstein, I.Vi.L. Costa and F.A. Oliveira, Mixing, ergodicity and fluctuation-dissipation theorem in complex systems, in Lecture notes in physics, Springer-Verlag, Berlin 2006.

ELECTRON-ELECTRON INTERACTION IN THE NON-RELATIVISTIC LIMIT

F. BARY MALIK

Department of Physics, Southern Illinois University,
Carbondale, IL, 62901, USA
and
Department of Physics, Washington University,
St. Louis, MO, 63130, USA
fbmalik@physics.siu.edu

Received 12 December 2009

The electron-electron potential in the one-photon exchange approximation with the omission of the spin-spin interaction, leads to the classical Coulomb interaction, but the inclusion of the latter results in the Møller interaction. Bethe and Fermi showed that the latter interaction leads to the Breit potential, if a few of the terms in the expansion of the retardation effect are considered. In this article, it is shown that the higher order terms omitted in the Bethe-Fermi treatment reduces to terms of the same order in Dirac's alpha-matrices considered by Bethe and Fermi. This raises questions whether the Breit interaction is the appropriate first order correction to the Coulomb potential in the non-relativistic limit. It is pointed out that the nature of the interaction between two bound (1s) electron derived by Brown using the Schwinger formalism of the quantum electrodynamics but proposed empirically in 1929 by Gaunt could be a better correction to the Coulomb potential for bound electrons in atoms. The calculated energies using these matrix elements plus the vacuum polarization energies are in reasonable agreement with the data. For comparison, calculated energies using the Breit interaction plus vacuum polarization energies are also presented.

Keywords: Møller and Breit interaction between electrons; applications to K-shell energies of W; Hg; Pb; Rn and Fm and K; LI; LII and LIII shells of Fm.

1. Introduction

The K- as well as L-shell ionization cross sections are reasonably accounted for in the plane wave Born approximation[1–5] with plane wave solutions of the Dirac equation and considering the Møller interaction (MI)[6] between two colliding electrons, whereas the calculations done with the simple consideration of only the Coulomb potential between two colliding electrons deviate from the data significantly already for about 2MeV incident electron energy[1–5] This is shown, for example, in Fig. 1 of Ref. 1 where the calculated cross sections in relativistic plane wave Born approximation are presented as a function of the ratio of incident energy, KE and ionization potential I, for:(a) Møller interaction for Au) and Hg, (b) Coulomb-only interaction for Hg and non-relativistic plane wave Born approximation with Coulomb interaction for Hg, and where these results are compared with the data of Ref. 17.

On the other hand, the corrections to the inner shell x-ray energies due to the first order deviation of electron-electron Coulomb potential are usually done using the Breit Interaction (BI)[7] derived from a semi-classical field theory. In an attempt to find a relation between the (MI) derived using the quantum electrodynamics and the (BI), Bethe and Fermi[8] expanded the retardation term of the MI in terms of the Dirac α-matrices of two-electrons and established the equivalence of the BI with that of MI by keeping only the first few terms in the expansion involving only α^2. In this article, we evaluate the next few terms and show that they also contain terms of the order of α^2 raising questions about their equivalence. Browns derivation[9] of the interaction between two s-electrons in atoms using Schwingers formalism of the quantum electrodynamics leads to a different interaction which is similar to the one proposed empirically by Gaunt.[10] The latter is basically equivalent to the (MI) without the retardation term. This is presented in the next section. The K-shell electron energies of W, Hg, Pb, Rn and Fm and K, LI, LII and LIII shell electrons in Fm using the MI without retardation along with the addition of vacuum polarization contribution using the relativistic Hartree-Fock orbitals discussed in the applications section are in good agreement with the data.

2. The Theory

In this article, we adopt the Bjorken-Drell-Schwaber representation of the gamma matrices, i.e. $\gamma^0 = \beta$ and $\gamma^i = \beta\alpha_i$ $(i = 1, 2, 3)$ and $\gamma^\mu\gamma^\nu + \gamma^\nu\gamma^\mu = 2g^{\mu\nu}$. The gamma matrices are related to the Dirac's α and β matrices as follows:

$$\beta = \begin{pmatrix} 1 & 0 & 0 & 0 \\ 0 & 1 & 0 & 0 \\ 0 & 0 & -1 & 0 \\ 0 & 0 & 0 & -1 \end{pmatrix} \equiv \begin{pmatrix} 1 & 0 \\ 0 & -1 \end{pmatrix}$$

$$\beta^2 = \begin{pmatrix} 1 & 0 & 0 & 0 \\ 0 & 1 & 0 & 0 \\ 0 & 0 & 1 & 0 \\ 0 & 0 & 0 & 1 \end{pmatrix}$$

$$\alpha = \begin{pmatrix} 0 & \sigma \\ \sigma & 0 \end{pmatrix}; \quad \sigma_x = \begin{pmatrix} 0 & 1 \\ 1 & 0 \end{pmatrix}; \quad \sigma_y = \begin{pmatrix} 0 & -i \\ i & 0 \end{pmatrix}; \quad \sigma_z = \begin{pmatrix} 1 & 0 \\ 0 & -1 \end{pmatrix}$$

Thus, $\alpha_i^2 = 1$, $i = x, y, z$, and $\beta^2 = 1$ The current, $\mathbf{j} = <c\alpha>$, is identified to be equivalent to velocity, $\mathbf{v}$. Hence usually $<\alpha>$ is taken to be a measure of $\mathbf{v}/c$.

Bethe and Fermi[8] considered the matrix-element, M.E., of the Møller operator in the following way:

$$M.E. = \left(\bar{\psi}_2(2)\bar{\psi}_1(1)|O_S(r_{21}) + O_V(r_{21})|\psi_2(2)\psi_1(1)\right) \tag{1}$$

where ψ is a four component wave function satisfying the Dirac equation in an external field and $\bar{\psi} = \psi^*\gamma^0$. Subscripts and numbers in parentheses refer, respectively, to all quantum numbers and coordinates of electrons 1 and 2. r_{21} is the distance between electrons 1 and 2, $r_{21} = |\vec{r}_2 \vec{r}_1|$. The scalar part of the Møller operator O_S is given by

$$O_S(r_{21}) = \frac{e^2 \gamma^0(1) \cdot \gamma^0(2)}{r_{21}} e^{(1/\hbar)[E_{2\prime} + E_{1\prime} - E_2 - E_1]t} e^{(i/\hbar c)[E_1 - E_{1\prime}] \cdot r_{21}} \tag{2}$$

where e is the electric charge of electrons 1 and 2, $\hbar$ is the Planck constant$/2\pi$, $E_{1\prime}$ and $E_{2\prime}$ are the final energies, and E_1 and E_2 are the initial energies of electrons 1 and 2, respectively.

The energy conservation requires that

$$E_{2\prime} + E_{1\prime} = E_2 + E_1 \tag{3}$$

Because of (3), we have

$$O_S(r_{21}) = \frac{e^2 \gamma^0(1) \cdot \gamma^0(2)}{r_{21}} e^{(i/\hbar c)[E_1 - E_{1\prime}] \cdot r_{21}} \tag{4}$$

Taking (3) into consideration, we can write $O_V(r_{21})$ in (1) as follows:

$$O_V(r_{21}) = \frac{e^2 \vec{\gamma}(1) \cdot \vec{\gamma}(2)}{r_{21}} e^{(i/\hbar c)[E_1 - E_{1\prime}] \cdot r_{21}} \tag{5}$$

Expanding the retarded part in (4) and (5) and denoting the successive terms as $W_i(s)$ and $W_i(\nu)$ $(i = 1, 2, 3, \ldots)$, respectively, we get

$$O_S(r_{21}) = \frac{e^2 \gamma^0(1) \cdot \gamma^0(2)}{r_{21}} [1 + (i/\hbar c))[E_1 - E_{1\prime}] \cdot r_{21}$$
$$+ (1/2)(i/\hbar c)^2 [E_1 - E_{1\prime}]^2 \cdot r_{21}^2 + (1/6)(i/\hbar c)^3 [E_1 - E_{1\prime}]^3 \cdot r_{21}^3$$
$$+ \cdots]$$
$$\equiv W_1(s) + W_2(s) + W_3(s) + W_4(s) + \cdots \tag{6}$$

$$O_V(r_{21}) = \frac{e^2 \vec{\gamma}(1) \cdot \vec{\gamma}(2)}{r_{21}} [1 + (i/\hbar c))[E_1 - E_{1\prime}] \cdot r_{21}$$
$$+ (1/2)(i/\hbar c)^2 [E_1 - E_{1\prime}]^2 \cdot r_{21}^2 + (1/6)(i/\hbar c)^3 [E_1 - E_{1\prime}]^3 \cdot r_{21}^3$$
$$+ \cdots]$$
$$\equiv W_1(\nu) + W_2(\nu) + W_3(\nu) + \cdots \tag{7}$$

Bethe and Fermi showed that the BI is given by $W_1(s) + W_2(s) + W_3(s) + W_1(\nu)$ (Sometimes the BI is referred to as the sum $W_2(s) + W_3(s) + W_1(\nu)$ only.)

In fact, the matrix element of $W_2(s)$ is zero because either $E_1 = E_{1\prime}$, or if $E_1 \neq E_{1\prime}$, the wave functions are orthogonal. Bethe and Fermi have shown that $W_3(s)$ can be rewritten as follows:

$$W_3(s) = (1/2r_{21})(\vec{\gamma}(1) \cdot \vec{\gamma}(2)) - (\vec{\gamma}(1) \cdot \vec{n})(\vec{\gamma}(2) \cdot \vec{n}) \tag{8}$$

where $\vec{n}$ is a unit vector defined as

$$\vec{n} = (\vec{r}_1 - \vec{r}_2)/|\vec{r}_1 - \vec{r}_2| \tag{9}$$

Thus

$$W_1(s)+W_2(s)+W_3(s) = (e^2/r_{12})(\gamma^0(1)\cdot\gamma^0(2)-(1/2)(\vec{\gamma}(1)\cdot\vec{\gamma}(2)+(\vec{\gamma}(1)\cdot\vec{n})(\vec{\gamma}(2)\cdot\vec{n}))) \tag{10}$$

The rationale for leaving out the next order terms is that they, *apparently* containing higher powers of γ, are expected to be of higher order than $(v/c)^2$. In the following, however, we show that this is not necessarily true by evaluating $W_4(s)$, $W_2(\nu)$ and $W_3(\nu)$:

$$W_4(s) = -(ie^2\gamma^0(1) \cdot \gamma^0(2)/6\hbar^3 c^3)(E_1 - E_{1\prime})^3 \cdot r_{21}^2$$

We can rewrite this using the energy conservation and replacing some of the energies by their corresponding Hamiltonian:

$$W_4(s) = -(ie^2\gamma^0(1) \cdot \gamma^0(2)/6\hbar^3 c^3)(E_1 - E_{1\prime})[H(2)F(12) - F(12)H(2)] \tag{11}$$

with

$$F(12) = H(1)r_{21}^2 - r_{21}^2 H(1) \tag{12}$$

In the calculation of the matrix element, we need

$$F(12)\psi_1(1) = +2i\hbar c\vec{\gamma}(1)(\vec{r}_2 - \vec{r}_1)\psi_1(1) \tag{13}$$

and

$$[H(2)F(12) - F(12)H(2)]\psi_2(2)\psi_1(1) = \frac{2\hbar^2 c^2 \vec{r}_2}{|r_2|}\psi_2(2)\psi_1(1) \tag{14}$$

Thus matrix elements of

$$W_4(s) = 0 \tag{15}$$

because either $E_1 = E_{1\prime}$ or ψ_1 is orthogonal to $\psi_{1\prime}(1)$.

In cases where we consider only matrix elements of $W_2(\nu)$ between states of the same energies, e.g. in calculating its contribution between two (1s) electrons in the first order perturbation theory, it is zero because $E_1 = E_{1\prime}$. Thus,

$$
\begin{aligned}
W_2(\nu) &= -\frac{e^2}{r_{21}}\vec{\gamma}(2) \cdot \vec{\gamma}(1)(i/\hbar c)(E_1 - E_{1\prime}) \cdot r_{21} \\
&= -(ie^2/\hbar c)\vec{\gamma}(2) \cdot \vec{\gamma}(1)(E_1 - E_{1\prime}) \\
&= 0 \quad \text{if} \quad E_1 = E_{1\prime} \\
&= -(ie^2/\hbar c)\vec{\gamma}(2) \cdot \vec{\gamma}(1)\Delta E_1 \quad \text{otherwise}
\end{aligned} \tag{16}
$$

ΔE_1 is energy difference of electron one before and after scattering or absorption/emission of photon

For small ΔE_1, $W_2(\nu) \simeq (\nu_1 \cdot \nu_2)/c^2$. Thus, for $E_1 \neq E_{1\prime}$, this omitted term in the Bethe- Fermi treatment contributes terms of the same order as those included in their calculation. We now consider

$$W_3(\nu) = -\frac{e^2}{2\hbar^2 c^2}\vec{\gamma}(2) \cdot \vec{\gamma}(1)(E_1 - E_{1\prime})^2 \cdot r_{21} \tag{17}$$

Because of the energy conservation, $(E_1 - E_{1\prime})^2 = -(E_1 - E_{1\prime})(E_2 - E_{2\prime})$ and we can rewrite $W_3(\nu)$ as

$$W_3(\nu) = -\frac{e^2}{2\hbar^2 c^2}\vec{\gamma}(2) \cdot \vec{\gamma}(1)[H(2)G(12) - G(12)H(2)] \tag{18}$$

with $G(12) = H(1)r_{21} - r_{21}H(1)$ Hence $G(12)\psi_1(1) = -(\hbar c/i)(\vec{\gamma}(1) \cdot \vec{n})\psi_1(1)$ and

$$W_3(\nu)\psi_2(2)\psi_1(1) = \frac{e^2}{2r_{21}}\vec{\gamma}(2) \cdot \vec{\gamma}(1)$$
$$\times (\vec{\gamma}(2) \cdot \vec{\gamma}(1) - (\vec{\gamma}(1) \cdot \vec{n})(\vec{\gamma}(2) \cdot \vec{n})\psi_2(2)\psi_1(1) \tag{19}$$

Thus $W_3(\nu)$ reduces to

$$W_3(\nu) = \frac{e^2}{2r_{21}}\vec{\gamma}(2) \cdot \vec{\gamma}(1)(\vec{\gamma}(2) \cdot \vec{\gamma}(1) - (\vec{\gamma}(1) \cdot \vec{n})(\vec{\gamma}(2) \cdot \vec{n}) \tag{20}$$

$$W_3(\nu) = \frac{e^2}{2r_{21}}\left[2 + \sum_{ij} A^{ij}(1)A^{ij}(2) - \vec{\gamma}(1) \cdot \vec{\gamma}(2) + (\vec{\gamma}(1) \cdot \vec{n})(\vec{\gamma}(2) \cdot \vec{n})\right] \tag{21}$$

where $A^{ij}(1) = (1/2)(\gamma^i(1)\gamma^j(1) - \gamma^j(1)\gamma^i(1))$ A comparison of (21) with the BI (10) reveals the very interesting fact that $W_3(\nu)$ is of the same order as (10), *irrespective of velocities and electrons*. In other words, the next order correction to the BI is of the same order as the BI itself! The derivation also shows that there is no guarantee that the successive terms in the expansion of O_V and O_S become less important. *In fact, the non- convergent nature of the successive terms in the expansion of O_V and O_S is obvious by noting that the diagonal part of a product containing two or more γs reduces to terms containing zero and linear powers of γ.* Møller already pointed out in 1932[11] that the interaction-form is valid for one bound and another free electron. Brown, using the Schwinger approach, indeed showed that the Møller interaction is valid between two bound electrons in atoms without the retardation term. O_M (for two bound electrons)

$$O_M = \frac{e^2}{r_{21}}(\gamma^0(1)\gamma^0(2) - \vec{\gamma}(1) \cdot \vec{\gamma}(2)) \tag{22}$$

This is actually proposed by Gaunt in 1929 in an adhoc way. Gaunt interaction

3. Applications

It is of interest to examine whether K and L-shell energies calculated using (22) as a perturbation are in accord with the observations. For this purpose, we use the Dirac-Hartree-Fock orbitals[12] as unperturbed states and the contributions of the Lamb shift and vacuum-polarization to the energies as determined by Desidario and Johnson.[13] The calculated K-shell energies for W, Hg, Pb, Rn and Fm and K, LI, LII and LIII energies of Fm are presented and compared to those observed in Tables 1 and 2, respectively. Our theoretical calculations are shown in the rows marked "Theo." in both tables.

For the comparison, we present in the row marked Theo(DJ) in Table 1, the results obtained using the BI plus vaccum polarization, as noted in.[13] Clearly, our results are in better agreement with the data. Similarly, the throretical results using the BI plus vaccum polarization from[17] for Fm are shown as Theo(FDW) in row 3 of Table 2. Once again our calculations compare more favorably with the data.

Table 1. Comparison of the theoretical K shell energies, in eV, using the Moller interaction in the first order perturbation theory plus vacuum polarization energy (VP), with those observed. Energies are in eV. Experimental value in parenthesis is from X-ray absorption energy. The data are from [14]. Row Theo(DJ) presents calculations using the BI plus VP as reported in [13].

	W	Hg	Pb	Rn
Exptl.	-69525±0.3	-83102.3±0.8	-88004.5±0.7	-98404±12
	(-69508)			
Theo.	-69512	-83086	-87986	-98380
Theo(DJ)	-69672	-83525	-88219	-98658

Table 2. Comparison of the theoretical energies, obtained from the Møller interaction in the first order perturbation theory plus vacuum polarization energy (VP), with observed ones for inner shells of Fm. Energies are in eV. The data are reported in [14]. Row Theo(EDW)re presents calculations using the BI plus VP as reported in [16].

	(1s)	(2s)	$(2p_{1/2})$	$(2p_{3/2})$	(3s)	$(3p_{1/2})$
Exptl.	-141963±13	-27573±8	-26644±7	-20868±7	-7200±9	-6779±7
Theo.	-141912	-27573	-26633	-20858	-7212	-6780
Theo(EDW)	-142410	-27677	-26655	-20866	-7238	-6786

4. Discussion and Conclusion

A number of attempts have been made to improve the remaining discrepancies between the theoretical calculations and the experimental results. For example, Mann and Johnson[18] tried to bridge the gap by using an average form of the BI arguing that the BI was originally determined for bare Coulomb field, whereas electrons in atoms perceive shielded Coulomb field. However, there are considerably uncertainties in determining the effect of shielding and their results differ from those of Kim.[19] Similarly attempts to include contributions of self-energy and vaccum fluctuations for *bound systems using the operators valid for scattering* states are fraught with uncertainties and the results[18] differ significantly from similar calculations of Brown *et al.*[20] The contribution from electron correlation energy is also difficult to determine, although they are likely to be very small for K-shell electrons. Another source of uncertainties is the exact determination of the nuclear charge distributions. A mere change of 0.1^{-15}m in either the nuclear half-density radius or surface thickness would result in a change of about 10eV in the binding energies of K-electrons.[21] Experimental data for W, Hg and Pb are for atoms in solids, whereas theoretical calculations are for a single atom in free space. Considering all these, it may be quite reasonable to look for an agreement between within 10 to 15eV, which can be achieved with the Møller interaction without the retardation term. This unretarded Møller interaction is clearly justified for s-electrons by the derivation of Brown, and the agreement between our calculations and the observed data.

For a colliding two-electrons, one expects a retardation term from the causality conditions to insure proper time difference between advanced and retarded wave propagation. For two bound electrons, one deals with the stationary standing wave and the system does not evolve in time, which corresponds to diagonal matrix elements of (22). A simple consequence of this is that energies of triplet states lie below those of singlet states for a given angular momentum, which is the case for He-like ions.

References

1. J.T. Ndefru and F.B. Malik, *Phys. Rev.* **25**, 2407 (1982).
2. J.T. Ndefru and F.B. Malik, *J. Phys. B: Atom. Mol. Phys.* **13**, 2117 (1980).
3. J.T. Ndefru, Doctoral Dissertation, Indiana University at Bloomington (Unpublished) (1978).
4. J.H. Scofield, *Phys. Rev.* **A 18**, 963 (1978).
5. J.T. Ndefru and F.B. Malik, *Condensed Matter Theories* **14**, 45 (2000).
6. C. Møller, *Zeit. Physik* **70**, 531 (1931); *ibid*, **70**, 786 (1931).
7. G. Breit, *Phys. Rev.* **34**, 553 (1929); *ibid*. **39**, 616 (1932).
8. H.A. Bethe and E. Fermi, *Zeit. Physik* **77**, 296 (1932).
9. G.E. Brown, *Phil. Mag.* **43**, 467 (1952).
10. J.A. Gaunt, *Proc. Roy. Soc. (London)* **A 122**, 513 (1929).
11. C. Møller, *Annalen d. Physik* **14**, 531 (1932).
12. J.P. Desclaux, *Atomic Data Nuclear Data* **12**, 311 (1973).
13. A.M. Desidario and W.R. Johnson, *Phys. Rev.* **A 3**, 1267 (1971).
14. J.A. Bearden and A.F. Burr, *Rev. Mod. Phys.* **39**, 125 (1967).
15. F.T. Porter and M.S. Friedman, *Phys. Rev. Lett.* **27**, 293 (1971).
16. D.V. Davies, V.D. Mistry and C.A. Quarles, *Phys. Rev. Lett.* **38A**, 169 (1972).
17. B. Fricke, J.P. Desclaux and J.T. Waber, *Phys. Rev. Lett.* **28**, 714 (1972).

18. J.B. Mann and W.R. Johnson, *Phys. Rev.* **A 4**, 41 (1971).

19. Y.K. Kim, *Phys. Rev.* **154**, 17 (1967).

20. G.E. Brown, D.F. Myers and E.A. Sanderson, *Phys. Rev. Lett.* **3**, 90 (1959).

21. T.A. Carlson, C.W. Nestor, Jr., F.B. Malik and T.C. Tucker, *Nucl. Phys.* **A 135**, 57 (1969).

AUTHOR INDEX

SUBJECT INDEX